# Technology of Breadmaking

Second Edition

Stanley P. Cauvain and Linda S. Young

# Technology of Breadmaking

## Second Edition

 Springer

Stanley P. Cauvain
BakeTran, UK
97 Guinions Road
High Wycombe Bucks HP13 7NU
UK
SPC@baketran.demon.co.uk

Linda S. Young
BakeTran, UK
39 Carpenters Wood Drive
Chorleywood Herts WD3 5RN
UK
LSY@baketran.demon.co.uk

*Cover Illustration:* Bread Varieties

Library of Congress Number: 2006932832

| ISBN-10: 0-387-38563-0 | e-ISBN-10: 0-387-38565-7 |
| ISBN-13: 978-0387-38563-1 | e-ISBN-13: 978-0-387-38565-5 |

Printed on acid-free paper.

9 8 7 6 5 4 3 2 1

springer.com

# Preface to the Second Edition

The manufacture of any processed food is constantly evolving and breadmaking is no exception. Even though bread has been made for thousands of years and its traditional forms remain as strong today as they did in past times, new ideas and new technologies are being developed and adapted to underpin modern production.

While new technologies will undoubtedly continue to drive developments in breadmaking, its traditional basis should not be neglected. Increasing diversity of products in part driven by consumer demand will contribute heavily to the future of breadmaking. More frequent travel and increased global communications expose many more people to the diversity of bread products. What is a traditional product in one part of the world is the novel product in another.

Since writing the the first edition of the *Technology of Breadmaking*, the knowledge base on which breadmaking is founded has undergone considerable enlargement and fuelled many of the more recent product and process developments. This second edition sets out not only to update the first edition but also seeks to identify and discuss the new knowledge that has become available since the mid-1990s or so.

We wish to thank those original authors who willingly gave their time to revise their original contributions.

Stanley P. Cauvain and Linda S. Young

# Preface to the First Edition

Not another book on breadmaking! A forgivable reaction given the length of time over which bread has been made and the number of texts which have been written about the subject.

To study breadmaking is to realize that, like many other food processes, it is constantly changing as processing methodologies become increasingly more sophisticated, yet at the same time we realize that we are dealing with a foodstuff, the forms of which are very traditional. We can, for example, look at ancient illustrations of breads in manuscripts and paintings and recognize products which we still make today. This contrast of ancient and modern embodied in a single processed foodstuff is part of what makes bread such a unique subject for study. We cannot, for example, say the same for a can of baked beans!

Another aspect of the uniqueness of breadmaking lies in the requirement for a thorough understanding of the link between raw materials and processing methods in order to make an edible product. This is mainly true because of the special properties of wheat proteins, aspects of which are explored in most of the chapters of this book. Wheat is a product of the natural environment, and while breeding and farming practices can modify aspects of wheat quality, we millers and bakers still have to respond to the strong influences of the environment.

The quality of the baker's main raw material, wheat flour, varies and so special knowledge is needed to ensure the right product qualities are formed in the bread for the consumer. Since some of the most significant changes in wheat quality are related to the environment in which it is grown, a most important tool for bakers is knowledge, without it they cannot adjust recipes or processing methods to ensure consistent product quality.

It is because breadmaking requires constant reaction to 'natural' changes and it has been the subject of scientific and technological study that there is room for another book on the subject. New ideas are being presented to bakers from wheat breeders, millers and ingredient and equipment suppliers, which are coupled with consumer and legislative pressures. These have to be integrated with 'natural' changes.

It is the purpose of this book to provide a useful tool to help bakers, scientists and technologists to cope with those changes. We hope that when you read through the contributions you will find something to make your particular job easier, or even something to enjoy.

As you read through the various chapters there will be occasions when you say to yourself 'I've read about that before'. When you get different authors to write about breadmaking, they have to consider the same common themes but they will approach them from their own special angles. The most common theme of course is the conversion of wheat to flour to bread. Each individual involved in that conversion process has a contribution to make, but in order for that contribution to be successful they must understand what part they play, and because of this they have different needs in their understanding. These different needs will be evident as they discuss common issues such as gluten development, so

- the cereal scientist seeks to understand the molecular reactions;
- the bakery technologist seeks to apply the understanding and solve bakers' problems;
- the flour miller seeks to ensure a consistent product by understanding the links between wheat, flour and bread quality;
- the ingredient suppliers seek to understand the contribution of their ingredients to bread quality;
- the equipment manufacturers seek to understand how dough behaviour interacts with their equipment; and
- bakers seek to make bread for their customers.

Editing a book of this type, just like breadmaking itself, is a team effort, and so we would like to thank the members of our team:

- The authors of the individual chapters who having agreed to write a contribution discovered like so many before that it is not as easy as it looks when you read a book written by someone else. We thank all of you for patience and perseverance.
- The publishers without whom this book would not have seen the light of day.
- Our many supporters, both moral and material.
- Our scientific mentors.

Why thirteen chapters? That is easy to answer – thirteen is the traditional 'bakers dozen'.

Stanley P. Cauvain and Linda S. Young

# Contents

Preface to the Second Edition ........................................... v

Preface to the First Edition ............................................. vii

List of Contributors ................................................... xxi

1. **BREAD – THE PRODUCT** ...................................... 1
   Stanley P. Cauvain
   1.1. Introduction ................................................. 1
   1.2. Quality Characteristics of Bread.................................. 3
   1.3. The Character of Bread.......................................... 5
   1.4. Bread Flavour................................................. 6
   1.5. Bread Types................................................... 8
   1.6. Assessing Bread Quality ........................................ 11
        1.6.1. External Character ....................................... 11
        1.6.2. Internal Character ....................................... 13
        1.6.3. Texture/Eating Quality and Flavour ......................... 14
   1.7. Nutritional Qualities of Bread and its Consumption.................. 15
   1.8. Conclusions .................................................. 18

2. **BREADMAKING PROCESSES**................................... 21
   Stanley P. Cauvain
   2.1. Functions of the Breadmaking Process ............................ 21
   2.2. Cell Creation and Control....................................... 25
   2.3. Major Breadmaking Process Groups ............................. 26
        2.3.1. Straight Dough Bulk Fermentation ........................... 28
               2.3.1.1. Yeast Level..................................... 28
               2.3.1.2. Flours......................................... 30
               2.3.1.3. Water Levels................................... 31
               2.3.1.4. Optional Ingredients ........................... 31
               2.3.1.5. Process Variations ............................. 32
               2.3.1.6. Creation of Bubble Structure .................... 32
        2.3.2. Sponge and Dough ........................................ 33
               2.3.2.1. Role of the Sponge............................. 33
               2.3.2.2. Formulations .................................. 34
               2.3.2.3. Improvers ..................................... 35

|   |   | 2.3.2.4. | Flours and Other Ingredients | 35 |
|   |   | 2.3.2.5. | Process Variations | 35 |
|   | 2.3.3. | Rapid Processing | | 36 |
|   |   | 2.3.3.1. | Activated Dough Development (ADD) | 36 |
|   |   | 2.3.3.2. | No-time Doughs with Spiral Mixers | 37 |
|   |   | 2.3.3.3. | The Dutch Green Dough Process | 37 |
|   |   | 2.3.3.4. | Role of Improvers and Other Ingredients in Rapid Processing | 38 |
|   | 2.3.4. | Mechanical Dough Development | | 38 |
|   |   | 2.3.4.1. | Chorleywood Bread Process | 39 |
|   |   | 2.3.4.2. | Role of Energy During Mixing | 40 |
|   |   | 2.3.4.3. | Energy and Dough Temperature Control | 42 |
|   |   | 2.3.4.4. | Flour Quality | 43 |
|   |   | 2.3.4.5. | Creation of Bubble Structure | 44 |
|   |   | 2.3.4.6. | Dough Rheology | 46 |
| 2.4. | Breadmaking Processes, Bread Variety and Bread Quality | | | 47 |

**3. FUNCTIONAL INGREDIENTS ... 51**
Tony Williams and Gordon Pullen

| 3.1. | Dough Conditioners and their Composition | | | 51 |
| 3.2. | Ingredients | | | 52 |
|   | 3.2.1. | Fats | | 52 |
|   |   | 3.2.1.1. | Summary of Fat Effects | 55 |
|   | 3.2.2. | Soya Flour | | 55 |
|   |   | 3.2.2.1. | Soya as a Flour Bleaching Agent | 55 |
|   |   | 3.2.2.2. | Gluten Oxidation by Soya Flour | 56 |
| 3.3. | Additives | | | 57 |
|   | 3.3.1. | Emulsifiers | | 57 |
|   |   | 3.3.1.1. | Diacetylated Tartaric Acid Esters of Mono- and Diglycerides of Fatty Acids [DATA Esters, E472(e)] | 57 |
|   |   | 3.3.1.2. | Use of DATA Esters in UK White Bread | 57 |
|   |   | 3.3.1.3. | Use of DATA Esters in High-Volume Bread and Morning Goods | 58 |
|   |   | 3.3.1.4. | Long Proof Processed Breads and Morning Goods | 58 |
|   |   | 3.3.1.5. | DATA Esters in Wholemeal and Seeded Breads | 58 |
|   |   | 3.3.1.6. | Sodium Stearoyl-2-Lactylate (SSL, E482) | 59 |
|   |   | 3.3.1.7. | Distilled Monoglyceride (E471) | 60 |
|   |   | 3.3.1.8. | Lecithins (E322) | 61 |
|   | 3.3.2. | Flour Treatment Agents | | 61 |
|   |   | 3.3.2.1. | Ascorbic Acid (Vitamin C, E300) | 61 |
|   |   | 3.3.2.2. | L-Cysteine (920) | 65 |
|   |   | 3.3.2.3. | Oxidizing Agents in the USA | 66 |
|   | 3.3.3. | Preservatives | | 66 |

3.4.    Processing Aids ................................................................ 68
        3.4.1.    Enzymes ............................................................ 68
                  3.4.1.1.    *Alpha*-Amylases ................................. 68
                  3.4.1.2.    Cereal *Alpha*-Amylase ........................ 71
                  3.4.1.3.    Bacterial *Alpha*-Amylase .................... 71
                  3.4.1.4.    *Alpha*-Amylase Which is a Product of
                              Modern Biotechnology ........................... 71
        3.4.2.    Hemicellulases .................................................. 72
        3.4.3.    Proteinases ....................................................... 73
        3.4.4.    Other Enzymes in Baking .................................. 74
        3.4.5.    Novel Enzyme Systems ..................................... 74
        3.4.6.    Health and Safety with Enzyme Usage ................ 75
3.5.    Summary of Small Ingredients ....................................... 75
3.6.    Liquid Improvers ........................................................... 76
3.7.    Bakers' Yeast ................................................................. 76
        3.7.1.    Where Does Yeast Come From? .......................... 76
        3.7.2.    Principal Forms of Yeast .................................... 77
                  3.7.2.1.    Compressed Yeast ................................ 77
                  3.7.2.2.    Granular Yeast .................................... 77
                  3.7.2.3.    Cream Yeast ....................................... 77
                  3.7.2.4.    Dried Pellet Yeast ............................... 78
                  3.7.2.5.    Instant Yeast ...................................... 78
                  3.7.2.6.    Encapsulated Yeast .............................. 78
                  3.7.2.7.    Frozen Yeast ....................................... 78
                  3.7.2.8.    Discussion .......................................... 78
        3.7.3.    Other Yeasts ..................................................... 79
        3.7.4.    Biology of Yeast Cells ....................................... 79
        3.7.5.    Overview of Commercial Yeast Production ............ 81
        3.7.6.    Baking with Yeast .............................................. 83
                  3.7.6.1.    Chorleywood Bread Process ................. 84
                  3.7.6.2.    Acidity or Alkalinity ........................... 84
                  3.7.6.3.    Storage Stability (Compressed Yeast) .... 85
                  3.7.6.4.    Effects of Mould and Rope Inhibitors .... 86
                  3.7.6.5.    Effect of Spices and Bun Spice ............ 86
                  3.7.6.6.    Yeast in Frozen Dough ........................ 86
                  3.7.6.7.    Salt .................................................... 87
                  3.7.6.8.    Sugar ................................................. 87
                  3.7.6.9.    Effect of Salt and Sugar on Yeast Activity 87
        3.7.7.    Conclusion ....................................................... 88
3.8.    Starter Doughs .............................................................. 89

4.  MIXING AND DOUGH PROCESSING ........................... 93
    David Marsh and Stanley P. Cauvain
    4.1.    Functions of Mixing ................................................. 93
    4.2.    Types of Mixer ........................................................ 94

|       | 4.2.1. | CBP-Compatible Mixers | 95 |
|       | 4.2.2. | Mixing Under Pressure and Vacuum with CBP-Compatible Mixers | 98 |
|       | 4.2.3. | Oxygen-Enrichment of the Mixer Headspace with CBP-Compatible Mixers | 99 |
|       | 4.2.4. | High-Speed and Twin-Spiral Mixers | 100 |
|       | 4.2.5. | Spiral Mixers | 102 |
|       | 4.2.6. | Low-Speed Mixers | 105 |
|       |        | 4.2.6.1. Twin-Arm Mixers | 105 |
|       |        | 4.2.6.2. Oblique-Axis Fork Mixers | 105 |
|       | 4.2.7. | Continuous Mixers | 106 |
| 4.3.  | Control of Dough Temperature and Energy Transfer | | 109 |
|       | 4.3.1. | Control of Dough Temperature | 109 |
|       | 4.3.2. | Energy Transfer | 110 |
| 4.4.  | Dough Transfer Systems | | 110 |
| 4.5.  | Dough Make-Up Plant | | 111 |
|       | 4.5.1. | Dividing | 112 |
|       |        | 4.5.1.1. Dough Damage During Dividing | 112 |
|       |        | 4.5.1.2. Two-Stage Oil Suction Divider | 113 |
|       |        | 4.5.1.3. Extrusion Dividers | 113 |
|       |        | 4.5.1.4. Single-Stage Vacuum Dividers | 115 |
| 4.6.  | Rounding and Pre-Moulding | | 115 |
| 4.7.  | Types and Shapes of Rounders | | 117 |
|       | 4.7.1. | Conical Rounders | 117 |
|       | 4.7.2. | Cylindrical Rounders | 118 |
|       | 4.7.3. | Rounding Belts | 118 |
|       | 4.7.4. | Reciprocating Rounders | 120 |
|       | 4.7.5. | Non-Spherical Pre-Moulding | 120 |
| 4.8.  | Intermediate or First Proving | | 120 |
|       | 4.8.1. | Pocket-Type Prover | 121 |
|       | 4.8.2. | First Prover Charging Methods | 122 |
|       | 4.8.3. | Indexing Conveyors | 123 |
|       | 4.8.4. | Pusher In-Feed Systems | 123 |
|       | 4.8.5. | Pallet In-Feed Systems | 124 |
|       | 4.8.6. | Discharging | 125 |
|       | 4.8.7. | Conveyorized First Provers | 125 |
| 4.9.  | Moulding | | 125 |
|       | 4.9.1. | Sheeting Action | 127 |
|       | 4.9.2. | Curling | 129 |
|       | 4.9.3. | Final Moulding | 129 |
|       | 4.9.4. | Four-Piecing | 131 |
|       | 4.9.5. | Cross-Grain Moulding | 132 |
|       | 4.9.6. | Other Sheeting and Moulding Systems | 132 |
| 4.10. | Modification of Gas Bubble Structures during Processing | | 133 |
| 4.11. | Sheet and Cut Dough Processing Systems | | 134 |

4.12.   Panning and Traying Methods................................. 135
4.13.   Equipment for Small Bread and Rolls....................... 135
    4.13.1.   Small Bun Divider Moulders ....................... 135
    4.13.2.   Integrated, Multi-Lane Roll Plants............... 137
4.14.   Combination Bread and Roll Plants ......................... 139

**5.   PROVING, BAKING AND COOLING** ......................... **141**
Chris Wiggins and Stanley P. Cauvain
5.1.   Introduction ...................................................... 141
5.2.   Psychrometry.................................................... 144
    5.2.1.   Definitions ................................................ 144
5.3.   The Proving Process........................................... 146
    5.3.1.   Practical Proving ...................................... 148
    5.3.2.   Prover Checklist ....................................... 149
5.4.   Modern Prover Design ........................................ 150
    5.4.1.   Airflow.................................................... 150
    5.4.2.   Ambient Conditions.................................... 150
    5.4.3.   Mechanical Handling................................. 152
5.5.   Developments in Proving..................................... 152
    5.5.1.   Higher Proof Temperatures......................... 152
    5.5.2.   Shorter Proof Time.................................... 153
5.6.   Prover to Oven ................................................. 153
5.7.   The Baking Process............................................ 153
    5.7.1.   Crumb Structure ....................................... 153
    5.7.2.   Yeast Activity and the Foam-to-Sponge Conversion... 154
    5.7.3.   Starch Gelatinization ................................. 155
    5.7.4.   Enzyme Activity........................................ 156
    5.7.5.   Baked Temperature ................................... 156
    5.7.6.   Crust Formation........................................ 156
    5.7.7.   Gloss Formation ....................................... 158
    5.7.8.   Crust Crispness ........................................ 160
    5.7.9.   Oven Break.............................................. 160
    5.7.10.   Practical Baking........................................ 161
    5.7.11.   Pan Strapping........................................... 162
    5.7.12.   Oven Design ............................................ 162
    5.7.13.   Developments in Baking ............................. 164
5.8.   Oven to Cooler................................................. 165
5.9.   The Cooling Process .......................................... 165
    5.9.1.   Practical Cooling ...................................... 167
    5.9.2.   Cooler Design........................................... 167
    5.9.3.   Developments in Cooling............................. 168
5.10.   Part-Baked Processes ........................................ 169
    5.10.1.   Frozen Processes ...................................... 170
    5.10.2.   Ambient Processing.................................... 170
5.11.   Process Economics............................................ 170

|       | 5.11.1. | Weight Loss | 170 |

5.11.1. Weight Loss ................................................................. 170
5.11.2. Life Cycle Costs ......................................................... 170

**6. DOUGH RETARDING AND FREEZING** ........................ **175**
Stanley P. Cauvain
6.1. Introduction ......................................................................... 175
6.2. Retarding Fermented Doughs ............................................. 176
    6.2.1. Suitability of Breadmaking Processes ....................... 176
    6.2.2. Recipe and Yeast Level ............................................. 177
    6.2.3. Retarding Temperature .............................................. 180
    6.2.4. Storage Time .............................................................. 182
    6.2.5. Proving and Baking ................................................... 183
    6.2.6. Guidelines for Retarded Dough Production ............... 186
6.3. Retarding Pizza Doughs ...................................................... 186
6.4. Freezing Fermented Doughs ............................................... 187
    6.4.1. Breadmaking Process, Recipe and Yeast ................... 187
    6.4.2. Processing and Freezing Doughs ............................... 190
    6.4.3. Defrosting and Proving ............................................. 194
6.5. Freezing Proved Doughs ..................................................... 195
6.6. Factors Affecting the Formation of White Spots
on Retarded and Frozen Doughs ......................................... 196
6.7. Causes of Quality Losses with Retarded and Frozen Doughs ..... 198
    6.7.1. Skinning ..................................................................... 198
    6.7.2. Crust Fissures ............................................................ 198
    6.7.3. Ragged Crust Breaks ................................................. 199
    6.7.4. Small Volume ............................................................ 199
    6.7.5. White Spots or Small Blisters ................................... 199
    6.7.6. Waxy Patches ............................................................ 200
    6.7.7. Black Spots ................................................................ 200
    6.7.8. Large Blisters ............................................................ 200
    6.7.9. Dark Crust Colour ..................................................... 200
    6.7.10. Uneven or Open Cell Structure ................................. 201
    6.7.11. Areas of Dense Crumb .............................................. 201
6.8. Principles of Refrigeration ................................................. 201
6.9. Retarder–Provers and Retarders ........................................ 202

**7. APPLICATION OF BAKING KNOWLEDGE IN SOFTWARE
SYSTEMS** ...................................................................... **207**
Linda S. Young
7.1. Introduction ......................................................................... 207
7.2. Examples of Systems for Use in Bread Technology ............ 209
    7.2.1. Bread advisor ............................................................. 209
    7.2.2. How the Bread Advisor is Used ................................. 210
    7.2.3. Fault Diagnosis or Quality Enhancement ................. 211
    7.2.4. Processing Details ...................................................... 214
    7.2.5. Questionioning Using 'What if?' ............................... 214

|  | 7.2.6. | Retarding Advisor | 215 |
|  | 7.2.7. | Baking Technology Toolkit | 217 |
| 7.3. | Software for Complex Logistical Bakery Challenges | | 218 |
|  | 7.3.1. | Rollout | 218 |
| 7.4. | Other Knowledge Sources | | 220 |
|  | 7.4.1. | Internet/Web Sources | 220 |
| 7.5. | Conclusion | | 220 |

**8. BAKING AROUND THE WORLD** .......................... **223**

John T. Gould

| 8.1. | Introduction | 223 |
| 8.2. | History | 223 |
| 8.3. | The Breadmaking Process | 227 |
| 8.4. | Flour and Dough Development | 228 |
| 8.5. | Water | 231 |
| 8.6. | Yeast | 232 |
| 8.7. | Salt | 232 |
| 8.8. | Other Improvers | 232 |
| 8.9. | Dividing | 233 |
| 8.10. | Resting | 233 |
| 8.11. | Moulding | 234 |
| 8.12. | Panning and Pans | 234 |
| 8.13. | Final Proof | 235 |
| 8.14. | Baking | 235 |
| 8.15. | Cooling | 236 |
| 8.16. | Slicing and Packing | 237 |
| 8.17. | Packaging | 237 |
| 8.18. | Bread and Sandwich Making | 238 |
| 8.19. | Crustless Breads | 239 |
| 8.20. | The International Market | 240 |

**9. SPECIALITY FERMENTED GOODS** ..................... **245**

Alan J. Bent

| 9.1. | Introduction | | | 245 |
| 9.2. | Hamburger Bun | | | 245 |
|  | 9.2.1. | Production Rates | | 245 |
|  | 9.2.2. | Formulation | | 246 |
|  | 9.2.3. | Liquid Brews and Fermentation | | 246 |
|  | 9.2.4. | Mixing | | 249 |
|  | 9.2.5. | Dough Transfer | | 249 |
|  | 9.2.6. | Do-Flow Unit | | 250 |
|  | 9.2.7. | Dividing | | 250 |
|  |  | 9.2.7.1. | Model K divider | 250 |
|  |  | 9.2.7.2. | Extrusion bun divider | 250 |
|  | 9.2.8. | AMF Pan-O-Mat | | 251 |
|  | 9.2.9. | Flour Recovery and Pan Shaker Units | | 251 |

9.2.10.    Proving.................................................................... 252
9.2.11.    Seed Application ..................................................... 252
9.2.12.    Baking..................................................................... 253
9.2.13.    Depanning and Cooling........................................... 253
9.3.    Part-Baked Breads........................................................... 255
9.3.1.    Introduction............................................................. 255
9.3.2.    Use of Part-Baked Bread.......................................... 256
9.3.3.    Manufacture of Part-Baked Breads............................ 257
9.3.4.    Baking Nets.............................................................. 257
9.3.5.    Proving and Baking................................................... 257
9.3.6.    Depanning and Cooling............................................ 258
9.3.7.    Storage of Part-Baked Breads ................................... 259
9.3.8.    Freezing of Part-Baked Bread................................... 260
9.3.9.    Second and Final Baking .......................................... 260
9.3.10.    Quality of the Final Product...................................... 260
9.3.11.    The Milton Keynes Process ...................................... 261
9.4.    Yeasted Laminated Products............................................ 263
9.4.1.    Introduction............................................................. 263
9.4.2.    Formulations ........................................................... 263
9.4.3.    Ingredients .............................................................. 264
9.4.4.    Flour Protein ........................................................... 264
9.4.5.    Fat Addition to the Base Dough ................................ 265
9.4.6.    Sugar Levels............................................................ 266
9.4.7.    Yeast Levels............................................................ 266
9.4.8.    Laminating or Roll-In Fat ......................................... 266
9.4.9.    Dough Mixing ......................................................... 269
9.4.10.    Methods of Adding the Roll-In or Laminating Fat
to the Dough................................................................ 270
9.4.11.    Lamination .............................................................. 271
9.5.    Frozen and Fully Proved Frozen Laminated Products................. 271

10. BREAD SPOILAGE AND STALING ........................... 275
Irene M.C. Pateras
10.1.    Introduction ................................................................. 275
10.2.    Microbiological Spoilage of Bread................................... 275
10.2.1.    Mould Spoilage ................................................. 275
10.2.2.    Bacterial Spoilage.............................................. 277
10.2.3.    Yeast Spoilage ................................................... 278
10.2.4.    Control of Microbiological Spoilage ..................... 278
10.2.4.1.    Preservatives................................. 278
10.2.4.2.    Modified Atmosphere Packaging ............ 279
10.2.4.3.    Irradiation ..................................... 282
10.2.4.4.    Ultraviolet Irradiation ...................... 282
10.2.4.5.    Microwave Radiation........................ 282
10.2.4.6.    Infrared Radiation ........................... 282

10.3.    Bread Staling .................................................................... 283
          10.3.1.    Crust Staling ...................................................... 283
          10.3.2.    Role of the Main Bread Components
                     in Crumb Staling ............................................... 283
                     10.3.2.1.    Starch.............................................. 283
                     10.3.2.2.    Starch Gelatinization................... 284
                     10.3.2.3.    Starch Retrogradation ............... 285
                     10.3.2.4.    Gluten............................................ 286
                     10.3.2.5.    Water Redistribution ................. 287
10.4.    Staling Inhibitors............................................................ 288
          10.4.1.    Enzymes.............................................................. 288
          10.4.2.    Emulsifiers ......................................................... 289
          10.4.3.    Pentosans .......................................................... 290
          10.4.4.    Alcohol ............................................................... 290
          10.4.5.    Sugars and Other Solutes ............................. 291
10.5.    Freezing of Bread............................................................ 292

11. PRINCIPLES OF DOUGH FORMATION ...................... 299
     Clyde E. Stauffer
     11.1.    Introduction .................................................................... 299
     11.2.    Flour and Dough Components........................................ 300
              11.2.1.    Starch ................................................................. 301
              11.2.2.    Gluten ................................................................. 301
              11.2.3.    Pentosans ........................................................... 302
              11.2.4.    Lipids .................................................................. 303
              11.2.5.    Water-soluble Proteins ................................... 304
              11.2.6.    Ash ....................................................................... 304
     11.3.    Flour Components and Water Absorption.................... 305
     11.4.    Wheat Gluten Proteins ................................................... 307
              11.4.1.    Amino Acid Composition ............................... 307
              11.4.2.    Gliadin ................................................................ 308
              11.4.3.    Glutenin .............................................................. 308
     11.5.    Stages in Dough Formation ........................................... 311
              11.5.1.    Hydration ............................................................ 312
              11.5.2.    Blending.............................................................. 312
              11.5.3.    Gluten Development........................................ 313
              11.5.4.    The Formation of Other Bonds..................... 316
              11.5.5.    Breakdown ......................................................... 317
              11.5.6.    Unmixing ............................................................ 317
              11.5.7.    Air Incorporation ............................................. 318
     11.6.    The Gluten Matrix.......................................................... 319
              11.6.1.    Dough Rheology................................................ 320
              11.6.2.    Gluten Structure................................................ 323
              11.6.3.    Bonding between Protein Chains ................. 324
              11.6.4.    Gluten Elasticity .............................................. 326

11.6.5.    Gluten Viscosity ........................................................ 326

11.6.6.    Extensibility .............................................................. 328

## 12. FLOUR MILLING ............................................... 333

Paul Catterall and Stanley P. Cauvain

12.1.    Introduction ..................................................................... 333

12.2.    In the Beginning................................................................ 333

12.3.    The Modern Flour-milling Process....................................... 335

    12.3.1.    Delivery of the Wheat................................................ 335

    12.3.2.    Wheat Testing........................................................... 336

        12.3.2.1.    Appearance, Off-odours and Taints ......... 336

        12.3.2.2.    Screenings (impurities) ........................... 336

        12.3.2.3.    Wheat Density........................................ 337

        12.3.2.4.    Protein Content ..................................... 337

        12.3.2.5.    Gluten Content ...................................... 338

        12.3.2.6.    Moisture ............................................... 338

        12.3.2.7.    Hagberg Falling Number ......................... 338

        12.3.2.8.    Hardness ............................................... 339

        12.3.2.9.    Electrophoresis ...................................... 339

    12.3.3.    Wheat Storage .......................................................... 340

    12.3.4.    The Mill Screenroom .................................................. 340

        12.3.4.1.    Size....................................................... 341

        12.3.4.2.    Specific Gravity ...................................... 341

        12.3.4.3.    Shape .................................................... 341

        12.3.4.4.    Magnetism............................................. 341

        12.3.4.5.    Air Resistance (Aspiration) ..................... 342

    12.3.5.    Conditioning ............................................................ 342

    12.3.6.    The Mill................................................................... 342

    12.3.7.    Break System............................................................ 343

    12.3.8.    Scratch System and Bran Finishers .............................. 344

    12.3.9.    Scalping, Grading and Dusting..................................... 344

    12.3.10.    Purifiers.................................................................. 345

    12.3.11.    Reduction System..................................................... 345

    12.3.12.    Flour Dressing ......................................................... 346

    12.3.13.    Storage and Packing.................................................. 347

12.4.    Food Safety and Product Protection ..................................... 347

    12.4.1.    Foreign Bodies.......................................................... 348

    12.4.2.    Chemical Contaminants............................................... 349

    12.4.3.    Biological Contaminants .............................................. 349

12.5.    De-branning and Flour Milling............................................ 351

12.6.    Controlling Flour Quality and Specification ........................... 351

    12.6.1.    Gristing Versus Blending............................................. 352

    12.6.2.    Additives.................................................................. 353

    12.6.3.    Ascorbic Acid........................................................... 354

    12.6.4.    Enzymes................................................................... 354

|          | 12.6.4.1. | Amylases | 354 |
|          | 12.6.4.2. | Protease | 355 |
|          | 12.6.4.3. | Hemicellulases | 355 |

12.7.   Nutritional Additions............................................................ 355
12.8.   Other Flour Types ................................................................. 356
    12.8.1.   Brown ....................................................................... 356
    12.8.2.   Wholemeal (Wholewheat)........................................... 356
    12.8.3.   Self-raising Flours ..................................................... 357
    12.8.4.   Malted Grain Flours ................................................... 357
12.9.   Flour Testing Methods ........................................................ 357
    12.9.1.   Protein and Moisture Content ..................................... 357
    12.9.2.   Flour Grade Colour .................................................... 359
    12.9.3.   Water Absorption ....................................................... 360
    12.9.4.   Hagberg Falling Number............................................. 361
    12.9.5.   Flour Rheology .......................................................... 362
    12.9.6.   Farinograph................................................................ 362
    12.9.7.   Extensograph ............................................................. 364
    12.9.8.   Alveograph ................................................................ 364
    12.9.9.   Other Rheological Testing Equipment......................... 365
    12.9.10.  The MixerLAB ........................................................... 365
    12.9.11.  Amylograph and Rapid Visco Analyser (RVA)........... 366
    12.9.12.  Using Testing Equipment............................................ 366
12.10.  Glossary of Milling Terms Used in this Chapter........................ 367

**13. OTHER CEREALS IN BREADMAKING ....................... 371**
Stanley P. Cauvain
13.1.   Introduction .......................................................................... 371
13.2.   Rye Bread .............................................................................. 371
    13.2.1.   Sour Dough Methods ................................................. 373
    13.2.2.   Doughmaking ............................................................ 373
    13.2.3.   Baking....................................................................... 374
    13.2.4.   American Rye Breads.................................................. 374
    13.2.5.   Keeping Qualities of Rye Breads................................. 375
13.3.   Triticale................................................................................. 375
13.4.   Other Grains and Seeds in Bread ............................................ 376
    13.4.1.   Multi-grain Breads..................................................... 376
    13.4.2.   Modifying Nutritional Properties with Non-wheat
           Sources.................................................................... 377
    13.4.3.   Malted Barley ............................................................ 378
13.5.   Wheatless Breads ................................................................... 379
    13.5.1.   Formation of Cake Batters .......................................... 379
    13.5.2.   Bread without Wheat................................................... 380
    13.5.3.   Cassava Bread ........................................................... 381
    13.5.4.   Rice or Maize Bread ................................................... 381
    13.5.5.   Sorghum Breads ........................................................ 381

              13.5.6.    Sorghum and Maize Bread Using a Continuous
                         Mixer .......................................................................... 382
              13.5.7.    Sorghum Flat Breads ................................................... 383
       13.6.   Unleavened Breads ........................................................................ 384
              13.6.1.    Recipes for Unleavened Breads .................................. 385
              13.6.2.    Balady ......................................................................... 385
              13.6.3.    Chapattis .................................................................... 385
              13.6.4.    Naan ........................................................................... 385
              13.6.5.    Papadams .................................................................... 386
              13.6.6.    Tortillas ...................................................................... 386
              13.6.7.    Flours for Unleavened Breads .................................... 386
       13.7.   Conclusion .................................................................................... 387

Index ............................................................................................ 389

# List of Contributors

| | |
|---|---|
| **Alan Bent** | International Centre for Baking Technology, South Bank University, 103 Borough Road, London SE1 0AA, UK |
| **Paul Catterall** | Campden and Chorleywood Food Research Association, Chipping Campden, Gloucestershire GL55 6LD, UK |
| **Stanley P. Cauvain** | BakeTran, 97 Guinions Road High Wycombe, Buckinghamshire, HP13 7NU, UK |
| **John Gould** | 106 Kopanga Road, Havelock North, New Zealand |
| **David Marsh** | Benier UK Ltd, 56 Alston Drive, Bradwell Abbey, Milton Keynes MK13 9HB, UK |
| **Irene M.C. Pateras** | Demokratias Avenue 18, Drosia, Athens, Greece |
| **Gordon Pullen** | retired from DCL Yeast Ltd, Lesaffre Group, 111 Epsom Road, Guilford, Surrey GU1 2LE, UK |
| **Clyde E. Stauffer** | TFC, 631 Christopal Drive, Cincinnati OH 45231, USA |
| **Chris Wiggins** | formerly with APV Bakery Ltd, Manor Drive, Paston Parkway, Peterborough PE4 7AP, UK |
| **Tony Williams** | retired from British Arkady, Skerton Road, Old Trafford, Manchester M16 0NJ, UK |
| **Linda S. Young** | BakeTran, 97 Guinions Road, High Wycombe, Buckinghamshire, HP13 7NU, UK |

# 1
# Bread – the Product

Stanley P. Cauvain

## 1.1. Introduction

Bread in its many forms is one of the most staple foods consumed by humanity. Traditionally bread is based on flour derived from the cereal wheat. Many other types of cereals, pulses and even legumes can be milled to give a 'flour' but the ability of the proteins present in wheat to transform a gruel of flour and water into a glutinous mass which becomes bread is currently limited to wheat and a few other commonly used cereal seeds. Genetic manipulation may yet combine the special protein characters of wheat with other more conveniently grown and processable seeds, for example to deliver no crease or a more rounded shape, but today we are still dealing with a cereal crop largely unchanged, in the genetic sense, from the time that humanity discovered its ability to make a special food many thousands of years ago.

Many consider bread to be one of the oldest, if not the oldest 'processed', food. We are unlikely ever to identify the moment when bread was 'discovered' though it is likely that the place of discovery was in the Middle East where the origins of cereal farming also lie in antiquity (Zohary, 1969). In its earliest forms, bread would have been very different from how we see it in industrialized countries today, and it would probably be closest in character to the modern flat breads of the Middle East. We will probably never know whether the gathering and cooking of wild grass seeds provided the spur to arable farming or whether the ability to grow and harvest the forerunners of modern wheats provided the impetus for breadmaking. Whichever way round the two events occurred, there is no doubt that one depends on the other and this simple relationship is the foundation of all modern breadmaking.

The move to improve the digestibility of the wild grass seed forerunners of early wheat types, by cooking or baking, represents a major step in the evolution of human food production. To make this step requires an appreciation, but not necessarily a scientific understanding, of the unique properties of the proteins in the grass seeds we call wheat, namely their ability to form a cohesive mass of dough once the grains have been crushed (milled) and the resultant product wetted (hydrated) and subjected to the energy of mixing, even by hand. This cohesive mass is the one we bakers call 'gluten', and once formed it has the ability to trap gases during resting (fermentation and proof) and baking and this allows the mass to expand to become a softer, lighter and even more palatable food after the final heat processing.

Another important event in the production of bread was the discovery that, if left long enough, the dough mass would increase in volume without being subjected to the high temperatures of cooking or baking. There is no doubt that the changes in the rheological character of the dough – the way in which it behaved on handling – would also have been keenly observed by those in charge of food production. The combined effect of these changes is for the subsequent baked mass to be further increased in volume and give a product with an even softer, more digestible character and different (improved?) flavour. Gradually the appreciation of the actions of wild yeasts and portions of old dough (e.g. starter doughs) were to lead to the transfer of fermentation technology from the brewing industry and eventually to the production of specialized bakers' yeasts.

Bread is a staple foodstuff and today there are few countries in the world where bread and fermented products are not made and eaten. Bread products have evolved to take many forms, each based on quite different and very distinctive characteristics. Over the centuries, craft bakers around the world have developed our traditional bread varieties using their accumulated knowledge on how to make the best use of their available raw materials to achieve the desired bread quality. Commonly this has been by adapting and changing pre-existing processing techniques and on occasions developing entirely new ones. Today, scientific study and technical development provide faster and more cost-effective ways of making bread, but even so bakers still have to use their collective knowledge, experience and craft skills to integrate the available raw materials and processing methods to satisfy their customer demands for fresh, wholesome and flavoursome fermented products. While our basic raw material, wheat, is generally much improved in quality and more consistent in performance, it is still a 'natural' material and as such is continually subjected to the influence of environmental factors during growth, at harvesting and during storage. All of these factors, and others, contribute both individually and collectively to variability in wheat during milling and flour performance during breadmaking.

In some countries the nature of breadmaking has retained its traditional form while in others it has changed dramatically. The flat breads of the Middle East and the steamed breads of China are examples of traditional bread forms which still remain an essential part of the culture of the countries where they are still produced in large quantities. On the other hand, in North America the arrival of wheats along with settlers and farmers from Western Europe was to lead eventually to production of new wheat varieties and the rapid industrialization of breadmaking in a country where the maize-based products of the native Americans had previously been the main cereal-based foods.

Today consumers are becoming increasingly cosmopolitan in their taste for bread as the influences of international travel and cultural exchange lead to a wider appreciation of the infinite variety of bread. In the UK, for instance,

Italian ciabatta, Indian chapattis and French baguette are all eaten along with UK-style sliced bread, and so for those of us who enjoy eating bread there is truly 'something for everyone'.

## 1.2.    Quality Characteristics of Bread

The product which we call bread today represents the progressive technical development and improvement of fermented wheat-based products over many thousands of years. In common with most, if not all products of modern life, the evolution of breadmaking processes has progressed further since the mid-1940s than in all of the preceding centuries and yet, because it is 'that most ancient of foods', it still evokes the most passionate of discussions about quality, taste and value for money. You have only to spend an hour or two in a room with bakers to appreciate just how emotive a subject breadmaking is, a strange mixture of craft, science, technology and to many 'love'.

The proliferation of bread varieties, a few of which are illustrated in Figure 1.1, derives from the unique properties of wheat proteins to form gluten and from the bakers' ingenuity in manipulating the gluten structures formed within the dough. The rubbery mass of gluten with its ability to deform, stretch, recover shape and trap gases is very important in the production of bread and all fermented products. Of all the cereals, wheat is almost unique in this respect. Some other cereal flours, such as those derived from rye and barley, can form gluten but

FIGURE 1.1. Bread varieties: rear left — mixed grain; rear right — sandwich; middle — baguette; front — ciabatta.

to a lesser extent than normally seen with wheat flours. It is possible to mimic some of the character of wheat-breads with products made from other cereals but if similar volume, crumb characteristics and flavour to wheat-based breads are required then any natural proteins present must be supplemented with other sources of gas-stabilizing ingredients, whether they are protein, carbohydrate or lipid based.

With such a long history of production and such diversity of form, bread-making is almost always an emotive subject. Whenever the subject of quality is raised amongst bakers and consumers, we can guarantee that there will be a diversity of opinion, with different bakers extolling the virtue of different breads, different processes, different doughmaking formulae and different ingredients. In these circumstances it is meaningless to describe bread as being 'good' or 'bad' (unless there is a genuine quality problem which renders it inedible), since the phrase 'good bread' will have a different meaning to each of us depending on our cultural background, our individual experiences and our personal likes and dislikes. For example, there are many supporters of the dense wholemeal (wholewheat) bread that one may typically make at home and they will find the large-volume, soft wholemeal of the plant bakery unacceptable, but the latter form will find greater acceptability in sandwich making for retail sale. It is interesting to note that while the quality characteristics of these two types of wholemeal bread may be very different, the nutritional and health benefits, if we ignore digestibility, remain largely the same.

We use the term 'bread' to describe a range of products of different shapes, sizes, textures, crusts, colours, softness, eating qualities and flavours. The characters of the products are diverse, and because of this the terms 'good' or 'bad' quality have no meaning, except to the individual making the assessment. A baguette is not a baguette without a crisp crust while the same crust formation would be unacceptable on North American pan bread. The fine cell structure of sandwich bread in the UK has no relevance to the flat breads of the Middle East. Clearly the 'ideal' loaf depends on who you are and where you are. Today we find bread made and eaten in parts of the world where wheat is not an indigenous crop and because of this some of the essential characters of bread have become universal, the form may be different but we all largely seek many of the same attributes from all of our fermented products.

Despite there being as many opinions on what makes 'good' bread as there are bakers and consumers, it is true to say that certain quality characteristics are required for different varieties to be acceptable to the widest cross section of consumers. For example, baguettes are characterized by a hard and crisp crust and without it we would reject the product, often describing a baguette with a soft crust as 'stale'. On the other hand, sliced pan breads in the USA, the UK and elsewhere are characterized by a thin but soft crust, and if the crust were thick and hard it would often be rejected by consumers, ironically also being described as 'stale'.

Loss of product freshness is as much about what we expect a product character to be as it is about its age since original manufacture. Whatever the criteria we use to judge bread staleness, it becomes clear that the single most common requirement of a fermented product is that it should ideally retain all of the attributes which it had when it left the oven; above all else we expect our bread to be 'fresh'. When we collect our bread from the baker and it is still warm to the touch we have no doubt as to its freshness but when we purchase it cold from the store shelf we need convincing of its freshness. The pursuit of fermented products which retain their 'oven-fresh' character for an extended period of time after they have left the oven has been one of the great challenges facing bakers, technologists and scientists for many years, and many different strategies have been evolved to meet this challenge. Whether they have been successful can really only be judged by consumers.

To be able to make our particular bread type, we must have an understanding of the complex interactions between our raw materials and the methods we will use in the conversion processes from ingredients to baked product. Our raw materials will change and our processes are time- and temperature-sensitive. Given the intricate nature of the process, it is a wonder that we manage to make bread at all. We do so because of accumulated knowledge – craft – augmented these days by scientific and technological understanding.

## 1.3.    The Character of Bread

What are the essential characters of breads? How do we distinguish them from cakes, pastries and biscuits? We have already considered that 'bread' requires wheat flour (mostly) and water to form gluten to trap the gas generated by the added yeast. We usually see at least one other ingredient used, namely salt, which is added to give more flavour to the baked product. Not that flour and water, when mixed and baked, with or without yeast, has no flavour since the very action of baking develops flavour compounds in the crust, and natural bacterial and wild yeast actions can also develop flavours within the bread crumb.

To some extent we can argue that a definition of the character of bread can be based on the ingredients used. To many, an essential difference between bread and other baked products is that cakes, some biscuits and some pastries contain sugar to confer sweetness (and other less obvious properties), but in the USA bread commonly contains added sugar, as do the sweet breads of India. In addition, where would we place fermented buns and rolls? The other ingredient commonly present in relatively high proportions in cakes, pastries and biscuits is fat. Most breads contain much lower proportions of fat than cakes, but where then do we fit fermented products such as croissant?

The fact is that we can never form a concise definition of bread since it is characterized by all of the ingredients we use to make it, and more besides. An essential component of bread is the formation of gluten, a process which does not occur in cakes to any significant degree and indeed is actively discouraged by the addition of sugar and, to a lesser extent, fat and the use of high level of water addition. Most biscuits and pastries also have limited gluten formation by comparison with most bread products. In laminated products, however, gluten formation is encouraged and bakers have evolved a specialized layering technique to allow the incorporation of fat for the modification of texture and eating quality but without much disruption of the gluten formation, and so we can say that in the case of croissant and Danish pastries, the addition of yeast places the products in the fermented goods category alongside bread.

In seeking a definition of bread we should not overlook the contribution of the water. Its addition in the 'right' quantity is essential for the formation of gluten and for modifying the rheology of the dough. The level of moisture remaining in the baked product is also a major contributor to the characters of breads. Too much or too little added water during mixing means that we cannot form the 'right' gluten qualities to trap the gases from yeast fermentation. Too little remaining in the baked product will result in the eating character being closer to that of pastries and biscuits.

The character of bread and other fermented products then depends very heavily on the formation of a gluten network in the dough, not just for trapping gas from yeast fermentation but also to make a direct contribution to the formation of a cellular crumb structure which after baking confers texture and eating qualities quite different from other baked products. Look closely at the crumb structures of most baked breads and you will see that the common linking theme is that they are formed of holes of differing shapes, sizes and distributions, each hole being embraced by a network of connected strands, coagulated gluten, in which starch granules and bran particles are firmly embedded. When this crumb is subjected to pressure with the fingers it deforms and when the force is removed, it springs back to assume its original shape, at least when the product is fresh. This combination of a cellular crumb with the ability to recover after being compressed largely distinguishes breads from other baked products and these are the very characteristics that bakers seek to achieve in most bread products.

## 1.4.  Bread Flavour

Nothing will provoke more debate in discussions on bread characteristics than that related to the flavour of fermented products. The judgement of what consti- tutes the 'right' flavour is another highly personal and emotionally charged issue. Sometimes bread products are eaten alone, but more often they will be eaten as an accompaniment to other foods in a meal or as part of a composite product, so that bread flavours tend to be more subtle than we would encounter in many other foods.

The development of flavour in fermented products comes from a number of different sources and includes contributions from the ingredients and the processing methods which are used. Many of the ingredients which are used in the production of fermented products make a significant contribution to the flavour of the product. Flour tends to have a fairly bland flavour with most of its contribution coming from the oils of the germ (embryo) and bran particles present. Since this is the case, we can reasonably expect that wholemeal, whole-wheat and bran- and germ-enriched white flours will yield bread with more flavour than white flours.

Breadmaking around the world has evolved many dough formulations which use ingredients to confer special flavours which have now become an essential part of that product character. The addition of salt (sodium chloride) to bread is the most obvious of those flavour modifiers, imparting both its own characteristic 'salty' taste and working in the mouth to increase our perception of other flavours which may be present. Since salt levels vary in many products so will our perception of flavour between products. Other common additions include fat, sugar, milk and malt products, each contributing its own special flavour. The level of yeast used in the recipe also makes its own unique contribution to bread flavour.

During the natural fermentation processes which occur in breadmaking, new flavour products are generated within the dough (Wirtz, 2003). Both the intensity of those flavours and the particular flavour 'notes' which are developed change with increasing fermentation time. The most commonly observed flavour changes are those associated with the development of acid flavours from microbial activity in the dough, which are readily detected in the flavour of the bread crumb. Not all of this flavour activity will come from the addition of bakers' yeast; some will come from wild yeasts and bacteria, especially lactic acid bacteria, which are present naturally in the flour. Usually several hours of fermentation are required before there are significant changes to the flavour profile of the bread crumb. Where the breadmaking process being used has no provision for lengthy fermentation times, it is often the practice to develop flavour in a 'pre-ferment', 'brew' or 'sponge' which is later mixed with the remaining ingredients to form the dough for final processing.

By far the most important contribution to bread flavour comes from the process of baking. During this heat-setting stage many of the flavour compounds present undergo major changes; some old ones are lost and many new ones are formed. We most readily see this phenomenon in the formation of a dark, mostly brown crust on the outer surfaces of the dough. These changes are associated with the complex processes commonly referred to as 'Maillard browning' and many of the compounds are highly flavoured. These compounds are very important to our perception of flavour in many baked foods, and views have been expressed that as much as 80% of bread flavour is derived from the product crust. In an interesting parallel in the bakery world, we have seen the introduction of the so-called 'high-bake' water biscuit, confirmation of the contribution that browning of the crust makes to flavour and quality.

The perception of flavour in bread is, then, no simple matter. It will, for example, be strongly influenced by the ratio of crust to crumb. The development of Maillard browning products during crust formation may help us understand why products as different as UK sandwich bread and French baguette have different flavour profiles. In the case of baguette, the proportion of crust to crumb is much higher, so that we will have a larger quantity of compounds which contribute to product flavour. The lower proportion of flavour compounds in UK sandwich bread may be seen by some as being to its detriment, but then the character of the bread is aimed at a completely different end use to that of the baguette.

## 1.5.   Bread Types

The development of particular types of bread has taken different directions in different parts of the world. As a consequence, not all of the terms used to describe breads and their quality attributes have the same meaning in all parts of the bakery world and this sometimes leads to misunderstandings between individuals regarding bread quality. The most obvious of examples are references to French breads which in the UK tend to be used to refer to baguette styles while in the USA may refer to pan breads with lean (i.e. low-fat) formulae, such as might be used in the production of French toast. In France the concept of 'French' bread as viewed in the UK would not be widely understood. Similarly, the term 'toast breads' will have different meanings in France, the USA and the UK. We must recognize that the use of bread-quality descriptors and product terminology will be strongly linked with local consumer preferences and traditions, even though in general the terms used to describe the quality attributes of fermented products can mostly be related to the categories of external and internal characters, eating quality and flavour as discussed below.

The characterization of a particular bread type will always include a description of its physical appearance, usually starting with its external form. Thus baguettes are likely to be described by their length and diameter, other bread forms by their pan shape, Middle Eastern and traditional Indian breads as 'flat' (Figure 1.2) and so on. Even markings on the surface may require definition, in the way that the number and direction of cuts on the dough surface may become an integral part of the traditional product character. A comment on the product crust colour may be included. Almost certainly a description of the interior appearance follows with references to the sizes, number and distribution of holes in the crumb and the colour of that crumb. Comments on crust hardness and the eating qualities of the crumb will almost certainly be made and there is likely to be some reference to the flavours present. To help with the characterization there may be references by the describer to other bread types, such as 'flatter than...' or 'with more holes than...'.

How often do we take a few moments to ask ourselves fundamental questions about the character of bread products such as 'Why is a baguette the length and shape it is'? The very attributes which characterize baguette or any other particular type of bread were identified many years ago and have become enshrined in our perception of the required product quality for a given loaf. For example, whoever heard of a baguette baked in a round pan? There is no reason why we should not make pan breads with crisp crusts and open-cell structures like those we normally see in baguette, but even if we do it is most

FIGURE 1.2. Naan bread from India.

unlikely that we would try to call it a baguette because most of our customers would not recognize it as such; essentially they would not view the product as 'authentic'. In discussing the technology of breadmaking we must recognize that most bread types are the product of long-forgotten traditions rather than the systematic development to give a specific product character.

Reference has already been made to the techniques of cutting the dough surface before baking. Such markings have become part of the traditional character of many breads, like baguette, coburgs and cottage loaves (in the UK). As well as providing a distinctive appearance they also play a significant role in forming many aspects of product character. In some products the cutting of the surface exposes a greater surface area to the heat of the prover and the oven and thereby improves expansion in both of these stages. In the oven the product may expand in a more controlled manner if the surface has been cut, and this may improve the overall product shape. This effect is readily observed in the inclination for the upper part of cottage loaves to become displaced during baking if the lower part has not been cut correctly before entering the oven.

The depth and direction of cutting contributes to the expansion and ultimate volume of a product, an effect best illustrated by once again considering the baguette. The 'traditional' French baguette is given a fixed number of shallow cuts, largely following the length of the dough piece (Figure 1.3), while in other parts of the world more numerous cuts may be made across the breadth of the dough piece. The first technique encourages the retention of much of the gas in the dough and confers greater expansion during baking, while the latter technique

FIGURE 1.3. Surface cuts on baguette.

is more readily inclined to release gas during baking and give a smaller volume in the final product, but of course calls for less skill on the part of the baker.

Such comments are not intended to denigrate or glorify any particular bread product, rather to show the traditional basis of the many bread types we encounter. There are a few relatively modern bread developments, but most of our bread types have a long period of development. However diverse our bread types are, they share a number of common elements largely based on the formation of gluten as discussed above and in more detail in later chapters. There are two main elements of bread: the crust and the crumb. The physical form of both crust and crumb and the ratio of one to the other are the very essence of what distinguishes one type of bread from another. The size and the shape of the loaf, together with the ingredients used, contribute to the overall quality but arguably such contributions are of lesser importance to bread character than those made by the nature and proportions of the crumb and the crust.

While we as scientists, technologists, bakers and consumers can earnestly debate the relative merits of crisp and soft bread crusts, we should not lose sight of the contribution that crumb structure makes towards these aspects of bread quality. The crisp baguette crust forms in part because of the open-cell structure created in the dough during processing and baking, while a fine, uniform cell structure is essential for the pan breads and must be created at the beginning of the process, principally in the mixer. Achieving the 'right' bread quality calls for the creation and control of gas bubbles in the doughmaking, moulding, proving and baking processes. These bubble structures eventually become the bread cell structure when the dough mass is heat set during baking.

The many different bread types which have been evolved with the passage of time all require their own individual bubble structure, processing techniques, processing equipment and process control mechanisms. It is an understanding of each of these factors against a background of an ever-changing raw material quality which allows bakers to maintain traditional bread qualities and to develop new ones in a changing market place. The challenges are many but the manufacture of 'good' bread still remains a pleasure for those of us involved in its production.

## 1.6.   Assessing Bread Quality

The process by which bread quality is determined still relies to a significant extent on subjective assessments by experts because of the difficulties associated with objective measurements of some highly 'personal' characters in breads. The most obvious examples of the assessment problems we face are those characters related to flavour and eating quality because of the diverse preferences of individual consumers. Nevertheless if we as bakers, technologists and scientists are to be in a position to assess the effects of new ingredients and processing methods, to match more closely bread quality with consumer requirements or to reduce product variability and limit quality defects, then we must have some basis on which to make our quality judgements. To state simply that a particular change of formulation has 'improved' bread quality is inadequate for others to judge the success of our efforts or for us to make longer-term assessments. Therefore we need to have objective criteria and in cases where this is not possible we need to standardize as much as possible the methods we use for our subjective assessments.

Various scoring techniques are usually employed to try and standardize subjective assessment (e.g. Kulp, 1991). The attribution of particular numbers will be based on individual requirements. Photographs or diagrams with attribute scores are often used as the basis for comparison between test and standard by the observer. In this way uniformity of scoring is improved.

The techniques for assessing bread quality usually fit into three broad categories: external, internal and texture/eating quality, which includes flavour.

### 1.6.1.   External Character

Among the characters we most often assess under this heading are product dimensions, volume, appearance, colour and crust formation.

The critical dimensions for most breads are their length and height, with breadth being of lesser importance. A large number of bread types are characterized by their length, for example baguette which should be 70 mm long in France (Collins, 1978). Devices for measuring product dimensions off-line can be simple and include graduated rulers and tapes. It is possible to measure product height and shape on-line using image analysis techniques. Measurement

FIGURE 1.4. Pan breads.

of height will often be used together with width (breadth) as a basis for an estimation of volume where the product shape makes such estimates meaningful, for example with rectangular pan breads (Figure 1.4).

The most common method of assessing whole product volume is by using a suitable seed displacement method. The apparatus concerned usually comprises a container of known volume, which has previously been calibrated with a suitable seed, usually rape seed or pearl barley (Cornford, 1969), into which the product is introduced. The seed is reintroduced and the product displaces a volume of seed equivalent to its own volume. It is important to keep such apparatus regularly calibrated with suitable 'dummy' products of known volume since the bulk density of seeds may change with time because of frictional attrition. More recently, infrared scanning of products has been introduced as a means of assessing product volume, for example the TexVol instruments (BVM-L series, www. Texvol.com).

Image analysis techniques have been applied to the measurement of product volume. One such method is based on the measurement of the cross-sectional areas of several slices from selected places along the length of the loaf and then integrating these with the known length of the product. This method may be preferable where uncut loaves are not available and samples are being taken straight from the production line.

The external appearance of the product will often be a major factor which attracts the eye of the consumer. To this end, cutting or marking of the surface, both in terms of the number and direction of the cuts, must be consistent with the product 'norm'. Any quality assessment of this character can be carried out by comparing the product with a standard illustration of the accepted product norm.

Often the contrast between the darker crust and the lighter areas which expand later in baking are seen as desirable quality attributes. This particular aspect is best seen with baguette and other cut and crusty breads. This expansion during baking is often referred to as 'oven spring' or 'oven jump', and in most products should be controlled and uniform, uncontrolled oven spring often being seen as a quality defect (e.g. 'flying' tops or 'capping' in pan breads). With pan breads, oven spring can be calculated directly by measuring dough piece height before entering the oven and loaf height after leaving the oven. In other products a subjective judgement will be used. A subjective judgment on the evenness of the break or 'shred' may also be made.

Crust colour is commonly assessed using descriptive techniques. Objective methods can be used based on comparison with standard colour charts, such as the Munsell system (Munsell, undated) or direct measurement with tristimulus type instruments (Anderson, 1995), but commonly crust colour is uneven on bread surfaces and this reduces the effectiveness of such measurements. The presence of unwanted surface blemishes may be noted and included as part of the loaf quality score.

The desired crust character varies with the product and is most often assessed subjectively. Instrumental methods based on puncture or snapping may be used, but are not common because the measurement so obtained will be strongly influenced by the complex architectures of the products.

## 1.6.2.  *Internal Character*

Our major concerns with internal character are normally limited to the sizes, number and distribution of cells in the crumb (crumb grain), the crumb colour and any major quality defects, such as unwanted holes or dense patches, visible in a cross section of the product. As discussed above, each bread type has its own special cell structure requirements and therefore there is no single standard which can be applied to all products. Because of this, subjective assessment of product crumb cell structure is still the most common method being used with some form of standard reference material, such as photographs. The assessment of crumb cell structure may well also include an evaluation of the thickness of cell wall material.

While being among the most important of bread characters, crumb cell structure remains the most difficult to quantify in a way which correlates well with the human perception of quality. Techniques employing photographs, or scanning with video cameras coupled with image analysis, may be used to provide more quantitative data on the cell structure and have been correlated with expert assessors (Jian and Coles, 1993; Rogers et al., 1995; C-Cell from Calibre Control International, *www.C-Cell.info*). The assessment of crumb colour may be subjective and descriptive or carried out by objective means using tristimulus colorimeters (Cauvain et al., 1983).

## *1.6.3. Texture/Eating Quality and Flavour*

Texture and eating quality are important properties of bread products and are different from one another. In assessing the texture of bread crumb we are concerned with its mechanical properties such as firmness and resiliency, and often we try to relate such properties to eating qualities by adaptation of more fundamental physical testing methods.

Crumb softness or firmness is the texture property which has attracted most attention in bread assessment because of its close association with human perception of freshness. In the subjective 'squeeze' test we simultaneously and subconsciously measure a number of product properties. The most obvious of these is the resistance of the product to deformation. Less obvious are how much the product recovers after the deforming force is removed and how much force we need to apply to compress the product to our 'standard squeeze'. Collectively, our subjective assessment will be recorded in degrees of softness or hardness, and sometimes as recovery or springiness.

There are numerous objective methods which have been applied to the assessment of bread texture (Cauvain, 1991). Compression tests form the bulk of such testing methods and mainly cover the measurement of softness or firmness (AACC, 1995a, b) and recovery or resiliency (Cornford, 1969). Usually force levels in such tests are insufficient to rupture the crumb matrix to any significant degree and the product will largely return to its uncompressed form after removal of the compressing force. Such tests mimic to a large extent the subjective assessment of crumb texture made by the fingers during squeezing or compression.

In the process of eating, the forces in the mouth will be large enough to cause irreversible changes to crumb properties. In addition to the strong compression and shearing forces, the action of saliva and the mechanical effects of chewing patterns form the basis of a method of assessment very different to that applied with the fingers and hand in the squeeze test. Subjective assessments of eating quality are carried out by individuals or groups with scoring based on descriptive terms (Amerine et al., 1965). The application of taste panel techniques for assessing eating quality has received much attention and details can be found in many standard works (e.g. Stetser, 1993). Objective methods for assessing eating quality in the mouth are clearly difficult, although electromyography (Eves and Kilcast, 1988) has been tried with some foods.

Texture profile analysis (TPA) is one technique which tries to use a common basis for both subjective and objective assessments of product eating qualities. TPA uses seven basic descriptors for eating quality derived from subjective assessment of a wide range of foods (Szczesniak, 1963a, b) and which have been subsequently translated to permit objective instrumental measurements (Szczesniak, 1972; Bourne, 1978). The objective technique uses two successive compressions on the same sample with forces which cause some irreversible changes to the food being tested. TPA has been applied to the assessment of changes in bread eating quality with varying degrees of success (Cauvain and Mitchell, 1986).

Evaluation of the flavour in bread products relies entirely on subjective assessments by individuals or groups (Stone and Sidel, 1985). The flavour of both crust and crumb may be assessed separately. Standard descriptors for flavour are commonly used but the provision of a standard against which to make such subjective flavour judgments will be difficult and it will be necessary to rely on comparisons or 'flavour memory', even with trained panels and agreed descriptors.

## 1.7.  Nutritional Qualities of Bread and its Consumption

Bread and other cereal-based products have become 'staple' foods throughout the world and are now established as an integral part of many modern diets. The nutritional qualities of cereals are well established, with most of the nutritional input from this category coming from wheat-based products. Although there will be some small changes in the nutritional qualities as a result of the milling and baking processes, wheat-based breads continue to provide significant sources of protein, complex carbohydrates (mainly starch), fibre, vitamins and minerals. The nutritional contributions are greatest in wholemeal (wholewheat) breads since they require conversion of 100% of the grain into flour. In lower-extraction white flours the removal of some of the bran and germ components from the wheat grain changes the overall nutritional qualities of the resultant product, although in spite of this, white breads continue to make significant contributions to the diet. Typical nutritional compositions for UK breads are given in Table 1.1.

Because of the significant role that bread products play in contributing nutrition to the diet, it has become the practice in many countries, for example in the USA and in the UK, to enrich white and some other flours with additional nutrients (Ranhotra, 1991; Rosell, 2003). The enrichment mainly comprises the addition of calcium, in some form, some of the essential vitamins and a readily assimilated form of iron, although other additions may be possible depending on the regulatory conditions in particular countries (e.g. folic acid). The other ingredients used in breadmaking formulations will also make contributions to the nutritional qualities of the products. Some, such as fat and to a lesser extent sugar, may not be considered to be adding to the nutritional value of

TABLE 1.1. Composition of bread (per 100g).

|               | White | Brown[a] | Wholemeal |
|---------------|-------|----------|-----------|
| Carbohydrate  | 49.3  | 44.3     | 41.6      |
| Protein       | 8.4   | 8.5      | 9.2       |
| Dietary fibre | 2.7   | 4.7      | 7.1       |
| Fat           | 1.9   | 2.0      | 2.5       |

[a] In the UK the term 'brown' denotes bread made from flour which consists of a mixture of white flour and a proportion of the bran component of the wheat.

the food, but in most bread products their levels of addition are modest and so the overall nutritional contribution is small. In some quarters the reduction of bread sodium levels has been advocated, the sodium coming mainly from the addition of salt (sodium chloride) which is used to improve the palatability of bread.

The level of bread consumption varies around the world and from region to region. In the so-called 'Western' diets of Europe and North America, bread consumption has been declining but not uniformly with bread type or region. The per capita consumption figures for bread products are published for many countries and tend to show the same gradual decline in consumption, even for traditionally 'large' consumers of bread product, such as Germany and France. In some countries, for example the UK, consumption statistics suggest that the rate of decline is slowing. A comparison of bread consumption in the UK for the years 1980–1995 is given in Table 1.2. The consumption of bread and other cereal-based products continues to make a significant contribution to human dietary needs throughout the world, and so its continued or increased consumption should be encouraged.

Increasingly the types of bread being manufactured and eaten are influenced by the medical professionals and nutrition lists. For example, wholemeal breads, including rye, are seen as a 'natural' source of folates in the diet while in other cases the enrichment of bread with selenium is seen as having positive health benefits. The latter arises because wheats grown in some parts of the world, for example in Canada, are naturally higher in selenium and much of the mineral finds its way through to the final flour. Thus in some geographical areas where the local soils are low in selenium, there is interest in the potential for fortification, whether at a voluntary or compulsory level. The decision to legislate for compulsory fortification of bread with any material is seldom straightforward. In part this arises from the emotive nature of the consumer to bread which is seen as naturally 'wholemesome' and should not therefore have lots of additives, however apparently beneficial they might be. The so-called 'food-journalists' in the 'popular' press in some parts of the world are never slow to draw attention to the long list of 'additives' used in 'mass produced' bread, and all the negative connotations associated with them without considering positive health benefits

TABLE 1.2. UK household consumption of bread (kg per capita per annum).

| Calendar year | Kg |
|:---:|:---:|
| 1985 | 45.86 |
| 1990 | 45.70 |
| 1995 | 39.36 |
| 2000 | 37.44 |

Source: UK National Food survey 2000; MAFF, 2001.

or that legislation compels millers and bakers to make some additions to the flour and bread. The other problem with additions of ingredients to promote health is that in many cases the medical benefits are only applicable to some sectors of consumers.

The baking industry has made positive reactions to consumer concerns to *trans* fats in the diet following the release of medical evidence which suggests that they should be reduced or even eliminated in the diet. Many bread products already have low levels of added fat in the recipe and in many cases the *trans* fat component has been eliminated from the formulation. Coupled with the reduction of *trans* fat in the diet has been interest in supplementing breads with *omega* 3 rich oils. The key to making changes to the levels and types of fats added in breadmaking is an understanding of their technical function and this is discussed in Chapter 3.

Organic farming and the sale of its products is a developing market and includes wheat-based products. The technology of bread production using organic wheats and flours remains very similar to that of the 'non-organic' equivalent though some adjustments to the formulation and processing methods may be required in order to achieve acceptable product quality. There is no consensus view on the nutritional or health benefits of organic products.

The concept of increasing fibre in the diet by consuming cereal-based foods has been in 'vogue' for many years and certainly influenced the move in some countries from white to wholemeal bread consumption. Increasing the consumption of fibre and whole grains is seen as a means of reducing a range of medical conditions including coronary heart disease, some forms of cancer and type 2 diabetes (Miller Jones, 2005). Increased whole grain consumption is seen as a positive contributory to reducing obesity. In this respect the increased consumption of bread flies in the face of the dieting trend of the 1980s and 1990s – the Atkins Diet which advocated a reduction in the consumption of carbohydrates. This particular dietary trend has passed but not without its impact on the bread industry, especially in the USA where bread consumption fell dramatically while the diet was in vogue. The knock-on effect of the Atkins diet was to generate a greater focus on the role of carbohydrates in the diet and along with it the role of bread in the diet. Attention has now shifted to the concept of the effect of carbohydrate on blood sugar levels and the potential impact on diabetes and obesity. The concepts are covered by the terms *Glycemic Index* (GI) and *Glycemic Load*. Much research is now focused on GI (Brennan et al., 2005) and new products have already reached market places around the world but it is too early to say if the introduction of the GI concept and these new products will result in an increase in the consumption of bread. While consumers require food to be healthy, it is an interesting question as to whether they will see breads as being a means to deliver 'medicines'. For the baker the challenge is to provide not only a nutritious and safe food but one which consumers will readily choose and enjoy. If bread is to be seen as a means of delivering 'positive' health benefits then the first task is to get consumers to purchase and eat the product.

## 1.8.   Conclusions

The manufacture and consumption of bread is global and products are even made for consumption in space. The use of wheat and other grains for the production of bread has a long history in many parts of the world. In some countries the use of wheat to provide a basic and staple food has a shorter history but has become quickly and clearly established. With a long history of production in diverse cultures, many different types of breads have been evolved and new variations continue to be developed to meet consumer demands for more varied and nutritious foods.

Bakers have developed an almost infinite variety of breads and production methods to meet consumers' needs and must achieve consistent product quality to maintain their markets in the face of increasing competition from fellow bakers and other forms of 'convenience' foods. In real terms, bread has always been a convenience product or part of one for the consumer. Because of the diverse nature of bread there is no right or wrong for product quality as each bread type requires its own suite of attributes to be authentic. Having identified the quality of the product which will most satisfy their customers' needs the challenge for bakers is to use their knowledge of their raw materials and processing methods to achieve that quality and to do so consistently.

## *References*

AACC (1995a) Method 74–09, Bread firmness by Universal Testing Machine, in *Approved Methods of the AACC*, 9th edn, Vol. II, American Association of Cereal Chemists, St. Paul, MN, USA.

AACC (1995b) Method 74–10, Staleness of bread, Compression test with Baker Compressimeter, in *Approved Methods of the AACC*, 9th edn, Vol. II, American Association of Cereal Chemists, St. Paul, MN, USA.

Amerine, M.A., Pangborn, R.M. and Roessler, E.B. (1965) *Principles of Sensory Evaluation of Food*, Academic Press, London.

Anderson, J. (1995) Crust colour assessment of bakery products. *AIB Technical Bulletin* XVIII, Issue 3, March.

Bourne, M.C. (1978) Texture profile analysis. *Food Technology*, **32**, July, 62–6, 72.

Brennan, C.S., Symons, L.J. and Tudorica, C.M. (2005) Low GI cereal foods: the role of dietary fibre and foods structure, in *Using Cereal Science and Technology for the Benefit of Consumers* (eds S.P. Cauvain, S.E. Salmon and L.S. Young), Woodhead Publishing Ltd., Cambridge, UK, pp. 95–101.

Cauvain, S.P. (1991) Evaluating the texture of baked products. *South African Food Review*, **22**(2), April/May, 51, 53.

Cauvain, S.P. and Mitchell, T.J. (1986) Effects of gluten and fungal alpha-amylase on CBP bread crumb properties. *FMBRA Report No. 134*, December, CCFRA, Chipping Campden, UK.

Cauvain, S.P., Chamberlain, N., Collins, T.H. and Davies, J.A. (1983) The distribution of dietary fibre and baking quality among mill fractions of CBP flour. *FMBRA Report No. 105*, July, CCFRA, Chipping Campden, UK.

Collins, T.H. (1978) The French baguette and pain Parisien. *FMBRA Bulletin No. 3*, June, 107–16.

Cornford, S.J. (1969) Volume and crumb firmness measurements in bread and cake. *FMBRA Report No. 25*, May, CCFRA, Chipping Campden, UK.

Eves, A. and Kilcast, D. (1988) Assessing food texture. *Food Processing*, **57**, January, 23–4.

Jian, W.J. and Coles, G.D. (1993) Image processing methods for determining bread texture, in *Image and Vision Computing NZ '93* (eds C. Bowman, R. Clist, O. Olsson and M. Rygol), Industrial Research Ltd., New Zealand, pp. 125–32.

Kulp, K. (1991) Breads and yeast-leavened bakery foods, in *Handbook of Cereal Science and Technology* (eds K.J. Lorenz and K. Kulp), Marcel Dekker, New York, pp. 639–82.

Miller Jones, J. (2005) Fibre and whole grains and their role in disease prevention, in *Using Cereal Science and Technology for the Benefit of Consumers* (eds S.P. Cauvain, S.E. Salmon and L.S. Young), Woodhead Publishing Ltd., Cambridge, UK, pp.110–17.

Ministry of Agriculture, Fisheries and Food (2001) *Household Consumption and Expenditure 2000*, Annual Report of the National Food Survey Committee, HMSO, London.

Munsell, A.H. (undated) *Munsell System of Colour Notation*, Macbeth, Baltimore, USA.

Ranhotra, G.S. (1991) Nutritional quality of cereals and cereal-based foods, in *Handbook of Cereal Science and Technology* (eds K.L. Lorenz and K. Kulp), Marcel Dekker, New York, pp. 845–62.

Rosell, C.M. (2003) The nutritional enhancement of wheat flour, in *Bread Making; Improving Quality* (ed. S.P. Cauvain), Woodhead Publishing Ltd., Cambridge, pp 253–70.

Rogers, D.E., Day, D.D. and Olewnik, M.C. (1995) Development of an objective crumb-grain measurement. *Cereal Foods World*, **40**(7), 498–501.

Stetser, C.S. (1993) Sensory evaluation, in *Advances in Breadmaking Technology* (eds B.J. Kamel and C.E. Stauffer), Chapman & Hall, London.

Stone, H. and Sidel, J.L. (1985) *Sensory Evaluation Practices*, Academic Press, Orlando, USA.

Szczesniak, A.S. (1963a) Classification of textural characteristics. *Journal of Food Science*, **28**, July–August, 385–89.

Szczesniak, A.S. (1963b) Objective measurement of food texture. *Journal of Food Science*, **28**, July–August, 410–20.

Szczesniak, A.S. (1972) Instrumental methods of texture measurement. *Food Technology*, **26**, January, 50–6, 63.

Wirtz, R.L. (2003) Improving the taste of bread, in *Bread Making; Improving Quality* (ed. S.P. Cauvain), Woodhead Publishing Ltd., Cambridge, UK, pp. 467–86.

Zohary, D. (1969) The progenitors of wheat and barley in relation to domestication and agricultural dispersion on the Old World, in *The Domestication and Exploitation of Plants and Animals* (eds P.J. Ucko and G.W. Dimbleby), Duckworth, London, pp. 47–66.

# 2
# Breadmaking Processes

Stanley P. Cauvain

In the same way that different bread varieties have evolved with the passage of time so have different methods which allow the conversion of flour and other ingredients into bread. In many cases the relationship between product and process is so strong that it may be wrong to consider them as separate issues. Just as there is no 'ideal' product so there is no 'ideal' breadmaking process. In reality each baker uses a breadmaking process which is unique, in that the combinations of ingredient qualities, formulations, processing conditions and equipment reflect the qualities of the products he or she is seeking to achieve. In practice the variations in such breadmaking processes are very small and usually consist of minor variations about a central 'standard' process, so that we are able to group many of the variations into a small number of more generic processes in order to consider the changes which occur within them and their contribution to final product quality.

## 2.1.  Functions of the Breadmaking Process

All of the processes which have evolved for the manufacture of bread have a single, common aim, namely to convert wheat flour into an aerated and palatable food. In achieving this conversion there are a number of largely common steps which are used.

- The mixing of flour (mainly wheat) and water, together with yeast and salt, and other specified ingredients in appropriate ratios.
- The development of a gluten structure (hydrated proteins) in the dough through the application of energy during mixing, often referred to as 'kneading'.
- The incorporation of air bubbles within the dough during mixing.
- The continued 'development' of the gluten structure created as the result of kneading, in order to modify the rheological properties of the dough and to improve its ability to expand when gas pressures increase because of the generation of carbon dioxide gas in the fermenting dough. This stage of dough development may also be referred to as 'ripening' or 'maturing' of the dough.

- The creation or modification of particular flavour compounds in the dough.
- The subdivision of the dough mass into unit pieces.
- A preliminary modification of the shape of the divided dough pieces.
- A short delay in processing to modify further the physical and rheological properties of the dough pieces.
- The shaping of the dough pieces to achieve their required configurations.
- The fermentation and expansion of the shaped dough pieces during 'proof'.[1]
- Further expansion of the dough pieces and fixation of the final bread structure during baking.

The main differences between individual or groups of breadmaking processes are usually associated with mixing and kneading, air incorporation, and the creation and development of the gluten structure; in summary, all of those operations which in practice deal with the formation of a large dough bulk. The subdivision of the bulk dough and the processing stages for individual dough pieces do contribute to the modification of product quality but tend to build on the dough development created before subdivision of the bulk dough. The processing stages at the end of the sequence, proving and baking, are common to most breadmaking processes and differences between individual bakeries tend to be in the type of equipment used and small variations in conditions which are applied in the bakery equipment, for example time and temperature.

Dough development is a relatively undefined term which covers a number of complex changes in bread ingredients which are set in motion when the ingredients first become mixed. The changes are associated first with the formation of gluten, which requires both the hydration of the proteins in the flour and the application of energy through the process of kneading. The role of energy in the formation of gluten is not always fully appreciated and is often erroneously associated with particular breadmaking processes, especially those which employ higher speed mixers and short processing methods (e.g. the Chorleywood Bread Process).

---

[1] The terminology used in baking can be confusing for many. This is especially true of two terms: fermentation and proof. Fermentation refers to the action of baker's yeast in the dough on the sugars which are present with the subsequent evolution of carbon dioxide gas and small quantities of alcohol. Fermentation will occur in the dough whenever the conditions are 'right' for the yeast (mainly the availability of food and an appropriate temperature). There are two main times when fermentation occurs: after the dough has been mixed and before it has been divided into unit pieces and after the dough has been finally shaped and before it enters the oven. The former is most commonly referred to as 'bulk fermentation' or simply fermentation, while the latter is most commonly referred to as 'proof'. This change in terminology logically allows the baker to understand which part of the process the dough has reached. In both cases 'fermentation' in the true sense occurs though the temperatures at which the stages carried out are different with proof commonly being carried out at a higher temperature. A further complication is that bakers may refer to 'first' or 'intermediate' proof to define a short rest period which occurs after first moulding and before final moulding.

Anyone doubting the validity of the need for energy in gluten formation should try a simple experiment which involves placing flour, water, yeast and salt together on a table and waiting for the gluten to form. They should then be encouraged to begin hand mixing of the ingredients to experience the transformation in the mixture which will occur. The best results in terms of improved bread volume and crumb softness will be achieved with vigorous and prolonged hand mixing and kneading. During the process of kneading, the dough, and more probably the 'baker', becomes warmer as energy is imparted to the dough. However, there is more to dough development than a simple kneading process.

In the process of developing a bread dough we bring about changes in the physical properties of the dough and in particular we improve its ability to retain the carbon dioxide gas which will later be generated by yeast fermentation. This improvement in gas retention ability is particularly important when the dough pieces reach the oven. In the early stages of baking, before the dough has set, yeast activity is at its greatest, and large quantities of carbon dioxide gas are being generated and released from solution in the aqueous phase of the dough. If the dough pieces are to continue to expand at this time, then the dough must be able to retain a large quantity of that gas being generated, and it can only do this if we have created a gluten structure with the correct physical properties.

Four physical properties of dough will concern us in breadmaking: resistance to deformation, extensibility, elasticity and stickiness. We can use the analogy of an elastic band to help understand the first three of these properties. When we stretch the band in our hands a degree of force is required to change its shape as it resists deformation. If we apply only a modest force and release one end of the band then, because it is an elastic material, it returns to its original shape. If we once again stretch the band and continue to apply force without releasing it we will eventually reach a point of extension when the band snaps, which we could take as a measure of its extensibility. The fourth physical property, stickiness, is largely self-explanatory. After some materials have been compressed they will stick to the surfaces within which they are in contact, so that when the direction of the compressing force is reversed they exert an adhesion force before parting from the surfaces concerned.

In practical doughmaking and dough handling, the distinction between one physical property and another is often hard to make because of the interaction of the dough with processing methods, such as moulding, as will be discussed in Chapter 4. During processing the dough is subjected to different magnitudes of force and those forces are applied at different rates. At the same time, the dough is subject to physical change as a result of chemical actions at the molecular level, along with further effects from the physical forces associated with gas production.

We can therefore see dough development as being the modification of some very important physical properties of bread doughs which make major contributions to the character of the final product. This modification of gluten structure can be achieved by a number of different physical and chemical processes, and various combinations of these form the basis of the different groups of breadmaking processes which are in common use.

Most of the desirable changes resulting from 'optimum' dough development, whatever the breadmaking process, are related to the ability of the dough to retain gas bubbles (air) and permit the uniform expansion of the dough piece under the influence of carbon dioxide gas from yeast fermentation during proof and baking. The creation of dough with a more extensible character is especially important for improved gas retention, while reductions in dough resistance and elasticity play a major role in the modification of bubble structures during processing, as will be discussed in more detail for some of the breadmaking process groups described below.

It is important to distinguish between gas production and gas retention in fermented doughs. Gas production refers to the generation of carbon dioxide gas as a natural consequence of yeast fermentation. Provided the yeast cells in the dough remain viable (alive) and sufficient substrate (food) for the yeast is available, gas production will continue, but expansion of the dough can only occur if that carbon dioxide gas is retained in the dough. Not all of the gas generated during the processing, proof and baking will be retained within the dough before it finally sets in the oven. The proportion that will be retained depends on the development of a suitable gluten matrix within which the expanding gas can be held. Gas retention in dough is therefore closely linked with the degree of dough development which occurs, and as such will be affected by a large number of ingredients and processing parameters which are not necessarily independent of one another.

A further distinction should be made between dough development and gas retention and the factors which affect both. Dough development is a poorly defined term used by bakers to indicate when they believe that the dough has all of the necessary physico-chemical properties it needs to deliver the required bread characteristics. This precise combination of properties varies with the breadmaking method employed, the equipment and the product type but is essentially based on the development and modification of the gluten network in the dough. Since gluten is the essential 'ingredient' in achieving the required dough development, it follows that the flour is the primary 'building block' in that process. Improved dough development leads to improved gas retention which is manifest through increased bread volume, crumb softness and changes in crumb texture.

The ingredient and process interactions which affect the formation of the gluten network in the dough will be discussed in detail in the following sections and chapters. It is recognised that the addition of many ingredients will impact on the gas retention properties of the dough and lead to increased product volume (and softness). However, not all of the additions which improve dough gas retention impact on product cell structure and this is because some of them do not make a direct contribution to dough development. Recent studies (Millar et al., 2005) of dough mixing using near infra red (NIR) technologies have shown no changes in NIR 'optimum' mixing (dough development) time with changing levels of ingredients such as emulsifiers and enzymes. The precise relationship between dough development and gas retention is not clear, but it is clearly wrong to assume that all improvements in gas retention are the direct result of improvements in dough development.

## 2.2.  Cell Creation and Control

The production of a defined cellular structure in the baked bread depends entirely on the creation and retention of gas bubbles in the dough. After mixing has been completed, the only 'new' gas which becomes available is the carbon dioxide gas generated by the yeast fermentation. Carbon dioxide gas has many special properties and at this point we are concerned with two: its high solubility and its relative inability to form gas bubbles. As the yeast produces carbon dioxide gas, the latter goes into solution in the aqueous phase within the dough. Eventually the solution becomes saturated and unable to hold any further carbon dioxide which may be produced. The rate at which saturation occurs depends on the fermentation conditions, but is fairly fast in all breadmaking processes, as shown by rapid dough expansion as the gas is retained within the developing or developed dough structure.

If the carbon dioxide does not form its own gas bubbles, how then does expansion of the dough through gas retention occur? Two other gases are available in significant quantities within the dough as a result of mixing oxygen and nitrogen, both of which are derived from any quantities of air trapped within the dough matrix as it forms. In the case of oxygen, its residence time within the dough is relatively short since it is quickly used up by the yeast cells within the dough (Chamberlain, 1979; Chamberlain and Collins, 1979). Indeed so successful is yeast at scavenging oxygen that in some breadmaking processes no oxygen remains in the dough by the end of the mixing cycle. The rapid loss of oxygen from mechanically developed doughs has been illustrated previously for a wide range of nitrogen to oxygen ratios (Collins, 1985).

With the removal of oxygen from the dough, the only gas which remains entrapped is nitrogen and this plays a major role by providing bubble nuclei into which the carbon dioxide gas can diffuse as the latter comes out of solution. The number and sizes of gas bubbles available in the dough at the end of mixing will be strongly influenced by the mechanism of dough formation and the mixing conditions in a particular machine. The influence of mixing action in each of the breadmaking processes will be further discussed below and the effects of mixer design in Chapter 4. At this stage it is only necessary to register the significant role that mixing will play in the creation of dough bubble structures.

It is now clear that for most breadmaking processes, particularly those which do not include a bulk dough resting time, the finest cell structure we can form is already in the dough by the end of the mixing process (Cauvain and Collins, 1995). During the processing stages subsequent to mixing, some modification of the bubble structure does occur which essentially comprises an expansion of the bubbles already created (Whitworth and Alava, 1999). The modification of bubble structures in the dough after mixing depends to a significant extent on the rheological qualities of the dough and we will be concerned with all three of the rheological properties described earlier.

## 2.3.   Major Breadmaking Process Groups

The sequences required for a complete breadmaking process have already been briefly described above. The processing stages which occur after dividing the bulk of the dough, such as shaping, proving and baking, are largely common to all breadmaking processes, and so when we discuss the different breadmaking processes we are mainly concerned with the methods which are used to produce a developed dough ready for dividing and further processing. In discussing the different processing methodologies we will also recognize the important contribution that different ingredient qualities and formulations play in determining dough development within a particular breadmaking process.

The methods by which dough development is achieved in the bakery may be fitted into four broad processing groups, although there are numerous variations and also elements of overlap between each of the individual groups. For discussion purposes we can name and characterize the groups as follows:

- *Straight dough bulk fermentation*, where resting periods (floor-time) for the dough in bulk after mixing and before dividing are the norm. For the purposes of discussion in this chapter, a minimum bulk resting period of 1 h will be required for a process to fit into this category.
- *Sponge and dough*, where a part of the dough formulation receives a prolonged fermentation period before being added back to the remainder of the ingredients for further mixing to form the final dough.
- *Rapid processing*, where either a very short ($< 1$ h) or no period of bulk fermentation is given to the dough after mixing and before dividing.
- *Mechanical dough development*, where a primary function of mixing is to impart significant quantities of energy to facilitate dough development, and the dough moves without delay from mixer to divider for further processing.

*Delayed Addition of Salt.* While strictly not a defined breadmaking process some bakers may delay the addition of salt until the later stages of mixing. The objective is to permit full hydration of the proteins in the flour and the optimisation of the gluten structure in the dough before the water-binding effects of salt (sodium chloride) are introduced. The salt is readily soluble in the dough and quickly disperses. However, the technique is most applicable to low speed mixing and with mixers which allow ready access to the dough during processing. Dosing salt late in short-time, high speed mixing processes (e.g. Mechanical Dough Development) present practical problems though technically it could be arranged.

The lack of inhibiting effect on the yeast by delaying salt addition is not commonly a problem because the delayed salt method is most commonly employed in breadmaking processes where the dough is rested after mixing and before dividing. The dough may have slightly more gas present by the end of the resting period but the impact will be small.

Each of the process groups identified above has a similar equipment requirement in that they all need some means of mixing the ingredients together to form a cohesive dough. The nature of that mixing equipment will make an important contribution to dough development, and so it is inevitable that in discussing the individual groups some consideration has to be given to the type of equipment used. To some extent, individual breadmaking processes have become synonymous with different mixers but in many cases that relationship is not absolute since several different mixer types may be capable of exploiting the principles of the same breadmaking process. The suitability of different mixers is most limited in the case of mechanical dough development processes where high mixing speeds and control of the atmosphere during the mixing cycle become very important in achieving the desired bread character (Cauvain and Young, 2006).

In commercial practice a close link has also developed between the type of breadmaking process and the scale of manufacture. Once again this link is not absolute with any of the breadmaking process groups being capable of exploitation by bakers of any size. Some breadmaking processes are more sensitive than others to variations in processing conditions, such as time and temperature, and consequently there has been a tendency in smaller-scale bakeries, where greater process flexibility is required, to use the more tolerant and less process sensitive of the breadmaking processes.

Polarization of products to different breadmaking methods has also tended to occur, in part because of the choices of equipment made by bakers, especially at the smaller end of the production scale. In seeking to use breadmaking methods in combination with equipment which give increased flexibility and 'tolerance' during processing, some bakers have limited their options with regard to the range of bread qualities which it is possible for them to make. At the other end of the scale, plant production has also tended to limit its options, although in this case it is because bakers have sought efficiencies of scale and close process control. In some cases, limited appreciation of the critical factors which affect bread quality for a particular process has resulted in particular bread types which have become synonymous with particular breadmaking processes. It is certainly true that not all of the breadmaking processes are equally capable of making optimum-quality bread over the full range of bread types we encounter, but often the possible range of qualities is greater than is appreciated or indeed exploited for any particular process.

It is a mute question as to whether the qualities of the flours which are available determine the breadmaking process to be used or whether the process to be used determines the flour qualities required. In fact, whichever way we answer the question we are correct. As breadmaking processes developed around the world, the 'strength' of the locally available flour had to be accommodated, but as our knowledge of the breadmaking process has increased we have come to learn that there are many different ways of achieving a particular bread quality, and in doing so we have learned that flour quality can be adjusted to achieve our desired aims. In modern breadmaking it is certainly true that the 'best' (usually taken to mean the strongest, the highest protein quantity or quality) wheats will make the 'best' flour for a given process. As we shall see, factors such as bread volume will increase with increasing protein content, but the price we may have to pay with such stronger flours is the adjustment of our preferred processing method. In some cases where we have a fixed processing method, we may not be able to accommodate 'improvements' in flour quality. This close relationship between flour properties and processing methods will be expanded upon in discussion of the individual processing method groups which follows.

## 2.3.1.  Straight Dough Bulk Fermentation

To many, the application of bulk fermentation for dough development is probably the most traditional and most 'natural' of the breadmaking processes. This process group is the most homogenous of all the groups we shall be discussing since the variations within it tend to be confined to different periods of bulk fermentation time, with variations in some other aspects of controlling fermentation, such as those associated with temperature or yeast level. There are only a few essential features of bulk fermentation processes and can be summed up as follows:

- mixing of the ingredients to form an homogeneous dough;
- resting of the dough so formed in bulk for a prescribed time (floor-time), depending on flour quality, yeast level, dough temperature and the bread variety being produced;
- part-way through the prescribed bulk fermentation period there may be a remixing of the dough (a 'knock-back' or 'punching down').

Dough formation for bulk fermentation is usually a low-speed affair carried out by hand or with low-speed mixing machines. Whether mixed by hand or by machine, the amount of energy which is imparted to the dough is very small by comparison with that experienced in other types of breadmaking processes. This is an important distinction because it shows that dough development is almost completely limited to that achieved in the fermentation period. This being the case, control of the factors which affect the bulk fermentation period and the quality of the ingredients used is especially important for optimum bread quality. The formulations for bulk fermentation need only contain a few ingredients as shown in Table 2.1.

2.3.1.1.  Yeast Level

The differences in yeast levels in the two examples of recipes given in Table 2.1 occur because more yeast is required with shorter bulk fermentation periods in

TABLE 2.1. Recipes for bulk-fermented doughs.

|  | 3 h (%) | 1 h (%) |
| --- | --- | --- |
| Flour | 100 | 100 |
| Yeast | 1 | 2 |
| Salt | 2 | 2 |
| Water | 57 | 58 |

order to achieve full dough development in the shorter time. This relationship between dough development time and yeast level probably comes from the contribution that enzymes present in the yeast cells, viable or dead, make to modification of the protein structures which are forming with increasing dough resting time. Of the enzymes present, the proteolytic enzymes and the natural reducing agent glutathione are likely to play the major roles. Flour too contains enzymes which can contribute to dough development.

If we take this relationship between yeast and bulk resting time to its ultimate conclusion we could continue to increase the yeast level and expect to eliminate bulk time altogether. Indeed it is possible to make a 'no-time' dough in this manner, but the resulting bread will be somewhat poorer in quality than we might expect from one or more hours of bulk time. This no-time doughmaking approach has been used by bakers and is often referred to as an 'emergency dough' to be made when there is insufficient time available to allow for a bulk rest of the dough (Ford, 1975). We can see then that while there is a working relationship between yeast level and bulk time the passage of a period of time is still very important for gluten modification and the production of suitable bread quality to occur, whatever the level of yeast added. For practical purposes a period of at least 1 h in bulk should be given to the dough.

Since the mechanism for dough development in bulk fermentation depends to a significant degree on yeast activity, we can also reasonably expect dough temperature to play a major role in determining the time at which full development is achieved for a recipe with a given yeast level. This is certainly the case and in bulk-fermented doughs it is normal to adjust the yeast level or bulk time, or both, with changes in dough temperature, whether the latter is deliberately introduced or occurs from some unintentional source. There are no 'hard and fast' rules for temperatures in the bulk dough at the end of mixing, but conventional practice places final dough temperatures in the region of 21–27 °C (70–80 °F). As a 'rule of thumb', a 4 °C rise in dough temperature can be offset by a reduction in yeast level by one-half. Conversely a 4 °C fall in dough temperature requires a doubling in added yeast. The practising baker will be familiar with such relationships and may make appropriate adjustments with rises and falls in ambient bakery temperatures in those countries which experience significant changes in daily and seasonal temperatures. In the context of dough temperature control, Calvel (2001) considered that the control

FIGURE 2.1. Effect of bulk fermentation time on bread quality.

of the final dough was one of the most significant factors in achieving consistent bread quality using fermentation systems.

### 2.3.1.2.    Flours

The 'strength' of the flour which can be used in bulk-fermented doughs is closely linked with the length of the bulk fermentation period which we employ. In general, the stronger the flour, the longer the fermentation period we will require in order to achieve optimum dough development (Figure 2.1), and the better the final bread quality will be (i.e. with a larger volume finer crumb structure and softer crumb). Flour strength is largely related to its protein content and quality, as will be discussed in Chapter 12, so that higher protein flours require longer bulk fermentation times than lower protein flours to deliver optimum bread quality. The level to which bran is present in the flour will also affect the length of bulk fermentation times, with wholemeal (wholewheat) flours requiring shorter bulk time than white flours. A typical white flour protein content for bulk fermentation would be 12% or greater (14% moisture basis).

Failure to match flour and bulk times will result in a number of quality defects in both the dough and the baked product. In the dough insufficient bulk time gives one which is 'under-fermented' and will exhibit a tough, rubbery gluten, not easily given to being moulded and which, in turn, will yield loaves of small volume, a dense cell structure and firm crumb. Too long a bulk time will result in the dough becoming 'over-fermented', readily giving up its gas at the slightest touch and liable to collapse under its own weight. If bread be made from such a dough it is likely to have a poor shape, although with adequate volume, and an irregular cell structure, frequently with large holes. In part, these changes occur because there is a progressive change in the physical properties in the gluten in the dough. Commonly referred to as 'relaxation', the changes are usually a loss of resistance to deformation which is accompanied by an increase in extensibility.

The supplementation of flours with dried, vital wheat gluten to raise the protein content of weaker base flours is a common practice in many parts of the world (Chamberlain, 1984; Cauvain, 2003). While such an approach works well in some breadmaking processes, gluten supplementation is less successful where mixing methods are of the lower speed, less intense form. This is sometimes the case in the production of bulk-fermented doughs and, even though a long resting period is available, the conditions are not suited to continuing gluten development. Some additional gluten development may be gained during remixing at knock-back but generally flours with higher levels of gluten supplementation do not perform as well as flours which contain the same level of indigenous protein.

Flours to be used for bulk fermentation processes are usually low in cereal *alpha*-amylase (high Falling Number) and will only be supplemented with low levels of fungal *alpha*-amylase, if at all, because of the potential softening effects on the dough handling character with extended bulk resting time.

### 2.3.1.3.   Water Levels

One of the most obvious manifestations of the changes taking place when the dough ferments in bulk is a progressive softening of the dough with increasing time. In breadmaking, bakers aim to achieve a 'standard' dough consistency for dividing and moulding. They accomplish this by adjusting the water level added during dough mixing according to the water absorption capacity of the flour (Chapter 12). During bulk fermentation progressive enzymic action is responsible for the softening of the dough which occurs. Since enzymic actions are time and temperature dependent, we can reasonably expect that dough softening will vary according to the bulk fermentation conditions, and in these circumstances adjustment of added water levels will have to be made to compensate for these changes. The recipes given in Table 2.1 show how a reduction in added water is required with longer bulk fermentation times in order to maintain a standard dough consistency for dividing.

### 2.3.1.4.   Optional Ingredients

While the only essential ingredients required are those given in Table 2.1, other ingredients are sometimes added for making bread by bulk fermentation. Typical rates of addition for these optional ingredients and the properties they confer to the dough and the bread are given in Table 2.2.

In addition to those optional ingredients identified in Table 2.2, 'improvers' may be added to bulk-fermented doughs. Usually the levels of addition are much lower than would be seen in no-time doughmaking processes. In some cases the 'improver' may consist of a small quantity of an oxidizing material added at the flour mill in order to assist in dough development (Chapters 3 and 12). Various flour-treatment agents are permitted for use around the world, although the numbers are becoming fewer.

TABLE 2.2. Optional ingredients in bulk fermentation.

|  | Percentage of flour weight | Improvement |
|---|---|---|
| Fat | 1.0–2.0 | Gas retention |
|  |  | Crumb softness |
| Emulsifiers | 0.1–0.3 | Gas retention |
|  |  | Crumb softness |
| Enzyme-active malt flour | 0.1–0.2 | Gas production |
|  |  | Gas retention |
|  |  | Crust colour |
| Enzyme-active soya flour | 0.2–0.5 | Crumb whiteness |
| Skimmed milk powders | up to 2.0 | Crust colour |
|  |  | Flavour |

### 2.3.1.5.  Process Variations

Reference has already been made to one process variation which may be encountered in bulk fermentation processes, namely the operation of 'knocking-back', 'punching down' or remixing the dough part way through the fermentation time. This operation tends to happen with doughs which are undergoing longer fermentation periods, such as greater than 1 h. A number of advantages are claimed for the operation, including equilibration of dough temperatures and the incorporation of more air into the dough to improve yeast activity.

Other variations include the delaying of the addition of salt and yeast to the later stages of mixing or, indeed, to the later stages of fermentation. Delaying the addition of these ingredients changes the manner and degree of the dough development process. For example, delaying the addition of salt until about two-thirds of the way through the bulk period increases the effects of fermentation without having to increase the bulk fermentation period. One of the common claims for delaying the addition of such ingredients is for the modification of flavours in the dough (Calvel, 2001). The development of bread flavour has already been discussed in Chapter 1.

### 2.3.1.6.  Creation of Bubble Structure

The basic elements of bubble structure creation have been described earlier in this chapter. As with all breadmaking processes, the gas bubbles in the dough at the end of mixing exhibit a range of sizes from a few micrometers to several millimeters. During the bulk fermentation period the evolution of carbon dioxide gas leads to the expansion of many of these bubbles. At the same time that the bubbles are being expanded, changes in the dough rheology are occurring which make it less resistant to deformation. Because of such changes it is possible to collapse many of the larger bubbles in the dough during the knock-back or moulding stages which leaves the many smaller bubbles which are subsequently be inflated by more carbon dioxide gas. Baker and Mize (1941) showed this to be the case for bulk-fermented doughs and considered that such events were major contributors to the formation of fine and uniform cell structure in bread made from bulk-fermented doughs. In the past, when bulk fermentation was the norm, craft bakers advocated the modification of bread cell structure using this principle of inflation, collapse, creation (more likely retention) of small bubbles and re-inflation for the production of the so-called 'competition breads' with a finer and more uniform cell structure (Horspool and Geary, 1985). Thus the creation of bread cell structures from bulk-fermented doughs clearly owes much to the manipulation of the dough during processing. This is not the case with no-time doughmaking processes as will be discussed below.

## 2.3.2.  Sponge and Dough

Elements of sponge and dough processes are similar to those for bulk fermentation in that a prolonged period of fermentation is required to effect physical and chemical changes in the dough. In sponge and dough this is achieved by the thorough fermentation of part of the ingredients rather than all of them. The duration of sponge fermentation may vary considerably, as may the composition of the sponge. In some cases the sponge component may be replaced with a flour brew in which the proportion of liquid is much higher than that used in a sponge.

The key features of sponge and dough processes are as follows:

- a two-stage process in which part of the total quantity of flour, water and other ingredients from the formulation are mixed to form a homogeneous soft dough – the sponge;
- resting of the sponge so formed, in bulk for a prescribed time (floor-time), mainly depending on flavour requirements;
- mixing of the sponge with the remainder of the ingredients to form a homogenous dough;
- immediate processing of the final dough, although a short period of bulk fermentation period may be given.

In the UK, sponge and dough formation tends to be a low-speed process carried out with low-speed mixing machines, while in North America more intense mixing is given to the sponge and the subsequent dough. On a small scale the mixer used to form the sponge may also provide the container in which to store it, provided it is not required for other uses. In large-scale production of sponges, separate containers which can be moved to temperature-controlled environments are needed in order to ensure uniformity of sponge development and to achieve the required scales of manufacture.

### 2.3.2.1.   Role of the Sponge

The main roles of the sponge are to modify the flavour and to contribute to the development of the final dough through the modification of its rheological properties. The process of flavour development in the sponge, though complex, is manifested in a relatively straightforward manner with an increase in the acidic flavour notes arising from the fermentation by the added yeast and other microorganisms naturally present in the flour. To maintain the right flavour profile in the finished product, the sponge fermentation conditions should be closely controlled and care should be taken to avoid a build-up of unwanted flavours by thorough cleaning of storage containers after use.

During the sponge fermentation period there will be a decided decrease in sponge pH with increasing fermentation (whether arising from changes in time, temperature or both). Under these conditions the rheological character of the gluten formed during initial sponge mixing will change, with the sponge becoming very soft and losing much of its elasticity. As standing time increases the condition of the sponge increasingly resembles an over-fermented dough. The low pH of the sponge and its unique rheological character are carried through to the dough where they have the effect of producing a softer and more extensible gluten network after the second mixing. In many cases the addition of the sponge changes the rheological character of the final dough sufficiently to render further bulk resting time unnecessary, so that dividing and moulding can proceed without further delay.

### 2.3.2.2.   Formulations

The main requirement for sponge and dough processes is to decide what proportion of the total flour is to be used in the production of the sponge. This proportion will vary according to individual taste and location. Two examples are given in Table 2.3, one for a typical 16 h (overnight) sponge in the UK and the other a 4 h example from North America. The contrast in the approaches is very evident.

TABLE 2.3. Examples of sponge and dough formulations (ingredient proportions expressed as percentage total flour weight).

|  | Sponge | Dough |
|---|---|---|
| *UK 16 h sponge* | | |
| Flour | 25.0 | 75.0 |
| Yeast | 0.18 | 1.75 |
| Salt | 0.25 | 1.75 |
| Water | 14.0 | 43.0 |
| Fat | 0.0 | 1.0 |
| *North American 4 h sponge* | | |
| Flour | 65.0 | 35.0 |
| Yeast | 2.4 | 0.0 |
| Salt | 0.0 | 2.3 |
| Water | 40.0 | 25.0 |
| Improver | 0.1 | 0.0 |
| Milk solids | 0.0 | 3.0 |
| Sugar | 0.0 | 6.0 |
| Fat | 0.0 | 3.0 |

### 2.3.2.3.  Improvers

Additions of improvers are not essential to the production of bread by sponge and dough methods since a contribution towards dough development is made directly by the sponge. However, as shown by the example of a North American recipe in Table 2.3, improver additions are common in some variations of the process. The choice of improver type and the timing of the addition, whether to the sponge or the dough, depend to a large extent on the bread variant being produced and traditional practices.

There will be different potential effects from the different oxidizing agents present if the improver is added to the sponge side of the process. Late-acting oxidizing agents have little or no effect until the dough reaches the prover (proofer) while faster-acting oxidizers, such as ascorbic acid and azodicarbonamide, will act in the sponge-mixing stage. In the case of ascorbic acid, oxygen is required for oxidation of the dough proteins to occur (Collins, 1994). Within the sponge the atmosphere will become anaerobic, and so opportunities exist for the ascorbic acid, in particular, to act as a reducing agent, its true chemical form, and to modify (weaken) gluten structures. The opportunities for enzymic action from the improver should also be considered, so that all in all there is a strong case for restricting improver additions to the dough side of the process where control of the changes which may occur is more readily achievable.

### 2.3.2.4.  Flours and Other Ingredients

Flours used in typical sponge and dough production will be at least as strong as those used in bulk-fermented doughs with protein contents not less than 12%. As with bulk-fermented doughs, flours for sponge and dough tend to have high Falling Numbers. High *alpha*-amylase activity could be a problem in the sponge, but is less likely to be a problem in the dough since the latter rarely has any floor-time after mixing.

### 2.3.2.5.  Process Variations

The most obvious of process variations encountered with sponge and dough systems will be variations in the sponge fermentation times. These will vary according to individual requirements for efficient processing, flavour development and available raw materials. Sponge temperatures will vary but are usually kept to maximum of 21 °C (70 °F). Final dough temperatures will fall into a similar range to those used in bulk fermentation, between 21 and 27 °C (70–80 °F).

In some cases the sponge may be incorporated into a dough which is then given a period of bulk fermentation or it may be added to doughs which are to be developed by a rapid processing method or by the Chorleywood Breadmaking Process (CBP), as will be discussed below.

Since a primary function of a sponge is to develop bread flavour, alternative liquid brew or ferment systems have developed to fulfil this function. In most cases there will be little or no dough development function from the brew, other than that which comes from the ingredients. The application of liquid brews to the production of hamburger buns is described in Chapter 9.

## 2.3.3.  Rapid Processing

This heading covers a multitude of slightly different breadmaking systems, each of which has evolved based on different combinations of active ingredients and processing methods. A common element to all breadmaking processes covered under this heading will be the inclusion of improvers to assist in dough development and the reduction of any individual fermentation period in bulk or as divided pieces (but not including proof) to less than 1 h. Processes which are covered by this heading include activated dough development, no-time doughs with spiral mixers and the Dutch green dough process.

### 2.3.3.1.   Activated Dough Development (ADD)

This process was developed in the USA during the early 1960s and became popular in smaller bakeries in the USA and the UK thereafter. Its essential features were as follows:

- the addition of a reducing agent, usually cysteine;
- the addition of oxidizing agents other than added at the flour mill;
- the addition of a fat or an emulsifier;
- extra water in the dough to compensate for the lack of natural softening;
- extra yeast to maintain normal proving times.

Since its first introduction Activated Dough Development (ADD) has undergone a number of changes and now probably no longer exists in its 'classic' form. When ADD was first introduced, potassium bromate was a common component in the added improver, together with ascorbic acid and L-cysteine hydrochloride. The increasing expense of L-cysteine and the withdrawal of potassium bromate from many permitted lists of breadmaking ingredients have both played a role in the demise of ADD certainly in the UK.

Since the dough development process in ADD was mostly chemically induced, low-speed mixers could be employed. This allowed craft bakers to continue using their existing low-speed mixers and eliminate bulk fermentation without purchasing the new high-speed mixers being developed for mechanical dough development processes in the 1950s and 1960s. With the passage of time many of the smaller bakers changed to spiral-type mixers which allowed them to move to improver formulations with fewer 'chemicals' at a time when consumer attitudes to 'additives' were changing.

A short period of bulk fermentation before dividing was beneficial for ADD product quality. Sponges could be added to change bread flavour if required. Final dough temperatures were in the region of 25–27 °C (76–80 °F).

### 2.3.3.2.   No-time Doughs with Spiral Mixers

In many bakeries, especially the smaller ones, the spiral mixer has taken over as the main type of mixer being used. Spiral mixers have a number of advantages for no-time doughmaking processes in smaller bakeries or where fine cell structures are not required in the baked product. These will be discussed in more detail in Chapter 4 but it is worth noting at this stage the input of higher work levels than those used with traditional low-speed mixers, with accompanying reductions of mixing times to achieve optimum dough development.

Although mainly used for no-time doughs some bakers may use short periods of bulk fermentation, usually 20–30 min, to assist with dough development after mixing. In these circumstances the control of final dough temperature is important in order to both control and optimize dough development. The additional gas generated during such bulk resting periods will place greater demands on divider weight control and yield products with a more open cell structure. Flavour development in the crumb is likely to be limited given the short time periods which are commonly used.

Some spiral mixers impart sufficient energy to raise dough temperatures above that expected from the ingredients. Final dough temperatures vary widely for no-time doughs with spiral mixers, and practical examples may be found from 21 (or lower, e.g. for frozen doughs) to 27 °C (70–80 °F). For many bakers the advantage of using lower dough temperatures lies in restricting yeast activity which comes with the usually higher levels of added yeast. A counter to this advantage is the reduction in chemical and enzymic activity which will occur at lower dough temperatures with a subsequent reduction in overall dough development.

### 2.3.3.3.   The Dutch Green Dough Process

This process was developed in the Netherlands, hence its name. It is included under this process group heading since the mixed dough passes without delay to dividing, although significant periods of resting are involved in the total process. The essential features of the process are as follows:

- mixing in a spiral-type mixer or extra mixing in a speeded-up conventional low-speed mixer;
- the dough is divided immediately after mixing;
- the divided dough is rounded and given a resting period of the order of 35–40 min;
- the dough is re-rounded and given a further resting period before final moulding.

The basis of the name 'green' refers to the fact that after the mixing the dough is considered to be underdeveloped or 'green' in classic bakery parlance. Dough development continues in the resting periods after each rounding. When first introduced, two or three resting periods were used; now it is more common to see one or, to a lesser extent, two.

2.3.3.4.  Role of Improvers and Other Ingredients in Rapid Processing

Although it is possible to make no-time doughs without additional ingredients, such as with the traditional Dutch green dough process, it is common for improvers to be added to assist with dough development in the absence of bulk fermentation time. The compositions of improvers which are used vary widely, although the most common ingredients are ascorbic acid, enzyme-active materials and emulsifiers. The degree of oxidation gained from the ascorbic acid depends in part on the level used and in part on the mixing machine and its ability to occlude air during the mixing operation. This latter aspect is discussed in more detail in Chapter 4.

Most no-time dough processes use flours of the stronger type with protein contents of 12% or more. Since there is no appreciable softening of the dough from fermentation before dividing, water additions will be higher than in bulk fermentation. The precise water level used will also be influenced by the type of mixer, with some doughs being softer and stickier when taken out of one machine compared with another. Often this initial stickiness is lost in the first few minutes after leaving the machine. The cause of this phenomenon is not clear but may well be related to the quantity of gas remaining in the dough at the end of mixing (see below).

## 2.3.4.  Mechanical Dough Development

The common elements of this group are that there is no fermentation period in bulk and dough development is largely, if not entirely, achieved in the mixing machine. In mechanical dough development the changes brought about by bulk fermentation periods are achieved in the mixer through the addition of improvers, extra water and a significant planned level of mechanical energy.

The principle of mechanical dough development was first successfully exploited in the Wallace and Tiernan 'Do-maker' in the 1950s. The 'Do-maker' loaf had a characteristically fine and uniform cell structure which eventually proved to be unpopular with many consumers, and today few installations remain in use. The 'Do-maker' used a continuous mixer and separate developer chamber. Other processes which exploited the same principles of mechanical dough development and continuous mixing were the Amflow process and the Oakes Special Bread Process. Like the 'Do-maker', few installations remain in use.

In 1958 the British Baking Industries Research Association at Chorleywood, UK (later merged into the Flour Milling and Baking Research Association and more recently into the Campden & Chorleywood Food Research Association) began to investigate the important factors in the mechanical development of dough. The work was to lead to the one mechanical dough development process which has stood the test of time – the CBP – and this process will serve as the basis for discussing and understanding the key issues in mechanical dough development. More detailed discussion of the CBP-compatible mixing equipment and its functions are given in Chapter 4.

2.3.4.1.   Chorleywood Bread Process

The basic principles involved in the production of bread and fermented goods by the CBP remain the same as those first published by the Chorleywood team in 1961, although the practices have changed with changes in ingredients and mixing equipment (Cauvain and Young, 2006). The essential features of the CBP are as follows:

- mixing and dough development in a single operation lasting between 2 and 5 min at a fixed energy input;
- the addition of an oxidizing improver above that added in the flour mill;
- the inclusion of a high melting point fat, emulsifier or fat and emulsifier combination;
- the addition of extra water to adjust dough consistency to be comparable with that from bulk fermentation;
- the addition of extra yeast to maintain final proof times comparable with those obtained with bulk fermentation;
- the control of mixer headspace atmosphere to achieve given bread cell structures.

The main difference between the CBP and the bulk fermentation processes lies in the rapid development of the dough in the mixer rather than through a prolonged resting period. The aim of both processes is to modify the protein structure in the dough to improve its ability to stretch and retain gas from yeast fermentation in the prover; in the case of the CBP this is achieved within 5 min of starting the mixing process. The advantages gained by changing from bulk fermentation to the CBP include the following:

- a reduction in processing time;
- space savings from the elimination of bowls of dough at different stages of bulk fermentation;
- improved process control and reduced wastage in the event of plant breakdowns;
- more consistent product quality;
- financial savings from higher dough yield through the addition of extra water and retention of flour solids which are normally fermented away.

Disadvantages include the following:

- faster working of the dough is required because of the higher dough temperatures used;
- a second mixing will be required for the incorporation of fruit into fruited breads and buns;
- in some views, a reduction of bread crumb flavour because of the shorter processing times.

The last disadvantage listed is one of continual debate which has been constantly fuelled by the detractors of the CBP without any real understanding of the processes by which bread flavour is developed. The basis of bread crumb flavour development was discussed in Chapter 1 and, while undoubtedly linked with the length of bulk fermentation time, in these days of predominantly no-time doughs, is probably more likely to be affected by ingredient additions and crust formation. If increased flavour is required in bread crumb made by the CBP, then the use of a sponge or a flour brew is recommended. Bulk fermentation after the completion of dough mixing in the CBP is not recommended because of the adverse changes which occur in the dough and the loss of subsequent bread quality.

### 2.3.4.2.   Role of Energy During Mixing

The role that energy plays in optimizing bread quality during mechanical dough development can be readily assessed by comparing the loaves illustrated in Figure 2.2. As the level of energy per kilogram of dough in the mixer increases, so bread volume increases, and with this comes a reduction in cell size and increased uniformity. With the range of flours studied when the CBP was first introduced into the UK, optimum energy levels quickly became standardized at 11 Wh/kg dough (5 Wh/lb) with little, if any, benefit being gained in varying from that level. Later work in New Zealand, the USA and eventually the UK was to show that optimum work input varied according to flour characteristics with those derived from 'extra strong' wheats requiring optimum energy inputs above the standard 11 Wh/kg. Despite the variation in total energy input, optimum bread quality with such flours in the UK was only gained by increasing the mixing speed in order to continue to deliver the energy within the specified 2–5 min time scale. An example of this relationship is illustrated in Figure 2.3 for a past UK wheat variety Festival.

The role of energy during CBP mixing has yet to be fully explained. It is very likely that the high energy inputs are capable of mechanically breaking the disulphide bonds holding the original protein configurations together since such processes are known to occur in the mechanical modification of other molecules.

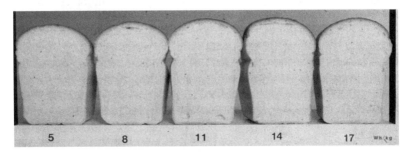

FIGURE 2.2. Effect of energy input during mixing.

FIGURE 2.3. Effect of energy input and mixing speed on bread quality. (a) 600 rpm and 5 Wh/kg, (b) 600 rpm and 17 Wh/kg, (c) 250 rpm and 5 Wh/kg, (d) 250 rpm and 17 Wh/kg).

The effect of mechanical energy might therefore be likened to the effects of natural or chemical reduction and, as such, will increase the sites available for oxidation. Chamberlain (1985) considered that only about 5% of the available energy was required to break the disulphide bonds with the rest being consumed by mixing of the ingredients and the breaking of weaker bonds. The mechanism of dough formation is discussed in more detail in Chapter 11.

The input of energy during mixing causes a considerable temperature rise to occur. Final dough temperatures are higher than those with other breadmaking processes and fall in the region of 27–32 °C(80–90 °F). Some bakers may see this as a disadvantage in trying to control yeast activity, but the very short processing times which are used after mixing should not give rise to undue problems in a well-controlled bakery. As with no-time doughs from rapid-processing techniques, the small advantages gained by reducing yeast activity are outweighed by the loss of dough development. Further, the higher dough temperatures give a dough which is more 'relaxed' (has less resistance to deformation) during moulding and is less susceptible to moulder damage of the type which will be discussed in Chapter 4.

### 2.3.4.3.  Energy and Dough Temperature Control

The transfer of energy to the dough during mixing causes the final dough temperature to be considerably greater than that would be expected from the simple prediction based on the knowledge of the temperatures and quantities of the ingredients used in the recipe. The control of dough temperature is vital if bread quality is to remain consistent.

The most common method for bakers to achieve a consistent final dough temperature is to adjust water temperature according to the temperature of the flour being used. The formula used is

$$\mathrm{T.water} = (2 \times \mathrm{T.dough}) - \mathrm{T.flour}$$

where the constant 2 is used because water has approximately twice the thermal capacity of flour and the level of added water is approximately half that of the flour weight, and T.dough is the dough temperature required at the end of mixing. Thus if T. dough $= 25$ °C and T.flour $= 20$ °C then T.water $= 30$ °C

This calculation works for hand and low-speed mixing and can be readily adjusted for changes in ambient bakery or equipment temperatures.

Mechanical mixing such as used in Mechanical Dough Development bread-making or with spiral mixed doughs complicates the relationship because of transfer of energy. The formula now becomes

$$\mathrm{T.water} = 2(\mathrm{T.dough} - \mathrm{T.rise}) - \mathrm{T.flour}$$

where T.rise is the difference between the final dough temperature if the ingredients where simply blended together and the actual temperature achieved in the dough by the end of mixing.

T.rise can be calculated from a few simple experiments starting with ingredients of known temperatures and masses.

For example, in Mechanical Dough Development the energy transferred to the dough during mixing to 11 Wh/kg resulted in the final dough temperature being 14 °C more than predicted from the simple relationship, and this value could be substituted in the equation as follows:

$$\mathrm{T.water} = 2(30 - 14) - 20$$

giving a required water temperature of 12 °C.

Because of the strong relationship between the temperature rise experienced by the ingredients in the dough formation process and the energy transferred during mixing it is possible to use the temperature data to 'cross-check' the energy balance during mixing. Such calculations are based on the specific heat capacities of the ingredients, their masses and temperatures, and the heat rise during mixing. For mixers running in a 'steady' state', that is the mixer bowl is losing heat to the bakery atmosphere as quickly as it gains it during mixing, the impact of the metal of the mixer bowl is limited. Examples of the method for calculating energy inputs during mixing may be obtained from equipment manufacturers or published literature (e.g. Cauvain and Young, 2006).

As mixing time or defined energy input changes, so will the temperature rise experienced by the dough. In practice it is advisable to carry out tests with a range of mixing conditions to cover the likely mixing scenarios.

In some cases the temperature of the ingredients (particularly the flour) and the heat rise experienced during mixing are so significant as to require the use of ice. The latent heat required to convert ice to water is significant and has a powerful cooling effect on the dough. However, gluten development depends on the presence of water and so the use of large quantities of ice may restrict the initial hydration of the gluten-forming proteins. The implications of restricting hydration in the early stages of mixing dough by the CBP are not clear but pre-hydration of flour before intense mixing is often considered to be beneficial.

### 2.3.4.4.  Flour Quality

The process of mechanical dough development has been shown to make better use of the flour protein, and in the early stages of the development of the CBP it was quickly recognized that a given bread volume could be achieved with a lower flour protein content in the CBP than with bulk fermentation. This finding was to lead to a gradual reduction in flour protein content, so that today a standard UK sandwich loaf can be made with a flour protein content of about 10.5% (14% moisture base).

In the CBP, more than with many other breadmaking processes, there are no disadvantages in supplementing the flour with added dried vital wheat gluten. The move to lower protein contents and supplementation of indigenous protein with dried gluten were important factors in the ability of the UK milling industry to reduce the importation of high-protein North American wheats and to attain near self-sufficiency using home-grown or EC-grown wheats.

Not all bread made using the CBP benefits from such a low protein content as the example given for the UK sandwich loaf. The quality characteristics demanded in the bread will largely dictate the flour specification and it is common for higher protein contents to be used for other bread varieties. For example, 'hearth' or 'oven-bottom' breads typically use protein contents of about 12%, although even this may represent a reduction from the level which would be used for these bread varieties in bulk fermentation, and for wholemeal (wholewheat) and mixed-grain breads flour protein contents may rise even higher (Figure 2.4).

FIGURE 2.4. Left to right, white, mixed grain and wholemeal breads made by the CBP.

### 2.3.4.5.  Creation of Bubble Structure

In contrast to the situation in bulk-fermented doughs the cell structure in the final bread does not become finer as the result of processing CBP doughs. In the case of CBP doughs, the final bread crumb cell structure is almost exclusively based on an expanded version of that created during the initial mixing process. The cell structure of UK sandwich breads made with dough taken straight from the mixer is contrasted with that which has received both the moulding stages in Figure 2.5. The loss of bread volume with the dough taken from the mixer can be attributed to the absence of intermediate proof during processing, an effect which will be discussed in Chapter 4. This illustration confirms the earlier work of Collins, (1983).

The creation of bubble structures in mechanically developed doughs, and indeed for many other no-time processes, depends on the occlusion and subdivision of air during mixing. The number, sizes and regularity of the gas bubbles depend in part on the mixing action, energy inputs and the control of atmospheric conditions in the mixer headspace. Collins (1983) illustrated how bread cell structure improved (in the sense of becoming finer and more uniform) with increasing energy input up to an optimum level with subsequent deterioration beyond that optimum. He also showed how different mechanical mixing actions yielded breads with varying degrees of crumb cell size.

More recent work to measure bubble distributions in CBP bread doughs (Cauvain et al., 1999) has confirmed that different mixing machines do yield different bubble sizes, numbers and distributions. However, in one CBP-compatible mixing machine, variation of impeller design had almost no effect

a                                    b

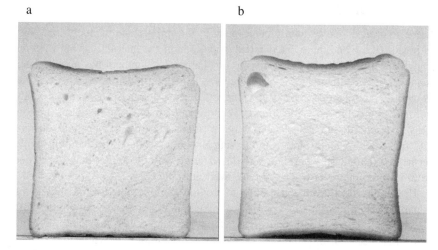

FIGURE 2.5. Comparison of bread cell structures from dough (a) ex-mixer and (b) ex-final moulder.

on the bubble population. The lack of differences in the characteristics of the various dough bubble populations was confirmed by the absence of discernible differences in the subsequent bread cell structures.

The modification of bubble populations through the control of atmospheric conditions in the mixer headspace has been known for many years, commonly through the application of partial vacuum to CBP-compatible mixers (Pickles, 1968). This control was useful in the creation of the fine and uniform cell structures typically required for UK sandwich breads, but was unsuited to the production of open cell structure breads. In the case of French baguette, an open and more random cell structure could be created by extending the intermediate proof time to 20 or 30 min (Collins, 1978), provided the delicate dough was given a gentle final moulding in order to preserve the large gas bubbles in the dough.

In a more recent development, CBP-compatible mixers have been invented which are able to work sequentially at pressures above and below atmospheric. When the dough is mixed under pressure larger quantities of air are occluded (Chapter 4), which gives improved ascorbic acid oxidation but more open cell structures. In contrast, dough bubble size becomes smaller as the pressure in the mixer headspace reduces and ascorbic acid oxidation decreases as the pressure decreases. The greater control of dough bubble populations realized in these mixers allows a wide range of bubble structures to be created in the dough. In addition to the fine and uniform structure created from the application of partial vacuum, an open cell structure for baguette and similar products can take place in the mixing bowl by mixing at above atmospheric pressure (Cauvain, 1994, 1995). Doughs produced from this type of mixer retain their larger gas bubbles in the dough without significant damage during processing, even when intermediate proof times were shortened from 20 to 6 min.

## 2.3.4.6.  Dough Rheology

Some references have already been made to the importance of dough rheology in breadmaking and there is a detailed discussion of the interactions between the dough and its processing after mixing in Chapter 4. Comment has already been made on the fact that water levels have to be adjusted (upwards) in most no-time doughmaking methods in order to achieve the same dough consistency as would normally be achieved with doughs at the end of a bulk fermentation period. This required dough consistency largely arises because the original designs and functions of many dough processing plants were based on handling bulk-fermented doughs and were not suited to dealing with the firmer doughs yielded from no-time methods when the latter were introduced.

The requirement to add extra water to provide a softer, more machinable dough is particularly true when the doughs are mixed under partial vacuum in the CBP. The lower the pressure during mixing, the 'drier' the dough feels and the more water that needs to be added to achieve the same dough consistency as that of the doughs at the end of a bulk fermentation period. This increased dryness with CBP doughs comes in part from the lower volume of gas occluded in the dough at the end of mixing (Chapter 4). If the dough is mixed at pressures greater than atmospheric, then the quantity of gas occluded during mixing increases and doughs become softer for a given water level. Practical limitations to the application of partial vacuum are the reduction of the amount of oxygen available for ascorbic acid conversion and the need for some air to be occluded to provide gas bubble nuclei (Baker and Mize, 1941). The degree to which mixer headspace can be lowered when mixing doughs in CBP-compatible mixers varies according to the machinery design and operational practices but commonly lies between 0.3 and 0.5 bar.[2]

The rheology of doughs from CBP-compatible mixers is also affected by changes in improver formulation, especially those related to dough oxidation. Cauvain et al. (1992) have shown that the resistance to deformation of CBP doughs was greater when ascorbic acid was compared with potassium bromate as the sole oxidizing agent (Figure 2.6). This effect is most likely to be accounted for by the difference in the rate of action between these two oxidizing materials, with that for ascorbic acid taking place in the mixer and for potassium bromate mainly in the later stages of proof and thereafter. Other increases in resistance to deformation are observed when the oxidation from ascorbic acid is increased by modifying the mixer headspace atmosphere. Once again, a reduction in resistance (or perhaps more correctly a restoration to standard) may be affected by changing

---

[2] Confusion over pressure units can exist because of the way in which they are expressed. In part this arises because gauges fitted to mixers often express atmospheric pressure as being 0. A partial vacuum may be given as 0.5 bar vacuum and positive pressure may be given as 0.5 bar pressure. In this discussion atmospheric pressure is taken as being equal to 1 bar (or full vacuum = 0). Thus, a figure of 0.5 bar is 0.5 bar below atmospheric pressure, 0.3 bar is 0.7 bar below atmospheric pressure, and 1.5 bar is 0.5 bar above atmospheric pressure.

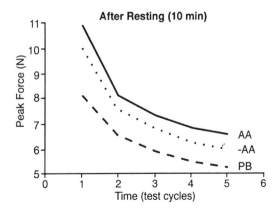

FIGURE 2.6. Comparison of effects of ascorbic acid and potassium bromate on dough resistance to deformation (AA = with ascorbic acid, —AA = without oxidant, PB = with potassium bromate).

the levels of added water, although in commercial practice the addition of other ingredients in improvers, for example enzyme-active materials, may act to reduce dough resistance and increase dough stickiness.

## 2.4.    Breadmaking Processes, Bread Variety and Bread Quality

Each of the breadmaking processes discussed above has particular advantages and disadvantages and almost all types of bread and fermented goods can be made with each of them. There are, however, some combinations of breadmaking process and product type which are more successful than others, and because of this successful 'partnership' there has been a narrowing of views, a closing of minds and a limiting of the potential for all breadmaking processes. It was never the intention of this chapter to review in detail the advantages and disadvantages of each of the breadmaking processes though some references to such matters have been made where deemed appropriate. It will, however, be useful to conclude by considering how product requirements might influence the choice of breadmaking process.

Reference has already been made to the link in many minds between the fermentation process and bread flavour. All current breadmaking processes involve at least one fermentation period, namely the one bakers call proof. Whether sufficient bread flavours are achieved within that one relatively short process will be endlessly debated because, as commented on earlier, bread flavour is a personal issue. For those who require more flavour in their bread crumb, the introduction of other fermentation stages is possible. With the introduction of a fermentation stage other than that required for proof come other factors which contribute to bread quality, the main one being an element of dough development, such as that seen in bulk development processes. Having chosen to make a 'flavourful' product, the baker must reconcile that requirement with a compatible breadmaking process. The choice of flavourful bread and no-time doughmaking process is not as incompatible as many would have us believe. There is absolutely no reason why we should not combine the benefits of better process control from no-time doughs with flavour, and several examples of using sponges to generate flavour before adding to no-time doughs have been discussed above.

Possibly more difficult to reconcile for the different breadmaking processes is the generation of the required cell structure for a particular product. Many bread products have distinctive crumb cell structures without which the product will simply not be authentic. The major difficulty lies not in making an open cell structure with a given breadmaking process since, as we have seen with the CBP, this may simply be a case of using a longer first resting period (intermediate proof), but rather in the creation of the fine cell structures that are required for many breads. The strong link between the formation of an open cell structure and a crispy crust must be recognized, but again there is no reason why this should limit the breadmaking process chosen to achieve these given aims.

No-time doughmaking methods offer the best opportunity of achieving the finer cell structures since little expansion of the gas cells normally occurs until after the dough has been moulded. Provided we have achieved 'optimum' dough development and do not treat the dough harshly during processing, we should retain all the necessary dough qualities to guarantee a fine crumb cell structure. Of the no-time doughmaking methods which have been discussed, the contribution of energy in mechanical dough development appears to offer the best opportunity for achieving this result.

In conclusion, we can see that it has been possible to study the underlying technology of breadmaking processes by considering the many variations which are used under four broad headings. Each breadmaking process offers unique advantages and disadvantages but there are few bread products that cannot be made with any one, once we as bakers have balanced our product requirements with our available raw materials and our process control needs.

## References

Baker, J.C. and Mize, M.D. (1941) The origin of the gas cell in bread dough. *Cereal Chemistry*, **18**, January, 19–34.

Calvel, R. (2001) *The Taste of Bread*, Aspen Publishers Inc., Gaithersburg, MA.

Cauvain, S.P. (1994) New mixer for variety bread production. *European Food and Drink Review*, Autumn, 51, 53.

Cauvain, S.P. (1995) Controlling the structure: the key to quality. *South African Food Review*, **22**, April/May, 33, 35, 37.

Cauvain, S.P. (2003) Dried gluten in breadmaking. *CCFRA Review No. 39*, Campden & Chorleywood Food Research Association, Chipping Campden, UK.

Cauvain, S.P. and Collins, T.H. (1995) Mixing, moulding and processing bread doughs, in *Baking Industry Europe* (ed. A. Gordon), Sterling Publications Ltd., London, pp. 41–43.

Cauvain, S.P. and Young, L.S. (2006) *The Chorleywood Bread Process*, Woodhead Publishing Ltd., Cambridge, UK.

Cauvain, S.P., Collins, T.H. and Pateras, I. (1992) Effects of ascorbic acid during processing. *Chorleywood Digest No. 121*, October/November, CCFRA, Chipping Campden, UK, pp. 111–114.

Cauvain, S.P., Whitworth, M.B. and Alava, J.M. (1999) The evolution of bubble structure in bread doughs and its effects on bread structure. In, *Bubbles in Food* (eds G.M. Campbell, C. Webb, S.S. Pandiella and K. Naranjan), Eagan Press, St. Paul, Minneapolis, USA, pp. 85–88.

Chamberlain, N. (1979) Gases – the neglected bread ingredients, in *Proceedings of the 49th Conference of the British Society of Baking*, pp. 12–17.

Chamberlain, N. (1984) Dried gluten in breadmaking – the new challenge. *British Society of Breadmaking, 30th Annual Meeting and 59th Conference Proceedings*, November, pp. 14–18.

Chamberlain, N. (1985) Dough formation and development, in *The Master Bakers Book of Breadmaking*, 2nd edn (ed. J. Brown), Turret-Wheatland Ltd., Rickmansworth, UK, pp. 47–57.

Chamberlain, N. and Collins, T.H. (1979) The Chorleywood Bread Process: the roles of oxygen and nitrogen. *Bakers' Digest*, **53**, 18–24.

Collins, T.H. (1978) Making French bread by CBP. *FMBRA Bulletin No. 6*, December, CCFRA, Chipping Campden, UK, pp. 193–201.

Collins, T.H. (1983) The creation and control of bread crumb cell structure. *FMBRA Report No. 104*, July, CCFRA, Chipping Campden, UK.

Collins, T.H. (1985) Breadmaking processes in *The Master Bakers Book of Breadmaking*, 2nd edn (ed. J. Brown), Turret-Wheatland Ltd., Rickmansworth, UK, pp. 1–46.

Collins, T.H. (1994) Mixing, moulding and processing of bread doughs in the UK, in *Breeding to Baking*, Proceedings of an International Conference at FMBRA, Chorleywood, CCFRA, Chipping Campden, 15–16 June, pp. 77–83.

Ford, W.P. (1975) Earlier dough development processes, in *Breadmaking: The modern revolution* (ed. A. Williams), Hutchinson Beenham, London, UK., pp. 13–24.

Horspool, J. and Geary, C. (1985) Competition breads, in *The Master Bakers Book of Breadmaking*, 2nd edn (ed. J. Brown), Turret-Wheatland Ltd, Rickmansworth, UK, pp. 400–424.

Millar, S.J., Bar L'Helgouac'h. C., Massin, C. and Alava, J.M. (2005) Flour quality and dough development interactions – the first crucial steps in bread production. In, *Using Cereal Science and Technology for the benefit of consumers* (eds S.P. Cauvain, S.E. Salmon, and L.S. Young), Woodhead Publishing Ltd., Cambridge, pp. 132–136.

Pickles, K. (1968) Tweedy (Chipping) Ltd. *Improvements in or Relating to Dough Production*, UK Patent No. 1 133 472, HMSO, London, UK.

Whitworth, M.B. and Alava, J.M. (1999) The imaging and measurement of bubbles in bread dough. In, *Bubbles in Food* (eds G.M. Campbell, C. Webb, S.S. Pandiella and K. Naranjan), Eagan Press, St. Paul, Minnepolis, USA, pp. 221–232.

# 3
# Functional Ingredients

Tony Williams* and Gordon Pullen[†]

## 3.1. Dough Conditioners and their Composition

The most basic bread dough one might use to produce a baked product would of necessity contain the following minimum ingredients: flour, water, yeast and salt. However, even those most skilled in the art of baking would agree that at the very least it would be difficult to make bread of a high, consistent quality from only these raw materials. The baker has always, where expedient, added small amounts of extra ingredients to enhance dough performance during processing or to improve the quality of finished bread. In the past these materials would usually be foodstuffs in their own right, such as fat, sugars, honey and malt flour. Although the principal benefits were probably considered to be related to the eating properties of the final baked article, it must have become apparent that it was possible to produce modifications to the dough itself during processing which might be equally beneficial in the finished product.

Virtually all of the early baking systems had one common factor: between the mixing stage and the final shaping of the dough prior to final proof there was a relatively long resting phase. This was dictated by the need to generate a sufficient rate of gas production by the yeast in the dough to ultimately produce a well-risen loaf of acceptable specific volume. Once compressed yeast became available in the mid-nineteenth century the rate of gas production was no longer the rate-determining factor in bread production. At this point it became clear to all that there were other benefits to be derived from a significant resting period between mixing and final moulding, and the concept of a bulk fermentation or floor time was evolved (Chapter 2). This fermentation stage not only enhanced gas production, but it also enhanced gas retention in the dough. Possibly it is for this reason that we have seen the name given to the group of small ingredients, additives and processing aids added to the dough to improve bread quality as, successively, yeast foods, flour improvers and currently as dough conditioners.

---

* All other sections
[†] Section on Baker's yeast

The term 'dough conditioner' as defined by the author is as follows: any material or combination of materials which are added to yeast-raised doughs to enhance and control gas production or gas retention, or both. The range and level of use of these substances is controlled by legislation in most countries. Over the last 15 years or so there have been a significant number of changes in both UK and EC regulations and the situation has reached a period of relative stability. It will, therefore, be assumed that at least for the foreseeable future the materials described will not disappear from the UK and EC permitted lists.

The functional components in bread doughs can be divided for convenience into three main groups: ingredients, additives and processing aids.

Ingredients may be described as foodstuffs in their own right, such as fats, soya flour, sugar and milk derivatives, although one might not usually consume them directly in the form that they are added to bread dough. It is not the intention in this section of the book to discuss ingredients which characterize particular varieties of bread, rather to examine in some detail a small number of key ingredients which demonstrate specific functional effects in dough processing and final bread quality.

Definitions of additives, legislation regarding their use and the labelling of such materials varies around the world. The definition of the range of additives or flour-treatment agents which might be used in bread and yeast-raised fine bakers' wares in the UK is contained within The Bread and Flour Regulations 1998 (SI 1998, No.141), and such materials are defined as any food additive which is added to flour or dough to improve its baking quality. In the European Union (EU), harmonization of additives has now largely been achieved. In theory the range of 'additives' which are now permitted for use in bread production has increased with the introduction of new forms of enzymes though in practice it is common to only use a small number of such materials to gain practical benefits.

The list of additives in bread improvers used throughout Europe is much shorter than comparable lists on products from the USA. However, the ingredient listings on packaging do not tell the full story. For example, in the EU some additives may be classified as 'processing aids' and as such need not be declared on product labels, currently! In contrast, the addition of *alpha*-amylase to bread dough in the USA would mean its appearance on the ingredient label. While non-declaration of enzyme additions is permitted under current EU regulations this may not be the case in the future. It is worth noting that many of the new enzymes which have been introduced as 'processing aids' in the EU will have been cleared through the 'Novel foods and novel food ingredients' regulations (EC, 2000).

## 3.2.   Ingredients

### 3.2.1.   Fats

The soft fats, those with melting ranges similar to those of lard, have been traditional components of white bread in the UK, the USA and many other countries for at least the last 100 years. With animal fats now being out of favour, lard and partially hydrogenated fish oils have largely been replaced by partially hydrogenated vegetable oils with melting characteristics similar to those of lard, or in other cases they have been replaced with liquid vegetable oils, such as rape or soya. More recently there has been a move away from partially hydrogenated fats because of concerns regarding the levels of *trans* fats in diets. As already commented in Chapter 1, it is worth noting that in many bread products total fat levels are low and even if *trans* fats are present the contribution that they will make to the diet is relatively small.

These types of fats may be described as 'enriching agents' in fermented products. They change the eating characteristics of bread and fermented goods giving a shorter, softer bite and at the same time a modest enhancement of the soft-eating shelf-life. The effects increase with increasing levels of addition. The level of use of such fats varies widely from zero in some breads to about 1% of flour weight for pan breads, increasing to levels of up to 10% of flour weight or even higher in rich morning goods, such as hot cross buns (a UK spiced and fruited bun traditionally associated with the Easter period).

The effects of high melting point fats on dough properties and final bread characteristics are far more significant than those of the softer fats. The importance of high melting point fats was dramatically demonstrated in the research which led to the eventual development of the Chorleywood Breadmaking Process (CBP) (Chapter 2) as a commercially viable bread manufacturing process but had been appreciated for some time (Baker and Mize, 1942). It was demonstrated that dough produced by the CBP which did not contain at least a prescribed minimum level of 'solid' fat at the end of final proof was highly unstable at this point, and gave bread with a blistered crust and little or no oven spring. The internal crumb structure was dark, open, coarse and the crumb was very firm, in fact all the symptoms of very poor gas retention (Chamberlain et al., 1965).

This effect can be graphically demonstrated by producing bread following the 'classic' CBP method and omitting fat from the recipe and then in successive doughs by adding controlled levels of various fats in a variety of dispersion systems. It was by empirical experimentation such as this in the 1960s that dough conditioner manufacturers devised, developed and refined their fat blends and at the same time established some of the basic rules regarding the nature of what came to be known as the 'fat effect' in CBP doughs.

The 'rules' for fat in the CBP and other no-time doughmaking systems can be summarized as follows.

- The ideal functional fat is a fully saturated one, with a chain length of $C_{16}$–$C_{18}$, that us tripalmin and tristearin with a melting point in the region of 55–60 °C (130–140 °F).
- Short-chain triglycerides, such as $C_{12}$ and $C_{14}$, are much less efficient on a weight-for-weight basis and require more rigid control of dough and proof temperatures.

- Fully hydrogenated fats based on fish and whale oils, even those with high melting points, are less efficient than fully hydrogenated animal and vegetable oils, presumably because the fatty acid chains are branched rather than straight.
- The fully saturated triglycerides must be in a highly dispersed form in the dough. A wide range of ingredients and techniques are capable of successfully dispersing the high melting point fat.
- The fat effect is not linear; too little fat produces bread of the same poor quality as bread without fat. There is then a dramatic increase in bread quality over a narrow range of increasing fat, after which there is little more to be gained in bread quality terms by further increases in the level of fat addition.
- The level of fat required to satisfy a CBP dough in terms of the fat effect is relatively low and may be in the region of only 0.1–0.2% of flour weight. It does, however, vary from flour to flour in response to factors which remain unclear and may rise to levels of about 1% in white breads or 4% in wholemeal (wholewheat) breads (Figure 3.1).

It is therefore essential to ensure that, in all circumstances where a high melting point fat is the main lipid source in the dough, a minimum level is added to allow for variations in flour characteristics.

The role of high melting point fat in the CBP is that of a key basic component in the dough and, once these finely dispersed fats had been developed for the CBP, their potential in other breadmaking systems was investigated. They have similar benefits for bread quality in other no-time systems, such as those based on spiral mixers (Chapter 2).

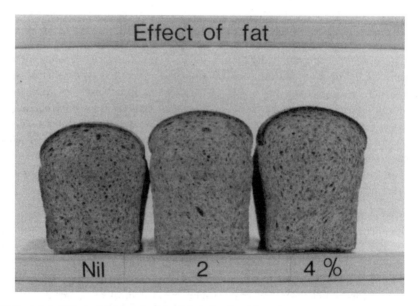

FIGURE 3.1. Effect of fat in wholemeal (wholewheat) bread.

The role of fat in long fermentation systems is more difficult to demonstrate. Although many traditional recipes contain significant levels of fat, it is possible to produce bread by bulk fermentation or sponge and dough methods with acceptable gas retention properties without any fat in the dough, given sufficient skill and appropriate flour properties, as seen, for example, in the production of 'traditional' baguette in France. However, even in these situations a level of high melting point fat equivalent to that used in the CBP will have a very significant improving effect on bread quality.

### 3.2.1.1.   Summary of Fat Effects

A quite specific type of fat in a finely dispersed form is of benefit in all types of bread dough. It is of particular importance in no-time dough breadmaking systems, such as CBP doughs. It is a highly functional, cost-efficient ingredient and can be used as an 'E-number-free' base for compound dough conditioners within the EC for a large number of bread products and manufacturing processes.

It is a source of great frustration to bakers and cereal scientists alike that having established the empirical rules by which the function of fat is determined in a wide range of production techniques and processes, the fundamental understanding of the nature of the 'fat effect' in breadmaking still remains unclear.

## 3.2.2.   Soya Flour

Full-fat enzyme-active soya flour as a functional dough ingredient was introduced into the UK in the 1930s. It was considered to have two principal beneficial functions, both being due to a particular enzyme system present in the uncooked bean. This enzyme system, lypoxygenase, catalyses the oxidation of unsaturated fats or lipids by atmospheric oxygen to lipid peroxidases via a number of intermediate oxidation compounds. It is these intermediate compounds that are the functional entities in the dough.

The two major functions of enzyme-active soya flour are as a flour bleaching agent and as a mild oxidizing agent.

### 3.2.2.1.   Soya as a Flour Bleaching Agent

The functional effect of soya flour as a bleaching agent is relatively simple and well understood. The intermediate oxidation compounds transfer oxygen from the atmosphere surrounding the dough to the yellow-coloured carotinoid pigment of flour. The oxidized products derived from these pigments are colourless, hence the dough is bleached – in much the same way as benzoyl peroxide functions in dough.

In the UK, where benzoyl peroxide and chlorine dioxide are no longer permitted, the value of soya flour in breadmaking should not be underestimated. However, even with high levels of soya it is not possible with most dough mixing systems to achieve the same efficiency of bleaching as with benzoyl peroxide or chlorine dioxide. This is due largely to the limited availability of atmospheric oxygen. It is therefore not surprising that spiral-mixed doughs are more readily bleached than CBP doughs mixed under partial vacuum. The effect of soya in a CBP dough mixed under vacuum used on unbleached flour is still significant, possibly because the intense shearing action during the high-speed mixing process (Chapter 4) exposes the dough surfaces repeatedly to the limited amount of available oxygen in the mixer.

It has also been demonstrated that by increasing the contact between the dough surface and oxygen the bleaching effect of soya can be greatly enhanced. For example, CBP doughs mixed either in oxygen-enriched atmospheres or under pressure/vacuum systems, where a flow of oxygen is maintained across the surface of the dough during mixing (Chapter 4), will produce dough and hence bread which is just as white as that produced by chemical bleaching, provided enzyme-active soya is included in the recipe.

### 3.2.2.2.  Gluten Oxidation by Soya Flour

The effect of soya flour as a mild oxidizing agent in dough is by no means as easy to demonstrate as its bleaching effect. The mechanism is complex. There may in fact be more than one route to the beneficial modifications of gluten via soya flour.

The 'direct' route has been demonstrated by Frazier et al. (1973). This suggests that the effect of soya is dependent on the presence of polyunsaturated free lipid and that, via a rather complex mechanism, bound lipid is freed from specific sections of the gluten protein thereby allowing this protein to become more hydrophillic and hence to be able to help form the viscoelastic surface of the gas bubbles in the dough.

It is also becoming clear that as the number of flour-treatment agents becomes ever more restricted, with the single remaining oxidant in many parts of the world being ascorbic acid, the role of soya flour in the efficient use of ascorbic acid is becoming more readily appreciated. In order for ascorbic acid to function in a dough it must first itself be oxidized to dehydroascorbic acid. Where the supply of oxygen is limited, for example in a CBP dough mixed under partial vacuum, or in a conventionally mixed dough where yeast activity begins to restrict the availability of oxygen, soya flour is a valuable adjunct to the oxidation system. This combination has proved to be beneficial in long fermentation processes.

Ascorbic acid is a quite rapidly acting oxidizing agent (see p. 61) and although it is difficult to over-treat a CBP dough with ascorbic acid this is not the case with bulk fermentation. Over-treatment results in 'bound', inelastic doughs which are difficult to process and can produce low volume bread with poor oven spring. Too low a level of ascorbic acid results in weak doughs with, once again, poor final volume and a coarse open texture. In the past many bakers had become used to flours containing a low level of potassium bromate, a slow-acting oxidant which functioned late in the process in the absence of oxygen. A combination of ascorbic acid and soya flour, while no means an ideal replacement, gives at least some tolerance and stability to long-processed doughs.

## 3.3.  Additives

### 3.3.1.  Emulsifiers

3.3.1.1.   Diacetylated Tartaric Acid Esters of Mono- and Diglycerides
of Fatty Acids [DATA Esters, E472(e)].

The legal status of DATA esters in the EC is that they are permitted at a *quantum
satis* (QS) level in all types of bread and fine bakers' wares.

The term 'DATA ester' (DATEM) is a generic one covering a range of similar
materials which vary due to the nature of the fatty acid and the ratio of mono-
to diglyceride. There are also further significant variations due to the level of
esterification by the diacetylated tartaric acid. The physical forms vary from an
oily liquid to a fairly free-flowing powder depending on the degree of saturation
of the base fat. In many commercial preparations there are at least five major and
many more minor components. There are a limited number of types of DATA
esters based on low melting point fats or oils which are used as components of
paste (fat-like) or fluid (pumpable) dough conditioners. In the UK it is now usual
for DATA esters to be based totally on non-animal raw materials, including
the glycerol used in the preparation of the monoglyceride prior to the final
esterification. By far the widest form of use is in powdered form in powder
dough conditioners.

The function of DATA esters in yeast-raised wheat flour doughs can be readily
demonstrated by practical baking tests and by a number of standard dough
rheology methods. The most basic description of the function of DATA esters is
that they enhance gas retention when incorporated into almost any yeast-raised
wheat flour-based dough. There is a fairly direct relationship between the level
of addition and the enhancement of gas retention up to an optimum level, after
which there is a plateau followed by a slight reduction of effect at excessive
levels of use. The final level of bread quality enhancement is far greater than
that of fat.

As to the mechanism, one can only select from the literature the one which
appears to fit most readily the empirical observations. There is strong evidence
that when DATA esters are incorporated into bread doughs they bond rapidly
and totally to the hydrated gluten strands. The resultant gluten network is not
only stronger but is more extensible and has a more resilient character. This
produces a dough which has a gas bubble network with small-sized, strong and
extensible gas cell walls. There are three major areas in which this property can
be exploited: in white bread, in morning goods (e.g. rolls) and in wholemeal
(wholewheat), multigrain and seeded breads.

3.3.1.2.   Use of DATA Esters in UK White Bread

When the flour used in such breads contains an inadequate amount or less than
ideal quality of protein, the inclusion of DATA esters assists in stabilizing the
dough at the end of normal proof and still provides a dough with reasonable oven

spring. The baked bread is of higher specific volume with a more symmetrical appearance. Internally it has a finer gas cell structure with thinner cell walls, and because of this the bread crumb appears whiter, it has a finer, more even texture, feels softer and is more resilient. Levels of addition will be less than 0.3% of the flour weight.

### 3.3.1.3.  Use of DATA Esters in High-Volume Bread and Morning Goods

The term 'high (specific) volume breads' also covers hard and soft rolls, baps and 'Danish' breads in the UK. The aim in their manufacture is to produce articles with a high specific volume whilst retaining a relatively fine, even structure. These properties are mainly obtained by increasing proof volume to a level which might be considered excessive by normal standards. The proof times for such products tend to be 50–60 min, but the recipes contain relatively higher yeast levels. Proof is followed by significant but controlled oven spring.

This type of product requires a good-quality bread flour as a base for the greater gas retention contributed by the emulsifier. The doughs benefit from the improved proof stability conferred by the emulsifier, especially when the low-density doughs are transferred from proof to oven and in the early stages of baking.

In the baked product the quality characteristics benefit in a manner similar to those described above for white bread and in addition with the better quality flour which is used (one hopes), the baseline quality is higher and hence the finished article is of enhanced quality.

### 3.3.1.4.  Long Proof Processed Breads and Morning Goods

This group covers products ranging from baguettes to Scottish morning rolls and includes overnight proved rolls. For such products DATA esters are required to provide proof stability over periods between 2 and 16 h at low proof temperatures. The overall flour quality varies in this product grouping between low protein flours for baguettes and high protein flours (in some cases with additional gluten) in overnight Scottish morning rolls.

All these products require a dough which is stable to extended and often variable proof conditions. For baguettes it must be possible to cut the dough at the end of proof without collapse. In all cases stability is required to ensure a lively movement in the oven to produce an article of high specific volume – baguettes in particular must demonstrate a good 'burst' at the cut (Collins, 1978).

### 3.3.1.5.  DATA Esters in Wholemeal and Seeded Breads

A major difficulty in the commercial production of these types of bread is that the particles of bran and flakes of wheat and seeds disrupt the gas cell network in the dough. One approach to overcoming this problem is the addition of extra wheat gluten to the recipe. An alternative is the use of DATA esters or a combination of emulsifier and wheat gluten, which is preferred. The DATA esters confer enhanced gas retention with a finer, more even texture and importantly once again, a thinner gas cell wall structure. The resulting bread not only has increased volume but it also has a much finer, more even crumb structure, a better internal appearance and a softer, less harsh eating characteristic.

DATA esters have no effect on the properties of starch–water mixtures and therefore will not directly modify or reduce staling or staling rates (Chapter 10). They do have a beneficial effect on soft-eating shelf-life due to their ability to increase finished product volume and to facilitate a fine, even, thin cell wall structure in a wide range of products.

### 3.3.1.6.   Sodium Stearoyl-2-Lactylate (SSL, E482)

In the UK, SSL is permitted for use in bread up to a level of 3 g/kg in bread and 5 g/kg in fine bakers' wares, based on finished product weight.

This is a less complex material than DATA esters, although the number of lactic acid residues may vary, usually between two and five per molecule. SSL is a white solid with a comparatively high melting point and can be added to doughs in powder form, either alone or as part of a compound dough conditioner. It is miscible with fat and therefore is an ideal component of fat-based concentrates, particularly for semi-rich and rich fine bakers' wares, including rich buns and doughnuts.

The SSL has some of the properties of DATA esters. It enhances gas retention in the dough but weight for weight it is less efficient in this function. At the same time it does demonstrate genuine soft-eating shelf-life extension. It is capable of binding to amylose in a manner similar to that of distilled monoglyceride, which must account for its crumb softening effect.

In practice in the UK, SSL is seldom used in the manufacture of standard bread or in lean or crusty recipes. Bakers tend to prefer DATA ester-based dough conditioners for maximum gas retention and add distilled monoglyceride at the desired level when extra softness is needed. However, in semi-rich and rich yeasted goods ranging from soft baps to hot cross buns, and in yeast-raised doughnuts, SSL is often the preferred emulsifier. It gives some of the proof stability of DATA esters, but with a more limited and controlled oven spring resulting in baked (or fried) items with adequate volume (in the case of doughnuts with less blistering in the frier) and a white, fairly dense crumb with a soft, moist-eating mouth feel and an extended soft-eating shelf-life.

The ideal mode of use for SSL is a part of a compound fat-based dough conditioner together with salt, sugar and flour-treatment agents. The range of end products to which it is best suited is that containing both fat and sugar. These two factors combine to make SSL an attractive emulsifier in quite a wide range of baked goods, which is why its use continues to grow in the UK market.

In the USA and elsewhere, SSL may be replaced with the calcium form, CSL, and used at levels similar to that for SSL. The overall effects of CSL in breadmaking are similar.

### 3.3.1.7.   Distilled Monoglyceride (E471)

In the EC this category of emulsifiers is permitted at a QS level in all breads and fine bakers' wares.

It is used as a crumb softener and functions by binding to the amylose fraction of the wheat starch at the elevated temperatures typical of baking. In doing so it slows down retrogradation of the starch during cooling and subsequent storage, hence it can be claimed that it actually retards staling (Chapter 10).

There is no doubt that incorporation of monoglyceride into a recipe does extend the softness of the crumb of yeast-raised products, particularly during the first 3 days after baking. The difference is sufficient to be appreciated by the average customer in a wide range of products. For this reason they may be used cost effectively as partial fat replacers. The distilled monoglycerides function extremely well in this role but there is a temptation, particularly in semi-rich and rich fermented goods, totally to replace the fat by what is considered to be an appropriate level of monoglyceride. Such actions can result in products in which the crumb is indeed very soft immediately after baking, but over the next 24 h becomes weak, dry and crumbly. In products other than standard UK white bread it is not advisable to replace more than 50% of the fat in the original recipe.

The most functional fatty acid base for crumb softening is stearic acid which for baking ingredients has a relatively high melting point, 55–65 °C (130–150 °F), and therefore to be functional it must be made water dispersible in some way and must be in the correct crystalline state. For these reasons, distilled monoglycerides are offered to bakers in two distinct forms: as hydrates in water and emulsions, and as water-dispersible powders.

Hydrates are oil in water emulsions produced by blending the melted emulsifier and hot water together with a low level of a stabilizer and then cooling, with stirring under carefully controlled conditions to form a stable emulsion. If formed correctly, this emulsion will remain in the correct crystal phase permanently. The emulsifier level in such hydrates is in the region of 20–40% by weight. It is also possible to combine high melting point fats into such products to give them a dual function.

There are no set rules for levels of addition of hydrates; however, the following can act as a basic guide:

- in standard UK 800 g white bread, up to 1% of hydrate by weight of flour should be sufficient to significantly extend shelf-life;
- in morning goods containing fat, add 50% of the original level of fat in combination with 25% of the original level of fat hydrate;
- if a water-dispersible powder is used then it should be added in the same way, but assume that one part of powder is roughly equivalent to four parts of hydrate.

On its own, distilled monoglyceride has only one function in yeast-raised doughs; it softens the crumb of the product post-baking and assists in the retention of the extra softness for up to about 3 days. It does not enhance gas retention and therefore does not improve proof stability, product volume, crumb colour or crumb texture. Those hydrates which exhibit some of these other properties do so only because of the high melting point fat which is contained within the emulsion.

### 3.3.1.8.  Lecithins (E322)

The legal status within the EC for lecithins is that they are permitted QS in all types of bread and fine bakers' wares.

The term lecithin covers a group of complex phospholipids found naturally in a wide range of animals and plants. The most usual source of lecithin used in the baking industry is soya. It is extracted as a viscous liquid which is approximately 65% phospholipid and 35% soya oil. The liquid is blended with gypsum or wheat flour to produce a free-flowing powder which can then form the base of a composite dough conditioner.

Lecithin is not widely used in the UK. It is, however, very popular in France as the principal lipid source in baguettes and other crusty breads. The reasons for its popularity are in part historical and in part technical. It was, and still is, the only legal source of lipid permitted for a traditional French baguette (as defined within the EC). It is capable of enhancing gas retention in a dough to some degree, although in this respect it is far less efficient than DATA esters or SSL. At the same time it has a very different effect on crust character. DATA esters in particular produce crusty baked goods with a thin, 'egg-shell' crust which can have an 'exhibition' appearance but which tends to become leathery during storage. Lecithins give a thicker denser crust which may not look as attractive but tends to retain its crispness qualities for longer.

In the UK, lecithin may only have limited use, but in crusty goods, either when used alone or in conjunction with DATA esters, it is of potential benefit and has possibly been neglected for too long, being described as 'old fashioned'.

## 3.3.2.  Flour Treatment Agents

In the UK, throughout the EU and most countries of the world this group of functional ingredients has been considerably reduced with the removal of potassium bromate in 1990 and azodicarbonamide in 1995. The other flour-treatment agent which has been retained is the reducing agent L-cysteine.

### 3.3.2.1.  Ascorbic Acid (Vitamin C, E300)

At present the level of use is limited in the UK to 200 ppm by weight of flour, although higher levels are used elsewhere.

To understand the beneficial effects of oxidants in the breadmaking process it is necessary to describe briefly the basic structure of a wheat flour–based dough. The key factor is the wheat protein. When mixed with water, wheat protein has a property unlike almost any other plant protein to form a viscoelastic sheet.

The hydrated wheat starch granules are embedded in this structure which forms at the surface of minute gas bubbles produced in the dough from occluded air present on the flour particles prior to mixing. The two major components of wheat protein are usually described as glutenin and gliadin. Glutenin consists of high molecular weight proteins in which individual polypeptide chains are cross-linked by the disulphide bonds of the amino acid cysteine. Gliadins are composed of lower molecular weight proteins in which the cysteine cross-links are intra- rather than inter-molecular (Figure 3.2).

Once hydrated both classes of protein are to a high degree in the form of either *alpha*-helix or random helix. In addition to the cross-linked cysteine present in wheat protein there are a number of cysteine amino acids present in the reduced form as shown in Figure 3.3. The ratio of —S—H groups to —S—S— bonds is approximately 1:20. The reactions which occur between —S—S— and —S—H links during mixing and bulk fermentation or mechanical dough development are varied and complex, and discussed in greater detail in Chapter 11.

At this stage we need to consider only the most basic change which occurs in the dough, which is a reduction in the stresses experienced in the gluten network and is commonly referred to as the disulphide–sulphydryl interchange. This exchange is illustrated in Figure 3.3. It occurs to some degree during conventional (low-speed) mixing, but at much more rapid rate during mechanical dough development or spiral mixing. Introducing an oxidizing agent into the system complicates matters further. It is now possible to oxidize the —S—H group,

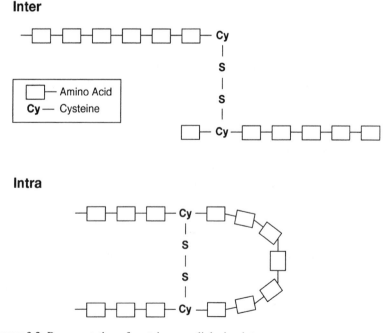

FIGURE 3.2. Representation of cysteine crosslinks in gluten.

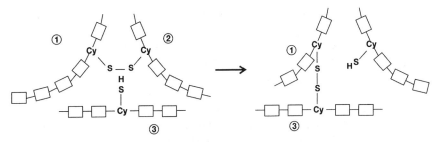

FIGURE 3.3. Representation of disulphide–sulphydryl interchange.

either to form new cross-links between protein chains or to oxidize the —S—H group to SO₃H, which is no longer capable of undergoing disulphide–sulphydryl interchange.

The possible effects of ascorbic acid in mechanically developed doughs can be listed as follows.

- Oxidation of water-soluble —S—H groups to remove them from the system. This would benefit the dough structure by preventing them from preferentially reacting with the —S—H groups of the glutenin molecules exposed during the development period.
- Causing an —S—S— bond to be formed between a water-soluble protein —S—H group and a glutenin —S—H group. This might well weaken the dough structure.
- Causing an —S—S— bond to be formed between two of the glutenin —S— H groups exposed during the development period. This would increase the elasticity of the dough structure.
- The direct oxidation of an —S—H group in a glutenin molecule to a stable form which is then unable to take part in further interchange reactions.

Thus the role of the oxidant is thought to be first to remove the water-soluble protein —S—H groups from the system. These groups will be the most readily reactable ones in the system, so that this must be the major reaction. The result is a shift of the balance of the reactions of glutenin molecules towards the formation of inter-glutenin —S—S— bonds rather than glutenin-soluble protein —S—S— bonds, thus producing a more elastic structure. It will also increase the degree of reaction of the 'masked' glutenin —S—H groups exposed during development, both to increase the number of —S—S— linkages and also to stabilize some as a non-reactive form. The overall effect is to produce a stable, stronger, more elastic gluten network capable of expanding without rupture, during the rapid growth of the gas cells in the early part of the baking process.

Bulk fermentation doughs are not subjected to the same conditions of stress as mechanically developed doughs since —S—H groups are not exposed to oxidation to the same degree. They are far less likely to react with small protein subunits to weaken the structure and so require far lower levels of oxidant to obtain optimum development. The optimum level of oxidant is therefore to a large extent determined by the dough processing system.

The use of ascorbic acid presents a further complication. Chemically it is a reducing agent and can only function as an oxidizing agent in a dough after it has been itself oxidized to dehydroascorbic acid, and to function efficiently requires the availability of atmospheric oxygen (Collins, 1994).

A breadmaking system which has significant needs for oxidation is the CBP which has traditionally been carried out using the application of partial vacuum during mixing to ensure a fine and uniform cell structure in some final products (Chapter 2). Coupled with this is the fact that in any yeasted dough the available oxygen is rapidly taken up by the yeast (Chamberlain and Collins, 1979) and it is easy to deduce that ascorbic acid is not the ideal choice of single, all-purpose oxidant for the UK and other baking industries using mechanical dough development processes like the CBP.

To summarize:

- Ascorbic acid is the only chemical oxidizing agent currently available in the European baking industry and many other parts of the world. This situation is very unlikely to change.
- It requires atmosphere oxygen and ascorbic acid esterase to function.
- It is a large molecule and may not be able to oxidize a number of —S—H groups which are 'protected' by the gluten molecule.
- In CBP and spiral systems a high level of addition is required (between 100 and 200 ppm by weight of flour).
- There is some evidence that, at levels higher than the currently permitted UK maximum of 200 ppm, it may be beneficial, for example in frozen doughs (Collins and Haley, 1992).
- In CBP doughs, particularly those mixed under partial vacuum, it very quickly becomes unable to access the atmospheric oxygen it requires to function effectively because of the competition from the yeast.
- In bulk fermentation systems it can only function efficiently in the mixer; later in the process it lacks the atmospheric oxygen to convert it to the dehydro-form.
- It is a relatively expensive material.

Against any of its shortcomings ascorbic acid does have a number of significant advantages.

- It is vitamin C and therefore it is almost inconceivable that it would under any circumstances be removed from the list of permitted additives for bread.
- The use of high levels of ascorbic acid in no-time doughs does not lead to the problems associated with over-treatment, such as those which occur with either azodicarbonamide or potassium bromate.
- The efficiency of ascorbic acid is greatly enhanced by the use of spiral mixers or by the newer CBP-type systems, such as oxygen enrichment or

pressure/vacuum mixers in which a supply of oxygen to the dough surface is maintained for part of the mixing cycle (Chapter 4).

It is possible to enhance dough characteristics using emulsifiers and enzymes to complement the action of ascorbic acid, as typically occurs with composite improvers. As discussed briefly above (see also Chapter 4), it is also possible to modify the mixing and dough development system to optimize the functions of ascorbic acid. In these circumstances the production of bread and other yeast-raised goods of high quality with ascorbic acid as the sole chemical oxidizing agent is possible. It may, however, require rebalancing of formulations or modification of production techniques, and more care and attention to detail on the part of bakers themselves to obtain optimum results; for example, through better control of dough processing times and temperatures.

### 3.3.2.2.   L-Cysteine (920)

This is an amino acid which due to its —S—H group can act as a reducing agent directly on the disulphide bonds in the gluten structure of the dough. It is commonly used in the hydrochloride form.

It was first used in the UK and Eire as a component of the Activated Dough Development (ADD) process as a mixed oxidant in conjunction with ascorbic acid and potassium bromate (Chapter 2). The ADD process allowed bakers to obtain some of the processing advantages of no-time doughs with low-speed mixers. L-cysteine relaxes the gluten structure during the mixing process, enhancing dough development. ADD was largely abandoned due to the loss of potassium bromate from the list of permitted additives in the UK and the progressive replacement of low-speed mixers by spiral mixer types.

L-cysteine does still have a role in the UK baking industry. It may be useful at a low level in single-piece 800 g CBP white bread to help to reduce (but not eliminate) streaks and swirls in the bread crumb cell structure generated at the moulding stage. In pinned-out morning goods, such as Scottish morning rolls and soft baps manufactured from CBP or spiral-mixed doughs using plants with little or no first proof, it can greatly ease stresses within the dough reducing misshapes, cores, streaks and swirls in the internal crumb structure. It can also reduce the incidence of holes under the top crust in such products.

It is beneficial in increasing flow in CBP-produced goods, such as hamburger buns which are processed in indented pans, often using short first proof systems usually designed for sponge and dough processes (Chapter 9). It can also be helpful in the manufacture of pizza bases, which must return to their original shape after sheeting or pressing (often such doughs contain a high level of rework which increases the resistance of dough to deformation).

Reducing agents are sometimes used to modify the behaviour of flours when the protein content is high or where the rheology of the doughs is unsuitable for the breadmaking process being used. In North America the treatment of strong flours which are said to be 'bucky' are commonly treated with a reducing agent can be relatively common. In such cases the addition of a reducing agent may be used to improve moulding behaviour.

### 3.3.2.3.   Oxidizing Agents in the USA

The list of permitted oxidizing agents which may be used in the USA is somewhat longer than that within the EC. At the time of writing, the use of potassium bromate is still permitted (and in Japan), although many bakers have stopped using it or do so at much reduced levels. Unlike the situation in Europe, North American bakers do not commonly use ascorbic acid, and salts of iodine, for example potassium iodate and azodicarbonamide are still in use. The latter are not limited to any great extent by available oxygen for their action and are suited to the sponge and dough processes in common use in North America (Chapter 2).

## 3.3.3.   Preservatives

The preservatives permitted for use in bread and fine bakers' wares in the UK are contained in The Miscellaneous Food Additives Regulations 1995. They are summarized in Table 3.1.

The permitted preservatives are intended to inhibit the growth of moulds and thermophilic bacteria (Chapter 10). Of the three groups of preservatives listed in Table 3.1, sorbic acid and its salts are of least value in bread and yeast-raised goods, because of their detrimental effects on dough characteristics rather than their efficiency in the inhibition of microbial growth. The levels of use required to give any significant increase in shelf-life result in doughs which are sticky and can be very difficult to process. The resultant baked product has a poor volume and a coarse, open cell structure.

If the customer is prepared to accept the presence of a preservative in the product, the propionates provide the most effective protection against mould growth. The usual choice from this group is calcium propionate.

The limits of use in the defined classes of baked goods are lower than those previously permitted in the UK regulations. They are, however, less restricting than they might at first appear. The levels refer to the free acid and they are based on the total weight of the baked product. There is a very significant loss in propionic acid level during baking. In practice the usual maximum levels of calcium propionate presently used in the UK, that is 3000 ppm by weight of flour, should give adequate protection and remain within the law for all classes of baked products. At this level the mould-free shelf-life can be extended by 2–3 days.

It is usual when adding calcium propionate to a dough to include it as part of a composite dough conditioner. This may be a little more expensive but it offers the following advantages.

- In the pure form, calcium propionate is not a pleasant material to handle. As part of a composite improver it is less likely to irritate the eyes and lungs.
- It can be incorporated into bulk handling systems for dough conditioners, so that the need for bakery staff to come into direct contact with it is virtually eliminated.

TABLE 3.1. Preservatives and their use.

| Product category | Maximum permitted level (ppm of finished product) | |
|---|---|---|
| All bread and fine bakers' wares | Acetic acid (E260)<br>Potassium acetate (E261)<br>Sodium diacetate (E262) | QS |
| Prepacked sliced and rye breads | Acetic acid (E260)<br>Potassium acetate (E261)<br>Sodium diacetate (E262) | QS |
| | Sorbic acid (E200)<br>Potassium sorbate (E202)<br>Calcium sorbate (E203) | 2000 |
| | Propionic acid (E280)<br>Sodium propionate (E281)<br>Calcium propionate (E282)<br>Potassium propionate (E283) | 3000 |
| Part-baked, prepacked and energy-reduced breads | Acetic acid (E260)<br>Potassium acetate (E261)<br>Sodium diacetate (E262) | QS |
| | Propionic acid (E280)<br>Sodium propionate (E281)<br>Calcium propionate (E282)<br>Potassium propionate (E283) | 2000 |
| Prepacked fine bakers' wares with $a_w > 0.65$ | Acetic acid (E260)<br>Potassium acetate (E261)<br>Sodium diacetate (E262) | QS |
| | Sorbic acid (E200)<br>Potassium sorbate (E202)<br>Calcium sorbate (E203) | 2000 |
| | Propionic acid (E280)<br>Sodium propionate (E281)<br>Calcium propionate (E282)<br>Potassium propionate (E283) | 3000 |
| Prepacked rolls buns and pitta | Acetic acid (E260)<br>Potassium acetate (E261)<br>Sodium diacetate (E262) | QS |
| | Propionic acid (E280)<br>Sodium propionate (E281)<br>Calcium propionate (E282)<br>Potassium propionate (E283) | 2000 |

- It minimizes the risk of errors in weighing, together with the possibility that it might be omitted totally in some dough or added at above the permitted levels in others.

Many producers do not wish to see preservatives included on the labels of their baked goods and the only material they will accept is acetic acid (vinegar). It is usually added in the form of a 12.5% solution at levels of about 0.6–0.9% based on flour weight. It is much less efficient in inhibiting mould growth than calcium propionate, although it does offer some limited protection. It has more benefit in inhibiting the growth of thermophilic bacteria provided that it is added at such a level that it lowers the pH of the dough to 5.2.

Where significant protection against microbiological spoilage is required it is best to use calcium propionate at a level of at least 2000 ppm by weight of flour and to use it all year round, since this offers protection against both moulds and thermophilic bacteria.

It should be noted that preservatives can only inhibit microbial spoilage; they do not destroy microorganisms. The essential basis for baked goods with an adequate shelf-life is a well-designed, clean, well-run bakery with a thorough, appropriate hygiene system correctly implemented and closely monitored.

## 3.4.    Processing Aids

### 3.4.1.    Enzymes

Enzymes are biological catalysts. They greatly accelerate the rates of reaction in natural plant and animal systems. In the past the use of enzymes in bread and flour has been strictly regulated in the UK to a limited positive list in successive Bread and Flour Regulations. Until recently the only two classes of permitted enzymes were *alpha*-amylases and proteinases. In 1995, after a long campaign by the whole of the baking industry, hemicellulases were added to that list. In 1996, much to their own surprise, the Bakery and Allied Traders Association succeeded in persuading the Ministry of Agriculture, Fisheries and Food in the UK to deregulate totally the use of enzymes in bread and flour in the UK. However, this act of deregulation was accompanied by the strong suggestion that any use of new enzymes in bread or flour should be preceded by an examination of each preparation by the Committee on Toxicity (COT) in the UK regarding all aspects pertaining to food safety.

#### 3.4.1.1.    *Alpha*-Amylases

The *alpha*-amylases are a range of enzymes which catalyse the same basic reaction, namely the breaking of hydrated starch molecules, both amylose and amylopectin un-branched and branched long-chain maltose polymers, into short-chain un-branched molecules known as dextrins. In combination with *beta*-amylase, an enzyme which attacks the ends of the amylose and amylopectin chains breaking off individual maltose sugar molecules, and given sufficient time and the right conditions, they are capable of converting starch almost totally to maltose (Figure 3.4). Each time a starch chain is broken by *alpha*-amylase two sites are created at which the *beta*-amylase can function. It is therefore the level of *alpha*-amylase which is the rate-determining enzyme in the system.

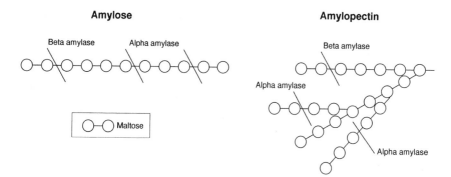

FIGURE 3.4. Representation of the actions of *alpha*- and *beta*-amylases on starch.

Most wheat flours contain an adequate level of *beta*-amylase, but usually only a low level of natural (cereal) *alpha*-amylase, and so it has become common practice to make adjustments to the amylase levels through additions of suitable materials. The effect of the addition of *alpha*-amylase to dough is dependent on at least three factors: the level of addition, the dough temperature and the heat stability profile of the particular type of *alpha*-amylase (Miller et al., 1953). Generalized heat and activity profiles are given for a range of types of *alpha*-amylases in Figure 3.5.

In the milling of wheat to flour for bulk-fermented dough, it was usual for the miller to add to the flour a low level of fungal *alpha*-amylase. This addition, in the region of 5–15 Farrand Units (FU) (Farrand, 1964), was to ensure the presence of an adequate amount of sugar for the yeast during final proof and in the early stages of baking. It was necessary to control both the damaged starch (Chapter 12) and the fungal *alpha*-amylase level to prevent excessive

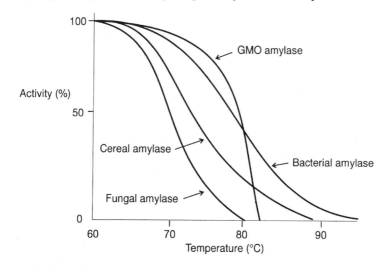

FIGURE 3.5. Generalized temperature–activity curves for *alpha*-amylases from different sources.

hydrolysis of starch, which leads to doughs becoming progressively softer and stickier throughout fermentation and subsequent processing.

In the UK, fungal *alpha*-amylase was preferred to cereal *alpha*-amylase derived from malted wheat or barley because of its lower heat inactivation temperature. This reduced the risk of the formation of an excessive level of dextrins in the bread during the later stages of the baking process. If there is an excess of dextrin formation, the baked loaf will have a dark crust and the internal cut surface will feel sticky to the touch. Problems even arise at relatively low levels of dextrin formation in sliced bread. The dextrins rapidly coat the slicer blades and it becomes progressively more difficult to achieve a clean cut. Eventually the bread becomes torn rather than cut and begins to compress and collapse at the in-feed to the slicer (Chamberlain et al., 1977). This difference is of much greater significance in no-time doughs, whether in spiral or high-speed mixers.

In many production units using the CBP, particularly when using flours which are lower in protein quantity and quality, it has been found beneficial to incorporate into the dough conditioner high levels of fungal *alpha*-amylase. Often these additions are in the region of 100–150 FU (Cauvain et al., 1985). In these circumstances the enzyme is not incorporated into the dough to provide a source of sugar for the yeast, but for its effect on loaf volume. The effect on dough characteristics and finished bread quality can be dramatic. Oven spring and hence final loaf volume can be considerably enhanced. The cut surface is whiter with a finer cell structure and a softer, more tender crumb.

It has been suggested that the increased oven spring is due to the action of fungal *alpha*-amylase on starch in the loaf in the temperature range of 55–60 °C (130–140 °F). At this point the hydrated starch begins to gelatinize. It is therefore very vulnerable to attack by *alpha*-amylase. If there is the correct balance in the system it is possible to delay the gelling of the starch within the baking loaf long enough for it to continue to grow in the oven for a little longer than would otherwise be possible. Observations on doughs containing high levels of *alpha*-amylase compared to a control of the same composition but containing little or no amylase confirm that the loaf containing the higher amylase levels does in fact continue to expand further into the baking process (Cauvain and Chamberlain, 1988).

The reason why cereal *alpha*-amylase represents a far greater risk than fungal *alpha*-amylase becomes apparent if one compares the heat inactivation curves of the two enzymes (Figure 3.5). Bearing in mind the additional factors that both yeast activity and *beta*-amylase activity cease by 55 °C, it is clear that in a baking loaf the cereal amylase will continue to produce dextrins during baking for far longer than fungal *alpha*-amylase, so that the final level of dextrins in the loaf will be higher.

The major risk associated with the use of fungal amylase is at the dough processing stage, and if care is not taken to determine the optimum level of addition, or if uncontrolled delays occur during processing, sticky un-processable doughs can result.

### 3.4.1.2.   Cereal *Alpha*-Amylase

The only sources of cereal *alpha*-amylase available to bakers are diastatic flours produced from malted barley and wheat. These flours contain complex enzyme cocktails which can be in some situations of great benefit. The two major components are *alpha*- and *beta*-amylases. In most but not all wheat flours there is an abundant supply of *beta*-amylase. Thus diastatic malt flour is regarded primarily as a source of cereal *alpha*-amylase.

As such, cereal *alpha*-amylase is not as widely used in the UK as fungal *alpha*-amylase. Its comparatively high heat inactivation temperature would make its use at levels of activity similar to those at which fungal *alpha*-amylase is currently used in the UK impractical for sliced breads. The level of dextrins produced within the bread would almost inevitably lead to slicing difficulties of the type discussed above.

In some circumstances diastatic malt flour does offer the baker specific advantages and opportunities, for example in the manufacture of unsliced crusty goods where a high specific volume with lively movement in the oven is desirable. The crust should be well coloured and crisp but not leathery. This is primarily achieved by the correct baking technique but can be facilitated by the addition of the correct level of diastatic malt flour. In Germany any compound dough conditioner designed for the production of small crusty rolls will contain diastatic malt flour rather than fungal *alpha*-amylase.

### 3.4.1.3.   Bacterial *Alpha*-Amylase

*Alpha*-amylases produced from bacteria are remarkable for their heat stability. They will in fact retain some of their activity throughout the baking process and can therefore continue to produce dextrins throughout the entire baking cycle and even during cooling. The result of adding even very low levels of bacterial *alpha*-amylase to doughs for the production of any sliced breads can be both spectacular and disastrous. The bread crumb becomes so sticky that it may even become impossible to slice the loaf. The only use for bacterial *alpha*-amylase in yeast-raised products is in the production of malt breads, where the incorporation of bacterial *alpha*-amylase into the recipe can produce such a loaf with a moist, sticky crumb which lasts for weeks rather than days.

### 3.4.1.4.   *Alpha*-Amylase Which is a Product of Modern Biotechnology

The ability of *alpha*-amylases to retard staling, that is the delaying of starch retrogradation has been recognized for some time. The practical difficulty has always been that any *alpha*-amylases which demonstrated this effect were relatively heat stable, that is they were of cereal or bacterial origin. Such amylases produced at the same time both extended softness and excessive stickiness in the baked product due to the formation of high levels of dextrins from both the amylose and the amylopectin fractions of the starch in the latter stages of baking.

If it were possible to fractionate *alpha*-amylases from existing cereal or bacterial sources to obtain a specific temperature stability profile and a controlled range of hydrolysis products, it might be possible to modify the amylopectin in the baked loaf to inhibit or delay starch retrogradation without the risk of causing excessive stickiness in the baked loaf.

Enzyme manufacturers have in fact been able to achieve this goal by producing *alpha*-amylase from a bacterium into which a gene for maltogenic amylase from a different bacterium has been introduced. The resulting *alpha*-amylase preparations do not contain enzymes which are themselves novel, nor do they contain genetically modified material. They do, however, have specific temperature stability profiles and produce a controlled range of break-down products from amylopectins, such that when used at low levels they are able to enhance soft-eating shelf-life without creating gumminess. Because of their low level of use they have no detrimental effects on the dough during processing.

In the UK, *alpha*-amylases have been specifically included in the National Bread and Flour Regulations since 1985. It is therefore considered that these new products are *alpha*-amylases and do not contain genetically modified material, and as such they should be permitted for use in bread (subject to general food safety regulations). It is also considered that since they are processing aids, they do not require to be labelled.

In practice the great advantage of these enzymes compared to traditional crumb-softening agents, such as distilled monoglycerides, is that they extend the soft-eating shelf-life for longer, that is they show a great advantage over the period of 3–6 days post-baking compared with monoglycerides, and are at least as good in the first 3 days. A significant further advantage is that this softness is not accompanied by either weakening or drying out of the crumb.

## 3.4.2.  Hemicellulases

The hemicellulases were added to the list of permitted enzymes in the UK prior to full deregulation, and therefore a significant number of commercial preparations have already been approved by the COT in the UK and are widely used in all sectors of the industry.

They can be quite complex enzyme systems, usually being derived from either fungal sources or from genetically modified organisms (GMOs). The function in their parent organisms is to break down plant cell wall material ultimately into its component sugars, mainly xylose and arabinose, to provide an energy source. The endosperm of wheat from which white flour is produced consists mainly of starch, protein and lipid (see also Chapter 11). It is subdivided in the grain into small cells, the cell wall material being composed of a complex mixture of long-chain molecules based primarily on the sugar xylose. There are a number of arabinose side chains, and in some cases the chains may be cross-linked via these arabinose sugars. In a limited number of molecules there is a direct bond between some of the material and protein via ferulic acid. This whole group of related materials is described as wheat pentosans and they are present in white flour at a level of about 2% of the total flour weight. They have historically been divided into two groups, the soluble and the insoluble pentosans, which are present in roughly equal weights.

The significance of the pentosans in dough structure becomes apparent if one examines the distribution of water in bread dough. Although representing only 2% of the total flour by weight (see Chapter 11), pentosans are associated with approximately 23% of the total dough water, that is they bind roughly seven times their own weight in water! Not surprisingly the pentosans have intrigued bakers and cereal scientists for decades. There are many reports of attempts to evaluate their function, which can be briefly summarized as follows: apparently, soluble pentosans enhance baking performance whilst insoluble pentosans detract from it.

The commercially available enzymes described as hemicellulases when added to dough have the effect of modifying wheat pentosans. It can be clearly demonstrated that a number of these enzyme systems (used at an appropriate level) have a considerable beneficial effect on the characteristics of a wide range of baked goods. At the dough stage their use results in dough with enhanced handling properties, and which is more extensible, without loss in strength or increase in stickiness. It may even result in a modest increase in water absorption. The dough has significantly better proof tolerance, and enhanced baking stability and oven spring. The baked product has greater volume together with a finer cell structure (Cauvain, 1985). The fine cell structure in turn yields better crumb colour, improved softness and resilience.

Bearing in mind that pentosans are a component of all wheat flours, it is not surprising that hemicellulases have been found to be beneficial across a wide spectrum of flour types and qualities and throughout the whole range of production systems.

The theories regarding the mode of action of hemicellulase in dough, as usual, lag behind the practical application. There is what is best described as a 'lively debate' as to its principal function. A number of suggestions are outlined below:

- by bringing about a controlled increase in the solubility of the pentosans, the rate of diffusion of carbon dioxide through the dough water is slowed down and gas retention is enhanced;
- by controlled breakdown of the most complex pentosans, their ability to interfere with the formation of the gluten network of the dough is reduced;
- bonds may be broken between pentosans and protein, both increasing the available protein and reducing potential disruption to the protein network;
- The release of water into the dough system during baking lowers its viscosity and permits greater and later expansion in a manner similar to that observed for *alpha*-amylase (S.P. Cauvain, personal communication).

## 3.4.3. Proteinases

This group of enzymes directly attacks protein chains, in doing so they increasingly and irreversibly weaken the gluten structure. They are therefore regarded with understandable suspicion by the UK baking industry, although they are more regularly used in North America to counteract the effects of 'bucky' flours, that is flours with elastic glutens (Kulp, 1993). 'Over-strong' gluten is not a common problem in most European bakeries. To achieve any discernible effect in the processing of no-time doughs, they need to be added at high levels. On occasions they can be of use in the production of pinned-out morning goods if used in conjunction with L-cysteine.

### 3.4.4. Other Enzymes in Baking

Oxidase enzymes find uses in breadmaking. Glucose oxidase is an enzyme which reacts with the *beta* form of glucose. In the presence of oxygen, glucose oxidase catalyses the oxidation of the *beta* glucose and in doing so produces hydrogen peroxide. The ability of hydrogen peroxide to oxidise thiol groups to form disulphide bonds often forms the basis of the claims for the improving effect of glucose oxidase in breadmaking (Wiksrom and Eliasson, 1998; Vemulapalli et al., 1998). However, in doughmaking systems where the availability of oxygen is restricted – usually by the yeast – the potential impact of glucose oxidase may be limited. For this reason glucose oxidase has found limited application in the CBP. The application of sulphydrl oxidase as a bread improver has been postulated (Kulp, 1993) and more recently transglutaminase (Bollain and Collar, 2005)

Lipase enzymes have been shown to be effective in improving bread quality (Si and Hassan, 1994). They have been shown to be particularly effective in the context of retarding staling (Poulson and Soe, 1996). The action of the lipase is to cleave the bond between the fatty acid esters and the glycerol on wheat flour lipids (Si, 1997), and the monoglycerides so formed may be involved in the reduction of the rate of staling in bread, much as the addition of glycerol mono-stearate does (see Chapter 10). Some forms of lipase will react with the phospholipids and glycolipids as well as the triglycerides in the dough system. The potential for lipases to replace DATA esters and SSL has been postulated (Rittig, 2005).

One enzyme which has attracted interest is asparaginase. Its use is not associated with improved bread volume, crumb softening or reduced staling. It has attracted attention because of its potential to reduce the formation of acrylamide during baking with reductions as high as 95% being claimed for some products into which the enzyme was incorporated (de Boer et al., 2005).

### 3.4.5. Novel Enzyme Systems

The recent deregulation of the use of enzymes in many baking industries has considerably broadened the range of enzymes which are now permitted for use. In a number of parts of the world, enzymes are considered to be 'processing aids' and as such do not need to appear on the ingredients listings. This is not a universal position with labelling of enzyme additions being required in the USA. Many new enzymes are derived from microbial fermentation. The microorganism may be found in nature or may come from a modified microorganism. In the EU the use of 'new' enzymes is covered by the Novel Food Regulations (EC, 2000) and will require clearance of the material if it does not come from an already permitted source.

### 3.4.6.  Health and Safety with Enzyme Usage

In recent year a numbers of concerns have been raised regarding the safety of use of enzymes in the bakery. These concerns are centered on the possible allergenicity of the materials for people working in manufacturing through exposure to dust in the atmosphere which may contain enzyme-active materials. In most bakeries, steps are regularly taken to reduce the presence of dust in the working atmosphere and the move in some bakeries to use liquid rather than powdered improvers (see below) is also contributing to the management of health and safety issues. There are no safety concerns for consumers eating products manufactured with enzyme additions. Most of the enzymes which are used in the manufacture of baked goods are inactivated by the heat of the oven. Some forms of bacterial *alpha*-amylase may not be inactivated during baking but these are not generally forms which find significant use in baking.

## 3.5.  Summary of Small Ingredients

The materials which have been discussed above constitute some of the principal functional 'small ingredients' currently found in bread and yeast-raised fine bakers' wares. When contemplating their use and potential benefits, it would be wise to bear in mind the following:

- their principal functions are to enhance and control the quality of baked goods;
- each one has a specific function and an optimum level of use in a particular system;
- they can only be used to best effect in carefully balanced combinations, so that each component complements the function of the others;
- they are able to compensate for minor variations and deficiencies in flour quality, and in plant and processing conditions;
- they are very definitely not intended to rescue the baker from incorrect or widely varying flour quality, nor from inappropriate or badly maintained plant;
- above all, they are no substitute for an adequate level of understanding of the science and craft of baking, conscientiously applied by the baker using the products.

It may appear that the range and permitted levels of such ingredients available to the baker are constantly being reduced. Sometimes it seems that the major customers, the consumer groups and regulatory bodies are in some dark conspiracy to reduce the formulation of bread back to the basics of flour, water, salt and yeast with which baking first began. An unbiased examination of the UK Miscellaneous Food Additives Regulations 1995 reveals a range and level of additives sufficient to produce the whole range of UK breads and fine bakers' wares of consistent high quality. In addition it is not likely that this list will be reduced in the foreseeable future. A further bonus is that a number of substances have been added to the list, for example gums and modified celluloses. When these changes are linked to the huge potential offered by the deregulation of enzymes, we may well be on the brink of fulfilling the old wish 'May you live in interesting times'.

## 3.6.  Liquid Improvers

Liquid dosing of ingredients to the mixer has been possible for many years. Indeed when first introduced the CBP was based on the delivery of the necessary ascorbic acid as a solution (Cauvain and Young, 2006). Liquid-based systems used for delivering functional ingredients are also found in the manufacture of hamburger buns (see Chapter 9). The increased automation of plant bakeries and improvements in the mechanical ability to meter in dosages of small ingredients has resulted in an increase in the use of liquid feed systems. Liquid yeast feeds have been used for some time (see below) but recent developments have led to the increased use of 'liquid' improvers. The new liquid improvers provide all of the necessary functional ingredients which are more commonly found in the powdered improver but they are oil-based rather than soya- or wheat flour–based.

In a liquid improver the necessary oxidant, enzymes and emulsifiers are suspended in a suitable vegetable oil. Solid fat may also be present, usually in small quantities and in a finely divided crystalline form. The absence of any water in the liquid improver ensures that there is no loss of functionality from active materials such as oxidants and enzymes. The technology used to manufacture and deliver the improvers ensures uniform dispersion of the low levels of functional ingredients that are present and therefore consistency between dough batches in the bakery.

Initially the use of liquid improvers was confined to the larger plant bakeries but increasingly they are becoming available for use in smaller bakeries using smaller pack sizes. One particular advantage to be gained from using liquid improvers is a reduction of dust in the bakery and this makes a significant contribution towards better working conditions for the operatives with reduced exposure to enzyme-active materials (see above).

## 3.7.  Bakers' Yeast

### 3.7.1.  Where Does Yeast Come From?

The use of yeast in breadmaking has at least 6000 years of history since fermentation of bread dough is thought to have started with the ancient Egyptians.

At that time, dough fermentation would probably have taken place by using a mixture of natural yeast and lactic acid bacteria. Bakers would save a portion of dough to seed subsequent doughs, and this method continued into the nineteenth century. In the Middle Ages, European bakers would produce a barm, a more liquid fermentation, which was often started with brewers' yeast from hops.

Louis Pasteur's research aided the understanding and development of yeast cultures. Commercial bakers' yeast started with the Vienna Process in the mid-nineteenth century. The basis of this process was to introduce air and a small amount of steam into the fermentation. The development of the Vienna Process increased the yield and enabled quality to be controlled. At this time, yeast was grown on grain. During 1915 (war-time in Europe) there was a shortage of grain and molasses started to be used commercially to grow yeast. By the 1950s, grain as a substrate was phased out, although today cereal is being used in areas where, for various reasons, it has become more economic to do so.

## 3.7.2.  Principal Forms of Yeast

Throughout the world today, bakers' yeast is produced in various forms which meet specific requirements of climate, technology, product, methodology, transportation and storage. Bakers' yeast comes in a number of different forms including compressed, granular, cream, dried pellet, instant, encapsulated and frozen. The variations are related to the physical form of the yeast, with the main differences being in the yeast moisture content. Various cultures are developed which depend on the properties required for the yeast's intended use; this will be discussed later.

### 3.7.2.1.  Compressed Yeast

This form is normally supplied in blocks and wrapped in waxed paper. A dry matter content of 28–30% is the standard for this product. In the UK, blocks are of 1 kg (2.2 lb); a 0.5 kg block is common in many other parts of the world. There is a small domestic market for 42 g (1.5 oz) blocks.

### 3.7.2.2.  Granular Yeast

This usually consists of small granules, has a dry matter content of 30–33%, and is supplied in laminated paper or plastic bags. Because it has a far greater surface area than compressed yeast, it is more vulnerable to temperature increases which impair the quality of the yeast. Granular yeast has in the main been superseded by cream yeast but has been used in brew processes and automated feed systems in breadmaking.

### 3.7.2.3.  Cream Yeast

Cream yeast is a pumpable form of yeast which has the consistency of cream. Cream yeast generally replaces compressed at a rate of 1.5:1, although some cream yeast is marketed at 1.7:1 replacement ratio, an important point for quality addition and costing. Also, when using cream yeast in the bakery, the extra water content should be compensated for in recipes.

### 3.7.2.4.   Dried Pellet Yeast

This was an early form of dried yeast, produced as small beige-coloured pellets with a very low moisture content. Its advantage over compressed yeast was that, when packed (and sometimes gas flushed), it could be transported and stored more easily, at ambient temperatures, and had a longer shelf-life. For use in the bakery, dried yeast needs to be reconstituted in five times its own weight of warm water. A variety of pack sizes are produced from 1 t packs for others to repack, through 12.5 kg (27.5 lb) packs to 20 g (0.7 oz) tins, in plastic containers and sachets. Dried pellet yeast has mainly been superseded by instant yeast for bakery and domestic uses.

### 3.7.2.5.   Instant Yeast

This form was developed in the 1960s; it has a very low moisture content and a fine particle size. The main advantage of using instant rather than dried yeast is that the yeast can de added directly to the flour. Instant yeast is popular on the islands around the UK and for premixes for bread and pizzas. The main use is for baking in parts of the world where compressed yeast is not available or where instant yeast is considered to be more convenient.

### 3.7.2.6.   Encapsulated Yeast

This is a specialist yeast produced for premixes where its use does not necessitate prior drying of the flour. Due to its high cost, encapsulated yeast is little used today.

### 3.7.2.7.   Frozen Yeast

This can be compressed yeast which has been frozen under special conditions; it needs to be defrosted slowly for use. There is also a specialist yeast produced for frozen dough which has a lower moisture content than compressed but higher than instant yeast, with the appearance of instant yeast and which, although frozen, is free flowing.

### 3.7.2.8.   Discussion

The above descriptions illustrate the various physical forms and appearances of yeast. Drying, freezing and encapsulation are all methods of yeast preservation. Different cultures and growth methods have also been developed by yeast manufacturers to produce yeasts for the different baking methods and fermented products. The cultures and growth methods will affect a number of yeast performance factors, such as activity, acid tolerance, osmotolerance, temperature stability, response to mould inhibitors and shelf-life. Some of these properties will be discussed later when considering the uses of yeast in baking.

## 3.7.3. Other Yeasts

Yeasts such as brewers' and distillers' yeasts may be used for bread type production but, as they are not specifically developed for that purpose, they are unsuitable because they have low activity and would not have the particular tolerances built into specialist bread yeasts. The following are examples of other yeast products.

- *Pizza yeasts* have low activity. They are an instant type of product with added benefits for pizza production, such as shrinkage reduction.
- *Deactivated yeasts* are another instant form of yeast product with additional properties such as characteristics to replace L-cysteine to assist in the relaxation of doughs.
- *Inactive yeasts* are produced to give flavour and as a flavour carrier, in products such as potato crisps.
- *Yeast extracts* are used in food spreads and soup flavourings.

## 3.7.4. Biology of Yeast Cells

Yeast is a living organism. It is found naturally, living around us, in the air, on the ground and on the fruits and leaves of many plants. A common example of a yeast is the bloom we can observe on grapes. There are about 500 known yeast species.

The scientific name for bakers' yeast is *Saccharomyces cerevisiae*, which means 'a mould which ferments the sugar in cereal (saccharo-mucus cerevisiae) to produce alcohol and carbon dioxide'.

The actions of yeast may be shown in a simplified form as follows:

$$\text{simple sugar} \Rightarrow \text{ethyl alcohol} + \text{carbon dioxide}$$

$$C_6H_{12}O_6 \Rightarrow 2C_2H_5OH + 2CO_2$$

Examining a yeast cell under a microscope will give a greater understanding of the composition and nature of yeast. The method for viewing a sample of yeast under a microscope is to disperse a small amount of yeast in water, causing the water to be slightly clouded, and then drop a spot of the liquor onto a glass slide. The drop is then covered and viewed with a $650\times$ magnification. The individual cells will take the general form illustrated in Figure 3.6.

If individual yeast cells were placed side by side, it would take approximately 1200 cells to measure 1 cm in length. Yeast is a unicellular organism (single cell) and 1 g of yeast contains around $15 \times 10^9$ cells. Individual cells will be seen to be round to oval in shape. Inside each cell are the following:

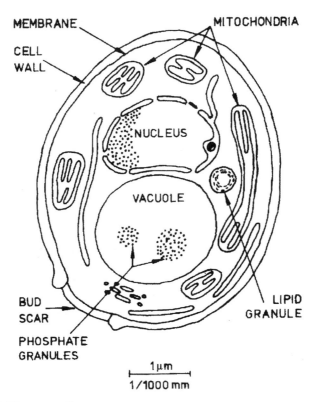

FIGURE 3.6. The yeast cell.

- A liquid solution of protoplasm, protein, fat and mineral matter.
- One or more dark patches called vacuoles.
- A darker spot which is the nucleus. This is where the cell's genetic information is stored as DNA which controls all the operations of the cell.

A yeast cell has 6000 different yeast genes. Yeast is made up of chromosomes, like any living thing; there are 16 different chromosomes in yeast compared with 23 in humans.

The double cell wall may have bud scars which are caused by budding, that is the cell reproducing itself. There can be up to ten such scars, which would cover the cell totally, after which the cell would expire.

The osmotic properties of a yeast cell are due to selective permeability of the cell wall with regard to solutions. This selectivity plays an important role in controlling the movement of nutrients into a cell. Nutrients are present in a medium in the form of ions, sugars and amino acids. The permeability of the cell wall also permits the release of alcohol and carbon dioxide from the cell during fermentation.

The yeast growth process is created by encouraging reproduction, by providing the correct conditions of warm water (30°C, 86°F) and nutrients (sugar). As was described earlier, yeast is asexual and reproduces by budding, the general form of which is shown in Figure 3.7. A parent cell grows a protuberance, this swells as the bud forms, a neck develops between the parent cell and the bud, and they separate. The process starts again and, in ideal conditions, a cell can reproduce itself in 20 min, so that numbers increase from one to two, then to four, to eight, to sixteen and so on. If the numbers are plotted on a graph, the line would take an exponential form.

Yeast can also reproduce by sporulation. This process occurs when cells are deprived of food. Each cell will produce up to four spores which, when the cell wall shrivels away, are released into the atmosphere. Yeast cells can survive in the form of spores for extremely long periods of time, and when later brought into contact with the right conditions the yeast will return to active life. Yeast microbiologists are able to use the state of yeast sporulation to interbreed yeasts in order to produce hybrid yeast with specific qualities. Today, genetic engineering of yeast cells in order to confer special properties is possible but, until it is more generally accepted, yeast producers consider that it would be unwise to risk a potentially devastating drop in bread sales.

### 3.7.5.  Overview of Commercial Yeast Production

The typical raw materials used for the commercial production of yeast growth are as follows:

- molasses cane/beet;
- ammonium hydroxide;
- magnesium sulphate;
- diammonium phosphate;
- air;
- sulphuric acid/sodium carbonate (processing aids for pH control).

For yeast growth by reproduction, a carbohydrate food source is required. This could be grain, but today molasses is a cheaper form of sugar and therefore is preferred. The vitamins and minerals in molasses support yeast growth. Molasses from beet, cane or a mixture of both can be used. As molasses consists of approximately 50% sugar, up to twice the weight of molasses is required to produce compressed yeast. Nitrogen is a major growth nutrient, which is added in the form of ammonia. Oxygen is provided in the form of filtered air.

A starter culture is grown under sterile conditions in a laboratory. A perfect, healthy cell is selected under a microscope, the culture is grown in a test tube with the required nutrients and is then transferred to flasks and larger vessels until it has grown to the required quantity for a commercial starter. The yeast culture is then constantly fed over a period of time and, as it grows and expands, it passes through approximately six stages of commercial propagators, as illustrated in Figure 3.8.

At this stage in the production cycle the yeast cells are suspended in a large amount of water. The concentrated yeast is extracted from the water by a

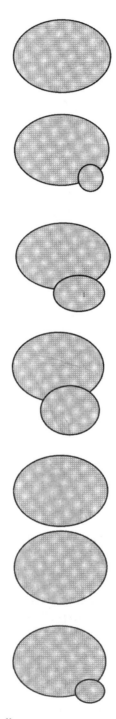

FIGURE 3.7. Budding in yeast cells.

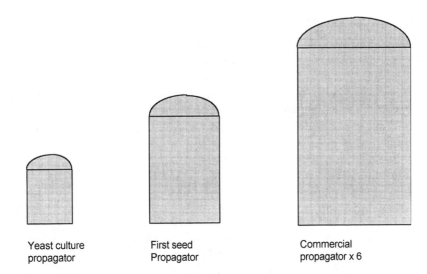

Yeast culture
propagator

First seed
Propagator

Commercial
propagator x 6

FIGURE 3.8. Commercial propagation of yeast.

centrifuging process in large separators, which provides yeast in a cream form and an effluent waste. The yeast cream is cooled and can be transported in that form without further processing to large bakeries for use.

The water content of the yeast cream is further reduced by filtering, passing through a filter press or rotary vacuum filtration unit. The next process is to prepare the yeast for its final form:

- extruded for compressed yeast;
- minced for granular yeast;
- dried, either by drum drier, spray drier or fluidized bed drier for pellet or instant yeast;
- freezing of compressed or semi-dried yeast.

The yeast is packed and then transported by road, rail or ship, depending on the destination, and in frozen, chilled or ambient forms, depending on the product.

### 3.7.6.  Baking with Yeast

Throughout the world there are many different types of bread, each of which requires different characteristics of yeast. Examples include products made in hot and humid climates; those requiring long product shelf-lives where additions of mould inhibitors at significant levels may be used; high-sugar sweetbreads; and those for slow or rapid fermentation processes.

### 3.7.6.1.  Chorleywood Bread Process

As a result of the introduction of the CBP in the early 1960s, a new type of yeast became necessary. The yeast commonly used in the industry had too slow a reaction for the new, faster process and had an inappropriate fermentation and gassing curve. Specifically, the yeast was lacking in gas production at the critical stages of proof and during the early stages of baking. Increasing the quantity used of the existing yeast type could not rectify the situation. The two graphs shown in Figure 3.9 illustrate the problems incurred with the existing yeast in the CBP and the effects of the new culture which was produced specifically to overcome these problems.

It will be seen from the results given in Figures 3.9a and b that when using the pre-1960s yeast in the CBP (Figure 3.9a), gas production dropped at the time that the bread was about to enter the oven, resulting in little or no oven spring. In the graph for the post-1960s yeasts (Figure 3.9b), there is no drop in gas production. Yeast for the CBP is used at a rate of 2% of flour weight, which is more than the percentage of yeast used in long-fermentation processes.

### 3.7.6.2.  Acidity or Alkalinity

The most favoured pH range for yeast is 4.5–6.0. Bread doughs are generally in the region of 5.5, so in normal breadmaking the effect of pH is not a particular consideration.

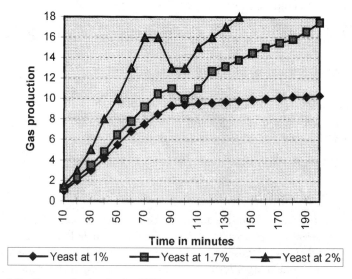

FIGURE 3.9a. Gas production in the CBP.

## Post 1960s yeast at varying levels

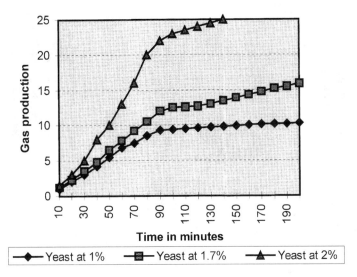

FIGURE 3.9b. (*Continued*)

3.7.6.3.   Storage Stability (Compressed Yeast)

Dough proof time is dramatically affected by the age and the temperature at which the yeast has been stored. The percentage difference in proof time of yeast stored at 4 °C (39 °F), 10 °C (50 °F) and 15 °C (59 °F) for both 7 days and 14 days is compared in Figure 3.10, which shows how proof time increases with the temperature at which the yeast has been stored.

## Yeast storage stability
## [high activity compressed yeast]

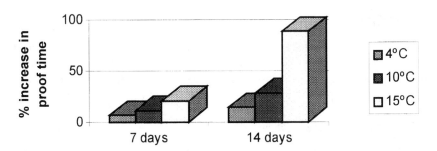

FIGURE 3.10. Yeast storage stability.

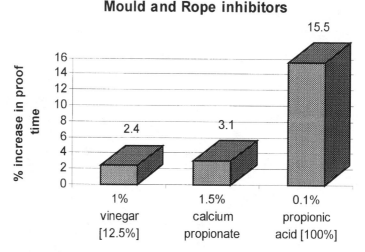

FIGURE 3.11. Effect of mould and rope inhibitors on yeast activity.

### 3.7.6.4.  Effects of Mould and Rope Inhibitors

Mould inhibitors, such as propionic acid and calcium propionate, have a retarding effect on yeast fermentation. Depending on the type of bread product, the required shelf-life and climatic conditions, the addition of inhibitors will vary. It will have the effect of slowing the yeast activity and extending the proof time. It is normal for bakers to add extra yeast to the mix when using mould inhibitors. The quantity will depend on which yeast is used, since acid resistance is built into yeasts which are normally sold for use in recipes and products which contain mould inhibitors.

To prevent rope occurring in bread, acetic acid or spirit vinegar (12.5%) may be added. This will also affect yeast fermentation but to a lesser degree than mould inhibitors. Since the introduction of the 'E'-number classification in Europe, there has been a preference by bakers to use spirit vinegar, a natural product, to avoid having an E-number and giving a 'clean label' to the bread product. A comparison of the effects of vinegar, calcium propionate and propionic acid on yeast activity (proof time) is shown in Figure 3.11.

### 3.7.6.5.  Effect of Spices and Bun Spice

Spices will affect the activity of yeast, the quantity of spice being determined by the required flavour of the finished product. It is normal for a higher yeast level to be used in such products, and bun spice is often added with fruit as late as possible in the process in order to reduce its effect on yeast activity.

### 3.7.6.6.  Yeast in Frozen Dough

As a generalization, a slower yeast is more robust and a faster yeast is more fragile, although yeast producers can take steps to overcome specific situations,

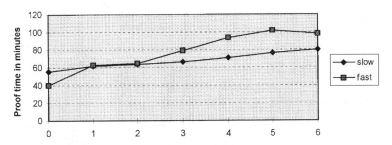

FIGURE 3.12. Yeast characteristics in frozen dough.

such as by selecting appropriate cultures and changing growth patterns. In Figure 3.12 the effects of different yeasts and their performance after being frozen and stored for up to 6 months at −18 °C (0 °F) are compared. The two yeasts shown are very similar at between 1 and 2 months storage. If the yeast was only frozen for up to 1 month, then the faster yeast would be better, whereas when frozen for between 3 and 6 months the slower yeast appears to be better.

There is a specialist frozen yeast which is produced, a free-flowing frozen yeast (F3). Freezing of this yeast takes place during its final stage of production, and it remains frozen during delivery and until it is used. This is considered to be the best yeast for producing frozen doughs as it has built-in tolerance to freezing.

### 3.7.6.7.  Salt

Salt can be used to aid the control of fermentation. There are a number of breadmaking methods, such as delayed salt (Chapter 2), which allow maximum fermentation before the salt is added to check the fermentation rate. Salt is also required to assist the flavour of bread; with no salt, bread is insipid (Chapter 1). The normal rate of salt addition is about 2% of flour weight, but when sugar is included, and especially with high sugar levels, this salt level can be reduced to 1%.

### 3.7.6.8.  Sugar

In the UK, little or no sugar is used in basic breads, although specialist breads and other fermented goods, such as morning goods, may have up to 15% sugar. Elsewhere, such as in African and Asian countries, it is not uncommon for as much as 30% sugar to be used. Special yeasts with high sugar tolerance, for high-sugar doughs, are produced for these countries.

### 3.7.6.9.  Effect of Salt and Sugar on Yeast Activity

Both salt (sodium chloride) and sugar affect the activity of yeast. The extent to which changes in osmotic pressure caused by increasing salt level affects the

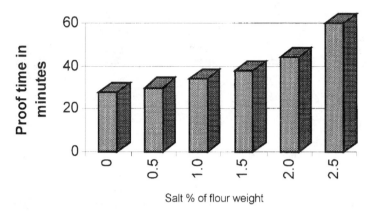

FIGURE 3.13. Effect of salt on yeast activity.

activity of yeast is shown in Figure 3.13. Weight for weight, salt has a far more dramatic effect on yeast than sugar, as can be seen by comparing the results illustrated in Figure 3.13 with those given in Figure 3.14. The quantities of salt and sugar are further adjusted in bread baking for reasons of flavour, texture and product shelf-life, and so the formulation of fermented products can have a significant influence on yeast activity.

## 3.7.7.  Conclusion

Throughout history, yeast cultures have been developed to meet market needs. The market for bakers' yeast has increased due to higher demand for fermented

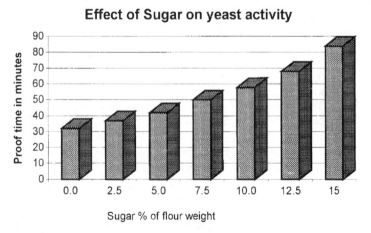

FIGURE 3.14. Effect of sugar on yeast activity.

goods and also due to the development of automated breadmaking processes where higher yeast levels are used. The requirements for yeast are diverse, covering local to worldwide use, from home baking once a week through hand and craft bakeries to fully automated bakeries using as much as 25 t of yeast per week. The finished products vary enormously according to local culture, the availability of grain and costs. Yeasts for these different needs may be single strain, hybrid or with mixed strains and propagation profiles. Their physical condition may be liquid, dried or compressed, with storage at ambient to frozen temperatures, either sealed airtight or inexpensively wrapped. Ultimately it is necessary for the baker to choose the most suitable yeast for the process, climate and recipe, according to commercial availability.

## 3.8.   Starter Doughs

Older methods of aerating bread dough were based on the use of portions of old dough which had fermented because of the presence of natural, wild yeasts and to a lesser extent bacteria. With the advent of the production of baker's yeast such methods became less popular, in part, because using modern baker's yeast provided a consistency of gassing performance which could not be readily achieved with 'natural' fermentation. However, in some parts of the world and for some products (e.g. rye bread, see Chapter 13) the principles of using starter or 'mother doughs' remains in practice.

In recent years there has been increased interest in the starter or 'mother doughs' as a means of changing bread flavour and to a lesser extent providing gas production. There are many variations on the starter dough concept but the basic principle is that only flour and water are mixed together, commonly in equal proportions and left to ferment naturally for 24 h or more. The starter is added to dough mix at levels between 10 and 25% of the flour weight depending on the degree of acidity that is required in the bread. The starter is perpetuated by retaining a portion of the original mix and replacing the mass that has been removed for baking with fresh flour and yeast. The starter dough may be based on baker's yeast or some other suitable strain.

If the starter dough principle is used it is important that the fermentation conditions used favour the yeasts and welcome bacteria. In the latter case the main organisms naturally found in flour will be Lactic acid bacteria and it is these that are largely responsible for the acidic bite that give sour dough products their distinctive flavour. It is important that unwanted organisms are not allowed to proliferate in the starter dough, otherwise off-flavours and odours may develop. In some cases the gas production potential of the starter may be impaired.

Control of the microflora in the starter culture is usually achieved through a combination of temperature and pH control. All microorganism have a combination of favoured temperature and pH conditions for growth. Conditions which favour the reproduction of one type of microorganism do not necessarily favour another and indeed may be used to suppress the activity of the undesired organism. In broad terms, yeasts prefer warmer conditions than bacteria. Thus, the functionality of lactic acid bacteria may be encouraged by using cooler fermentation or proof conditions than would normally be used, for example, in bulk fermentation. This difference contributes to the more acidic flavour that is often observed when dough is retarded or proved at low temperatures (see Chapter 6). There is some evidence that lactic acid bacterial activity in sours and sponges (or through cool proof) contributes to product shelf-life through an anti-fungal effect.

The sponge of the sponge and dough process (see Chapter 2) differs in concept to that of the starter or mother dough in that the sponge is made, fermented and completely used in the subsequent manufacture of bread. Even if the sponge is used subsequently to make more than one dough, it is necessary to make new sponges from all fresh ingredients. Nevertheless as discussed previously using a sponge can introduce different flavour into the subsequent products.

## References

Baker, J.C and Mize, M.D. (1942) The relation of fats to texture, crumb and volume of bread. *Cereal Chemistry*, **19**, 84–94.

de Boer, L., Heeremans, C.E.M. and Meima, R.B. (2005) Reduction of acrylamide formation in bakery products by application of *Aspergillus niger* asparaginase, in (eds S.P. Cauavin, S.E. Salmon and L.S. Young) *Using Cereal Science and Technology for the Benefit of Consumers*, Woodhead Publishing Ltd., Cambridge, pp. 401–405.

Bollian, C. and Collar, C. (2005) Impact of microbial transglutaminase on the fresh quality and keepability of enzyme supplemented pan breads, in (eds S.P. Cauavin, S.E. Salmon and L.S. Young) *Using Cereal Science and Technology for the Benefit of Consumers*, Woodhead Publishing Ltd., Cambridge, pp. 152–157.

The Bread and Flour Regulations (1998) UK SI 1998, No.141 as Amended by SI 1999, No. 1136, HMSO, London.

Cauvain, S.P. (1985) Effects of some enzymes on loaf volume in the CBP. *FMBRA Bulletin*, No. 1, February, CCFRA, Chipping Campden, UK, pp. 11–17.

Cauvain, S.P. and Chamberlain, N. (1988) The bread improving effect of fungal *alpha*-amylase. *Journal of Cereal Science*, **8**, November, 239–248.

Cauvain, S.P., Davies, J.A.D. and Fearn, T. (1985) Flour characteristics and fungal *alpha*-amylase in the Chorleywood Bread Process. *FMBRA Report No. 121*, March, CCFRA, Chipping Campden, UK.

Cauvain, S.P. and Young, L.S. (2006) *The Chorleywood Bread Process*, Woodehead Publishing, Cambridge, UK.

Chamberlain, N. and Collins, T.H.. (1979) The Chorleywood Bread Process: the roles of oxygen and nitrogen. *Bakers' Digest*, **53**, 18–24.

Chamberlain, N., Collins, T.H. and Elton, G.A.H. (1965) The Chorleywood Bread Process: improving effects of fat. *Cereal Science Today*, **10**(8).

Chamberlain, N., Collins, T.H. and McDermott, E.E. (1977) The Chorleywood Bread Process: the effects of *alpha*-amylase activity on commercial bread. *FMBRA Report No. 73*, June, CCFRA, Chipping Campden, UK.

Collins, T.H. (1978) The French baguette and pain Parisien. *FMBRA Bulletin*, April, CCFRA, Chipping Campden, pp. 107–116.

Collins, T.H. (1994) Mixing, moulding and processing of bread doughs in the UK, in *Breeding to Baking*, Proceedings of an International Conference at FMBRA, Chorleywood, 15–16 June, CCFRA, Chipping Campden, pp. 77–83.

Collins, T.H. and Haley, S. (1992) Frozen bread doughs: effect of ascorbic acid addition and dough mixing temperature on loaf properties. *FMBRA Digest 114*, February, CCFRA, Chipping Campden, pp. 21–3.

EC (2000) *Regulation (EC) No. 258/97 of the European Parliament and the Concil concerning noven foods and novel food ingredients.*

Kulp, K. (1993) Enzymes as dough improvers, in (eds B.S. Kamel and C.E. Stauffer) *Advances in Baking Technology*, Blackie Academic & Professional, London, UK, pp 152–176.

Farrand, E.A. (1964) Flour properties in relation to modern breadmaking processes in the United Kingdom, with special reference to *alpha*-amylase and damaged starch. *Cereal Chemistry*, **41**, March, 98–111.

Frazier, P.J., Leigh-Dugmore, F.A., Daniels, N.W.R. et al. (1973) The effect of lipoxygenase action on the mechanical development of wheat flour doughs. *Journal of Science, Food and Agriculture*, **24**(4), 421–436.

Miller, B.S., Johnson, J.A. and Palmer, D.L. (1953) A comparison of cereal, fungal and bacterial *alpha*-amylase as supplements in breadmaking. *Food Technology*, **7**, 38.

The Miscellaneous Food Additive Regulations, UK SI3187 (1995) HMSO, London.

Poulson, C.H. and Soe, J.B. (1996) Effect and functionality of lipase in dough and bread, in (eds S.A.G.F. Angelino, R.J. Hamer, W. Van Hartingsveldt, F. Heidekamp and J.P. Van Der Lugt), First Symposium on Enzymes and Grain Processing, Nutrition and Food Research Institute, Wageningen, The Netherlands, pp. 204–214.

Rittig, F.T. (2005) Lipopan F BG – unlocking the natural strengthening potential in dough, in (eds S.P. Cauavin, S.E. Salmon and L.S. Young) *Using Cereal Science and Technology for the Benefit of Consumers*, Woodhead Publishing Ltd., Cambridge, pp. 147–151.

Si, J.Q. (1997) Synregistic effects of enzymes for breadmaking. *Cereal Foods World*, **42**(10), 802–803.

Si, J.Q. and Hassan, T.T. (1994) Effect of lipase on breadmaking in correlation with their effects on dough rheology and wheat lipds. *Proceedings of International Symposium AACC/ICC/CCOA*, AACC, St. Paul, Minneapolis, USA.

Vemulapalli, V., Miller, R.A. and Hoseney, R.D. (1998) Glucose oxidase in breadmaking systems. *Cereal Chemistry*, **75**, 439–442.

Wiksrom, K. and Eliasson, A.C. (1998) Effects of enzymes and oxidising agents on shear stress relaxation of wheat flour dough: additions of protease, glucose oxidase, ascorbic acid and potassium bromate. *Cereal Chemistry*, **75**, 331–337.

# 4
# Mixing and Dough Processing

David Marsh and Stanley P. Cauvain

An integral part of all breadmaking is the formation of a smooth and homogeneous dough with a developed gluten structure. As discussed in an earlier chapter, in some breadmaking processes dough development continues during resting after mixing, while in others full development is achieved during the mixing process itself. Whatever the method by which dough development is achieved the next stage in bread manufacture is the subdivision of the bulk dough (dividing) and the shaping of the individual dough pieces (moulding) to conform to the requirements of the bread variety being made. Shaping may be a multi-stage operation and may involve a further resting period between moulding stages (intermediate or first proof). Once finally formed the dough pieces commonly pass on to be proved before baking.

Before the introduction of machinery, dough, all over the world, was made by hand mixing of the ingredients and then by kneading the mixture until a dough was created. The processes of mixing, dividing and moulding can be carried out by hand; indeed this is still the case in many bakeries, for example in India where the production of loaves in Mumbai is still based on hand mixing. Increasingly the operations of mixing, dividing and moulding are becoming mechanized. The purpose of this chapter is to discuss the essential elements of dough mixing and processing, to consider how they are achieved and to consider how equipment design can impact on final product quality.

## 4.1. Functions of Mixing

All mixing machines available today are designed to incorporate both the mixing and the kneading characteristics of the manual process. In essence, mixing is simply the homogenization of the ingredients, whereas kneading is the development of the dough (gluten) structure by 'work done' after the initial mixing. In mixing machines today, this 'work' is carried out by a variety of methods, each suiting the output capacity required, the type of dough required for the final product specification and its use in subsequent processing.

Some of the basic requirements for dough mixing have been introduced in previous chapters, but it is worthwhile to summarize them again before considering the different types of mixing machines which are available and how they may or may not meet the basic requirements of dough mixing.

We can summarize mixing requirements as the following:

- to disperse uniformly the recipe ingredients;
- to encourage the dissolution and hydration of those ingredients, in particular the flour proteins;
- to contribute energy to the development of a gluten (hydrated flour protein) structure in the dough;
- to incorporate air bubbles within the dough to provide gas nuclei for the carbon dioxide generated from yeast fermentation and oxygen for oxidation and yeast activity;
- to provide a dough in a suitable form for subsequent processing.

## 4.2.  Types of Mixer

We will see that mixing machines vary widely from those that virtually mimic a hand mixing action to high-speed machines which are able to work the mix intensively to the required dough condition within a few minutes. Many mixing machines still work the dough as originally done by hand through a series of compressing and stretching operations (kneading), while others use a high speed and intensive mechanical shearing action to impart the necessary work to the dough.

In many mixing processes the velocity of the dough being flung around within the mixing chamber is used to incorporate the full volume of ingredients into the mix and to impart energy to the dough from the mixing tool during kneading. Where mixing systems rely more heavily on this effect, they tend to require a higher minimum mixing capacity for a given mixing chamber capacity in order to remain efficient because the mixing tool does not come into intimate contact with every ingredient molecule. This practical effect tends to limit the higher-speed mixers to the large-scale bakeries where bread plants are running at near maximum capacities and variations in batch mixing sizes are not common. In smaller-scale production greater versatility of batch size may be required from the mixers, and so lower mixing speeds and more intimate contact between the mixing tool and the dough are an advantage.

In order to describe the most common variants of mixing machines and their applications, they may be divided into five common groupings (the first four being based on batch mixing) as follows:

- Chorleywood Bread Process (CBP) compatible, where the essential features are high mixing speeds and high-energy input, to develop the dough rapidly, and control of the mixer atmosphere;

- high speed and twin spiral, where a high level of work can be input to the dough in a short time;
- spiral, in which a spiral-shaped mixing tool rotates on a vertical axis;
- low speed, where mixing is carried out over an extended period of time;
- continuous where the dough leaves the mixer in a continuous flow.

## 4.2.1.  CBP-Compatible Mixers

The essential features of the CBP have been described in Chapter 2. For a mixer to be compatible with the CBP it must be capable of delivering a fixed amount of energy in a short space of time, usually 2–5 min. The required energy will vary according to the properties of the flour and the product being made. In the UK, energy levels of about 11 Wh/kg (5 Wh/lb) of dough in the mixer are common, while in other parts of the world or with products such as breads in the USA, this may rise to as much as 20 Wh/kg (9 Wh/lb) of dough (Tweedy of Burnley Ltd., 1982). Whatever the absolute energy level to be used, the short mixing time is very important in achieving the correct dough development and bubble structure formation in the dough.

Because of the CBP requirements, motor power levels will be large. The most common CBP-compatible mixers consist of a powerful vertically mounted motor drive, directly coupled through a belt system to a mixing blade (impeller) mounted vertically in a fixed bowl or tub (Figure 4.1). The high velocity of dough being flung off the impeller sweeps the bowl walls clean during mixing and subsequent mechanical development. The mixing bowl is mounted on horizontal pivots and is capable of being tilted to receive ingredients and for dough discharge into a trough (Figure 4.2). Other systems are available using a horizontal motor drive directly coupled to the mixing tool via a straight coupling or gearbox. Some versions have additional motor-driven bowl scrapers to encourage dough

FIGURE 4.1. Mixing chamber for CBP-compatible mixer shown with tool for removing mixing blade (courtesy ECS Engineering).

FIGURE 4.2. Mixer and ingredient feed system for CBP doughs (courtesy APV Baker Ltd.).

into the mixing area. In these mixers discharge is through an end door or through a bottom-mounted gate.

There are many variations to the design of the impeller blades used within CBP-compatible mixers. The primary function of the impeller is to aid dough development and it does this by interaction with a series of projections fitted on the inside of the bowl. As the dough impact on the bowl wall the projections turn it back towards the impeller blade and gravity pulls the dough downwards. In addition to this tumbling action the action of the impeller blades sweeping past the internal projections stretches the part of the dough which is momentarily trapped in the narrow space between projection and impeller plate. The number and positioning of the internal projections can have an effect on the rate of energy transfer during mixing, as can the design of the impeller or impact plate. However, measurements of gas bubble populations and bread crumb cell structure suggest that the design of the impeller has a limited effect on these dough and bread properties (Cauvain, 1999).

In many CBP-compatible mixers, control of the headspace atmosphere is incorporated into the mixing arrangements. In its 'classic' form this consisted of a vacuum pump capable of reducing the headspace pressure to 0.5 bar (15 in of mercury). This arrangement permitted the addition of extra water to the dough, provided a denser dough at the end of mixing (Collins, 1983) and reduced variations in dough divider weights. It also yielded a finer, more uniform cell structure, a softer crumb and a brighter crumb appearance. The application of partial vacuum can continue throughout the mixing period or may be delayed to the latter part of the mixing cycle. The advantages in delaying the application of partial vacuum are that better oxidation via ascorbic acid can be achieved in the early part of the mixing cycle before oxygen levels are depleted by gas volume reduction and yeast activity.

In the production of US style-breads, where fine cell structures and higher energy inputs are required to achieve optimum dough development, CBP-compatible mixers may be fitted with a cooling jacket to maintain control of final dough temperatures (French and Fisher, 1981).

Additional features of CBP-compatible mixers include the following:

- measurement of energy input to permit mixing to a defined dough energy input;
- automatic control of the mixing cycle and changes in mixer headspace pressure;
- automatic ingredient feed systems;
- programmable logic controls integrated with a preselectable recipe menu and fault diagnostic system;
- an integrated washing and cleaning system.

The most common applications of CBP-compatible mixers are the high-capacity production of bread and rolls with production lines rated for continuous 24 h output. Typical mixing plants are available as single or duplex mixers with outputs from 2,000 (single mixer) to 10,000 (larger twin mixers) kg dough/h (4,400–22,000 lb/h). More bread is produced from dough mixed with CBP-compatible mixers in the UK than from any other mixing system, and they can be found in use in many other countries around the world, including Australia, New Zealand, South Africa, India, the USA and more recently Germany, Spain and France. A wide range of bread products are manufactured with CBP-compatible mixers and include pan breads, rolls and hamburger buns (Cauvain and Young, 2006).

## 4.2.2. Mixing Under Pressure and Vacuum with CBP-Compatible Mixers

The very strong link between the gas bubble structure in the dough during mixing and the crumb cell structure in the baked product has been commented on above. Similarly the oxygen dependency of ascorbic acid and its contribution to dough development has been discussed. With the loss of potassium bromate as a permitted oxidizing agent in UK breadmaking (and elsewhere), the relationship between mixer headspace atmosphere and ascorbic acid became more critical. If ascorbic acid was the sole oxidant and the whole of the (short) dough mixing cycle was carried out under a partial vacuum, the resultant bread lacked volume and had a coarse crumb cell structure and a dark-coloured crumb. Delaying the introduction of partial vacuum to the later stages of mixing brought about improvements in final product quality. However, a key requirement of the successful application of partial vacuum was that the dough should be subjected to the reduced pressure setting for a reasonable period of time during mixing. In other words, it was not simply a case of achieving the final reduced pressure level at the end of the mixing cycle, the reduced pressure had to be achieved before mixing was completed. Typically the length of time required for the reduced pressure to be effective in delivering the required cell structure in the final product was in the order of 15–30 seconds. Thus, in a 3- or 4-minute mixing cycle the efficiency of the vacuum pump in lowering the mixer headspace pressure to the required level becomes critical. In practice these requirements tended to limit the introduction of pressure reduction to about half-way through the mixing cycle and deficiencies in bread quality could still arise.

In response to the deficiencies in product quality of some breads produced using the CBP with only ascorbic acid, a new CBP-compatible mixer was developed in which mixer headspace pressures could be varied both above and below atmospheric (APV Corporation Ltd., 1992). This mixer is most commonly referred to as a 'pressure-vacuum' mixer. It utilizes many of the basic principles of the CBP-compatible mixer based on the original design of Tweedy mixer but has a mixer bowl which is capable of withstanding positive pressures as well as operating at negative pressures. In the base of the mixer an inlet device allows for the movement of air through the mixer which ensures improved ascorbic acid–assisted oxidation. In some versions of the mixer the running speed of the motor may be varied. The mixer headspace pressure may be changed during the mixing cycle and so it is possible to start at one mixer headspace pressure and move sequentially to another. This arrangement is similar to the delayed application of partial vacuum commonly used with CBP-compatible mixer but differs in that pressures greater than atmospheric may be applied.

The versatility of control of mixer headspace atmosphere pressures with the pressure–vacuum mixer provides a mixer capable of producing fine-structured sandwich breads or open-structured French baguette simply by varying the pressure combinations applied during the mixing cycle (Cauvain, 1994; 1995). As discussed above, it is necessary for the dough to be subjected to the pre-set pressure for a short period of time before mixing is completed. This is true whether the required mixing pressure is above or below atmospheric pressure. The changes introduced in the pressure–vacuum mixers have a direct impact on the gas bubble populations which are formed in the dough during mixing and in doing so directly impact the final cell structure in the baked product. The control of mixing conditions is critical because of the strong link between crumb cell structure and the softness and eating qualities in bread and fermented products (Cauvain, 2004). The versatility of the pressure–vacuum mixer with its ability to more closely control the final crumb structure provides a unique opportunity for bakers.

## 4.2.3.  Oxygen-Enrichment of the Mixer Headspace with CBP-Compatible Mixers

The critical relationship between oxygen and ascorbic acid in the development of a suitable gluten structure in modern bread dough has been discussed above. In the case of the pressure–vacuum mixer, pressures greater than atmospheric can be used to deliver more oxygen to the dough via increased air flow. An earlier alternative to the use of pressure to deliver increased oxygen levels to the dough during mixing was developed for CBP-compatible mixers based on oxygen-enrichment of the gases in the mixer headspace. This method of mixing was based on the study of the role of gases in the CBP by Chamberlain and Collins (1979) who showed that a mixture of 60% oxygen and 40% nitrogen in the mixer headspace would yield bread with improved volume and finer cell structure than could be obtained from doughs mixed in air. The other advantage of using an oxygen-enriched gaseous mixture was that the application of partial vacuum was not required to yield fine and uniform cell structure in the product.

The concept of oxygen-enrichment of the mixer headspace was developed to a commercial-scale based on a mixture of oxygen and air rather than oxygen and nitrogen. All of the necessary safety features were developed and applied in the manufacturing environment. The quality of the bread was considered acceptable but the process was discontinued in the UK because of concerns about the 'legality' of the process. The concerns revolved around whether the use of oxygen in this way determined that it should be classed as an 'additive' rather than be caught up in a protracted and potentially expensive investigation the commercial bakeries concerned stopped using the concept.

## 4.2.4. High-Speed and Twin-Spiral Mixers

This category includes the widest variation in mixing design. It may be defined by the single ability of the mixer to impart a high level of mechanical work to the dough in a short period of time. Effectively, any mixer which can fully develop a dough within 5 min could be termed a high-speed mixer but the absence (mostly) of mixer headspace control separates this category of mixers from the CBP-compatible types.

Mixing criteria vary from one mixer design to another. Mixing to fixed time is the most common element, but some manufacturers offer alternative mixing controls based on dough temperature, energy consumption or even a combination of two different criteria (e.g. mix for $y$ min unless dough temperature exceeds $x$ °C). Not all high-speed mixers achieve mechanical dough development through an intense shearing and tearing action as experienced in most fixed-bowl machines with high-speed impellers. Manufacturers of twin-spiral and the so-called 'Wendal' mixing systems (Figure 4.3) claim that mechanical development is achieved through the stretching and folding action induced during the kneading stage (Figure 4.4). Whatever the mechanism of dough development, the absence of mixer headspace atmosphere control in this category of mixers does not permit as wide a range of cell structures in the final product as can be

FIGURE 4.3. Wendal kneader (courtesy Dierkes and Sohne Gmbh).

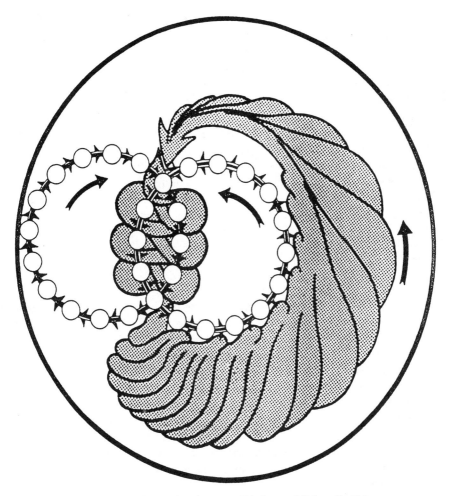

FIGURE 4.4. Wendal kneading action (courtesy Dierkes and Sohne Gmbh).

obtained in the CBP-compatible style. The absence, in some cases, of mixing to a fixed energy may lead to variations in the final bread qualities.

High-speed mixers come in a multiplicity of designs, many of which were not originally born from the needs of bread manufacture. Usually the bowl or mixing tub is static during mixing. In some versions the mixing bowl is removable to enable dough to be handled across the bakery, reducing the time for the dough to be transferred from one container to another. Twin spiral mixers are a development of the single spiral mixer where the dough is brought by the rotation of the driven bowl through two mixing tools, instead of one as seen on conventional single spiral mixers. A variation of this form, which is able to generate a very high intensity of mixing action, is where the dough is drawn by the action and design of the mixing tool between two high-speed integrated tools (e.g. Wendal). Most twin-spiral mixers have removable bowls, with the bowl driven during the mixing cycle and offer two speeds for mixing (slow) and kneading (fast). Most commonly, dough outputs from such mixing arrangements are 1,500–3,000 kg dough/h (3,300–6,600 lb/h).

High-speed mixer capacities vary from 50 to 300 kg dough/h (110–6,60 lb/h). The largest high-speed mixers can be found in the USA and Japan with batch capacities up to 1,000 kg (450 lb). Such machines are used on high-output production lines where very highly (by UK standards) developed doughs are produced on horizontal or 'Z-blade' mixers. This type of mixer usually consists of twin horizontally turning mixing tools contra-rotating in a drum-shaped mixing tub which can be tilted about its horizontal axis for discharging. The energy imparted to the dough during mixing is so great that pre-cooling of the ingredients or additional cooling from a refrigerated mixer jacket are required to maintain the required final dough temperatures.

## 4.2.5.  Spiral Mixers

The spiral mixer in its many subtle variations has become the most common batch mixer throughout the baking industry with mix sizes ranging from 10 to 300 kg (22–660 lb). Production capacities may rise to 2,300 kg/h (5,000 lb/h), or higher when spiral mixers are combined into integrated mixing systems.

A basic definition for this ubiquitous type is that the mixing machine is equipped with a spiral-shaped mixing tool (Figure 4.5) rotating on a vertical axis against the inner circumference of a bowl which is also rotating about its vertical axis. The mixing criteria are usually based on mixing time with most mixers having two speeds, slow for mixing and fast for kneading. Some mixers are available with control criteria based on dough temperature or energy consumption, and one example is available with a form of dough viscosity measurement.

The speed of the fast mixing setting on spiral mixers in this category is often slower than that typically produced by high-speed or CBP-compatible mixers, and energy input into the dough is much less, typically half that required for CBP in 5 min of mixing. As a result of lower energy inputs, the temperature rises experienced during spiral mixing are lower than with high-speed and CBP-compatible types.

The rotation of the mixing tool with respect to bowl rotation and the ratio of spiral blade diameter to bowl diameter vary from one manufacturer to another. Mixing tool designs and speeds also vary, although the basic mixing action is for the mixing blade to generate a downward force on the dough. Ingredients are mixed and later the dough is kneaded by the action of the spiral blade stretching the dough against the bowl wall and folding the dough on itself repeatedly. The intensity of the mixing action varies with the different velocities and surface areas of different spiral mixers. In each case the rotation of the bowl is used to re-circulate the mix/dough back to the mixing tool for more work.

FIGURE 4.5. Spiral mixer (courtesy Dierkes and Sohne Gmbh).

Many claims are made with respect to the advantages of different designs to ensure that all of the dough is mixed in the most beneficial way. Some manufacturers provide a large, high-powered mixing tool which sweeps an area, the diameter of which is greater than the bowl radius in order to eliminate 'dead spots' in the centre of the bowl. Others engineer central posts to guide the dough into a smaller mixing tool. These central posts are also engineered to achieve different effects on dough development; some are large central spigots which increase the effective kneading surface, while others are blade-like to generate a shearing action between them and the mixing tool. All such designs may have advantages of one type or another. The test of the complete mixing system, however, depends on how well the dough is homogenized, its structure, the time taken to accomplish mixing and the energy lost as heat during mixing.

The variety of dough structures required from various ingredients for various bread products also means that there is no perfect design for all doughs. In reality, since most spiral mixers come equipped with two speeds, the ratio of the slow to fast periods is varied to accommodate the different kneading intensities required. Typically the slow period is extended for 'weaker' doughs with a corresponding reduction in the length of the fast period.

Reference has already been made to the lower energy input from spiral mixers compared with CBP-compatible types, and the effect that this will have on dough development has been discussed in an earlier chapter. There are two other important differences which require comment, and both are related to gas occlusion in the dough during spiral mixing. Compared with CBP-compatible mixers, spiral mixers are more effective in occluding air into the dough during the mixing cycle. A comparison of typical gas occlusion values is given in Table 4.1.

The occlusion of a greater volume of air during spiral mixing increases the quantity of oxygen available for ascorbic acid conversion, so that the potential oxidizing effects are greater. However, the mixing action involved in a typical spiral mixer generates a gas bubble size range which is considerably greater, and with a larger 'average' size than would occur with CBP-compatible mixers (see below). These two factors, perhaps more than any others, explain why spiral mixers have become so commonly used in the production of breads with an open cell structure, such as French baguette.

Dough handling for spiral mixers is typically either by hand from machines where the bowl is fixed or by a bowl discharge system, usually tipping the bowl above a hopper for subsequent handling. In the latter case the bowl is usually removable from the machine. The rotating bowl is particularly suited to automated scraping of sticky doughs from its walls during discharge by means of tipping. There are variations on this theme with examples where the whole mixer is tipped up for discharging, where the bowl discharges through an orifice in its base to the elevating systems and where the bowl and its drive are attached to a high tip which is an integral part of the machine.

The flexibility provided by the removable but interchangeable mixing bowl equips this style of mixer for use in automated mixing systems incorporating many ingredient stations, mixing stations, emptying stations and resting/fermentation stations. Earlier designs for such operations were based on the so-called 'carousel' arrangement where several bowls rotate in a frame

TABLE 4.1. Gas occlusion in different mixers.

| Mixer type | Proportion of gas by volume (%) |
| --- | --- |
| Spiral | 12–15 |
| CBP-compatible | 8 |
| CBP+partial vacuum | 4 |
| CBP+pressure | 20+ |
| Low speed | 3–5 |

around a central axis, moving from one station to another. More recent systems incorporating large numbers of resting/fermentation stations are based on linear arrangements where the bowl is handled from rails (mounted above or below) between stations positioned either side of the rail system.

## 4.2.6.  Low-Speed Mixers

The first development of mixing machines for bread doughs was what we would now describe as slow-mixing systems. This was due to the requirement to mimic the mixing process in practice, rather than to a limitation of engineering capability. Low-speed mixers are still used today for some products where they are still the most appropriate mixing system for the particular ingredients used and the specification of the dough required. The most common slow-mixing systems are the twin reciprocating arm mixer and the oblique axis fork mixer or 'wishbone', and less commonly the single-arm reciprocating mixer. All feature a gentle mixing action and consequently a low rate of work input. The low level of mechanical development and comparatively low rate of air occlusion are the main reasons why today these mixers are most commonly linked with bulk fermentation processes, as they were before the advent of mechanically and chemically developed doughs.

### 4.2.6.1.   Twin-Arm Mixers

In a direct mimic of hand mixing, two linked arms are driven in a symmetrical reciprocating action such that the mixing tools mounted on the end of the arms fold ingredients from the centre to the outside of the mixing bowl during mixing (Figure 4.6). The arms also lift, stretch and fold the dough during kneading. Ingredients are returned to the mixing tools by rotation of the mixing bowl, which also aids stretching during kneading. Unlike other mixing systems very little kneading takes place against the bowl wall and the mixer usually has only one speed.

Typical mixing times are between 15 and 25 min and the mixing time is dependent on machine capacity, ingredient specification and the dough character required. Capacities for this type of mixer typically range from 50 to 350 kg (110–770 lb) dough weight. This style of mixer is particularly effective for the incorporation of delicate fruits without damage or the mixing of doughs with a weak or delicate gluten structure.

### 4.2.6.2.   Oblique-Axis Fork Mixers

This type of mixer has a single mixing tool shaped like a wishbone, with profiled ends, mounted obliquely to the axis of the bowl (Figure 4.7). The mixing bowl typically has a centre boss such that the tool action takes place between the boss and the bowl wall. Most mixers of this type have no bowl drive but allow the bowl to rotate against a friction clutch, the drive force being provided by the action of the tool against the outer bowl wall as the tool rotates. Adjustment of the clutch

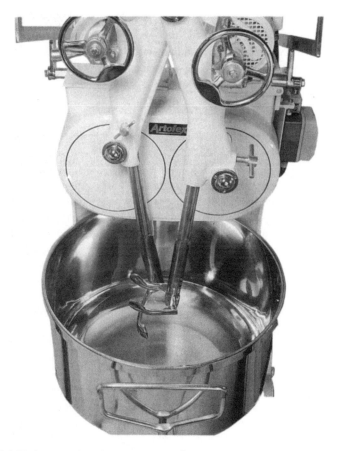

FIGURE 4.6. Twin-arm mixer (courtesy Artofex).

therefore has a direct influence on the kneading characteristics of the mixer. If a mixer of this type has a driven bowl it will have a fixed kneading characteristic.

Initial mixing is a function of folding the ingredients into each other as the mixing tool rotates. Later kneading is with respect to the squeezing action between the mixing tool and the bowl. The profiled ends of the tool act as plough shares to incorporate dough from between the forks into the kneading zone. Typical mixing times are between 15 and 20 min. Capacities for this type of mixer typically range from 50 to 350 kg (110–770 lb) dough weight. Larger mixers are available with removable bowls.

## 4.2.7. Continuous Mixers

These have been developed to meet the needs of dough make-up processes that are particularly sensitive to changes in dough consistency and density arising from variations in processing time between mixer and divider which occur with

FIGURE 4.7. Oblique-axis or fork-type mixer (courtesy VMI).

some batch mixing processes. Where bakers produce large quantities of dough to a narrow specification, the continuous mixer also offers advantages in terms of operator requirements because some versions can run continuously without operator supervision.

The most common form of continuous mixer in use for bread production today is the two-stage mixing system. As with all such systems it is the integration of an ingredient feed system with a flow-through mixing system. Dry ingredients are stored locally to the mixer, in bins which can be discharged at a controlled rate. The reader will understand that proper control over the discharge rate is essential, and so some systems continuously check this by placing the storage bins on weighing load cells so that the 'loss in weight rate' can be checked against the required recipe. Discharge is typically achieved by properly sized spiral or screw conveyors which have variable-speed drives to enable the rate of loss in weight to be adjusted by the control system.

Some systems feed directly from the dry ingredient bins to the primary mixing chamber, while others feed into a transport auger which conveys and blends the different dry ingredients. At the primary mixing chamber, water is added along with other liquid ingredients such as cream yeast or pumpable (fluid) fats. If fresh yeast is used, it is often added after the primary mixing chamber. Various designs of primary mixing chambers are offered by manufacturers. Important aspects of this part of the system are that the ingredients are uniformly distributed and the mix achieves homogeneity. The primary mixing chamber design is such that the mix is 'pumped' from this chamber into the secondary mixing or kneading chamber.

The common kneading action at this stage is similar to that of a horizontal mixer described earlier. However, the mixing tool is placed in a tightly configured trough with an opening at the discharge end, so that the mix and dough fed into one end of the kneading trough displace fully developed dough at the discharge end by virtue of the fact that the total dough quantity is greater and the level of dough in the trough is higher.

The work done during kneading is a function of the flow rate and tool speed. Some manufacturers include variable-speed tool drives in order to adjust the dough development with respect to recipe, throughput and product requirements. While variations in mixing speeds in order to match changes in plant speeds can be readily accomplished, this may lead to variations in dough development unless a compensatory change can be made. Such a change could be a variation in the speed of the mixing tool, such that more or less work is imparted to the dough as the throughput rate changes. An adjustment in ingredient temperatures or the effectiveness of the cooling system might also be required. Such considerations explain why continuous mixers are best suited to production lines running dedicated or very limited product ranges.

The kneading trough is commonly provided with a refrigerated cooling jacket to maintain and control dough temperature during the mixing process. The use of an integrated control system is essential with this type of mixing plant, and so a computer or PLC-based system is used to enable the operator to choose different recipes and mixing parameters without having to set up complex machine functions.

It is necessary for the system to be full of ingredients and product for it to work effectively and efficiently. Hence, some product is lost at the beginning and end of production shifts, if the mixer is stopped, during product changes and during cleaning periods. Some of the potential drawbacks of continuous mix systems have been the difficulties associated with reconciling raw material in and dough product out, that is yield, and potential losses from plant interruptions.

The creation of bubble structure and dough development during mixing follow similar lines to those discussed previously for batch mixers and no-time doughs. In general, bubble distributions from continuous mixers are uniform. Some opportunities exist for modifying mixer atmospheres, but the relatively 'open' nature of the mixer limits the potential for oxygen enrichment. Because of these practical considerations the range of bread structures which may be created during continuous mixing are mostly limited.

More recently there have been attempts to adapt continuous mixing to permit the control of pressure in the mixing chamber. In the COVAD project (Alava et al., 2005), a prototype continuous mixer was developed to provide sections capable of operating at both above and below atmospheric pressure and thus was able to achieve some of the advantages of the batch pressure–vacuum mixer, namely to increase ascorbic acid–assisted oxidation and to provide a range of cell structures in the final product.

# 4.3.    Control of Dough Temperature and Energy Transfer

## 4.3.1.    Control of Dough Temperature

During mixing, the temperature of the mixture of ingredients which constitute the developing dough begins to rise as a direct consequence of energy being transferred to the dough. It is important for bakers to produce dough with a consistent final dough temperature in order to ensure uniform processing after mixing and to optimise final product quality. The heat rise which typically occurs during mixing is compensated for through the adjustment of ingredient temperatures, most notably the temperature of the water. The principles which can be used for calculating the required water temperature for a given final dough temperature and a given mixer are discussed in Chapter 2.

The availability of sufficient chilled water is critical to the delivery of a dough at a consistent temperature at the end of mixing. This means that there must be a sufficient capacity of chilled water and the refrigeration equipment must have the capability of delivering the chilled water at the required rate. The calculation of the required capacity of the refrigeration plant is a relatively straightforward calculation but should be based on realistic conditions which take into account flour and water temperatures in the warmest conditions likely to be experienced in the bakery.

In some cases the mixing environment will dictate that the temperature of the dough water required will be 0 °C or even lower. Clearly this is not realistic for water which at such temperatures will form as ice. It is possible to use ice to aid the control of dough temperature, not least because of the high latent heat required to convert ice to water. As commented in Chapter 2 the addition of ice may have an inhibitory effect on hydration of the gluten-forming proteins, never-the-less the cooling advantages to be gained cannot be overlooked in those situations where temperature control could not otherwise be achieved. In practice, the use on an ice 'slush', a mixture of finely divided ice and water, is possible.

An alternative to the ice slush is the use of a chilled salt solution which has the advantage that the solution will remain liquid a few degrees below zero (32 °F) because the salt depresses the freezing point of water. A disadvantage of using a salt solution in this way is that variations in the level of liquid addition to compensate for any variations in the water absorption capacity of the flour or the dough handling requirement of the plant result in small variations in the level of salt in the dough which, in turn, leads to potential variations in yeast activity.

Other proposed means of combating the heat rise during mixing have been the chilling of the flour and the use of carbon dioxide snow. The problems and expense associated with the chilling of flour are considerable and are not really practicable. The delivery of carbon dioxide snow directly to the mixer does have significant potential for compensating for the heat rise of mixing. However, the use of carbon dioxide in this way will impact on the gas composition in the mixer headspace. In particular it will increase the carbon dioxide concentration and most critically reduce the oxygen (from the air) concentration with the potential for limiting the effectiveness of ascorbic acid additions.

### 4.3.2.  Energy Transfer

Mixer design and operating speed have significant impacts on the transfer of energy to the dough during mixing. A key element is the interaction between the mixing tool and the dough as it moves around the mixing bowl. In most mixing actions the dough is squeezed through a relatively narrow space which stretches the dough in a manner similar to that achieved with hand mixing. If the mixing speed is low then the heat which is transferred to the dough may quickly dissipate and there may be no sign of a rise in dough temperature. As the speed of mixing increases then there is less opportunity for the energy to be dissipated and it is stored as heat in the dough. This is commonly the case with all mechanical mixers.

In some cases the rate of transfer of energy to the dough is increased through the use of internal projections in the mixing bowl while in others the mixing tools are designed to 'screw' the dough more vigorously. A number of such variations have been discussed above. In all cases it appears that the basic principle illustrated in Figure 2.3 applies, namely that an increase in the rate at which energy is transferred to the dough will give increased dough gas retention for a given work input. In other words, the faster that the dough is mixed (the lower limit seems to be around 90 to 120 seconds) and therefore the more rapid the development of the dough, the greater will be the bread volume for the optimum work input of the flour being used.

## 4.4.  Dough Transfer Systems

As discussed above, mixing systems can discharge dough in a variety of ways to the next dough processing stage. The most common is batch handling the full mixing capacity of the mixer using either a mobile bowl or receiving dough tub (trough) from the mixer to a receiving hopper feeding the dough divider or extruder. Alternative mixing systems provide for a continuous flow of dough either directly to the divider or via a conveyor system. Some equipment combinations require the dough to be pre-divided prior to the divider to allow a divider hopper of smaller capacity than that of the batch mixer.

Ideally all such transfer equipment should be minimized by choosing compatible mixing and dough processing batch equipment and arranging the equipment such that transfer distances are as short as possible. When dough transfer equipment is required, the following points should be borne in mind.

- Whenever we work the dough, we change (usually detrimentally) the dough structure to a greater or lesser extent.
- Dough transfer equipment usually has to handle several varieties of dough through one plant. The potential for cross-contamination should therefore be minimized by making the equipment as resistant as possible to 'dough pickup'. There are also obvious hygiene considerations.
- When handling doughs which are particularly difficult, care should be taken to avoid modifying the dough to suit the handling system. Usually the most difficult doughs are those which are wet and sticky, and the temptation is to use excessive dusting flour, oiling of the conveyors and hoppers, or to skin the dough excessively by warm air circulation systems used to 'dry' conveyor belts.
- Many doughs are sensitive to transfer times between mixer and divider or the time between mixer and moulder.
- Hoppers should be designed so that dough flow is even across their section without the risk of dough 'eddy' causing some dough to age excessively in the hopper.

## 4.5.   Dough Make-Up Plant

As described earlier, dough is delivered to the divider with most of its structural properties and rheological character already determined by the ingredients and formulation, and the bubble structure created during mixing. In these respects, further mechanical handling after mixing can only alter the outward size and shape of dough; it cannot improve the dough's structural properties but it can 'damage' them. The Dutch 'green dough' process (Chapter 2) is an exception in that it incorporates a fermentation stage after dividing and initial rounding, which changes dough rheology in such a way as to permit advantageous modification of bread cell structure.

The critical issues for the dough as it is processed revolve around the degree to which the gas bubble structure created in the mixer is modified during the collective and individual processing stages before it reaches the prover. In most no-time doughmaking the ultimate bread cell structure is largely created in the mixer and during subsequent processing the bubble structure in the dough undergoes little beneficial modification. The absolute gas volume in the dough as it reaches the divider depends on the type of mixer and the breadmaking process being used, as discussed above. Generally, gas volumes in no-time doughmaking processes are much lower than in sponge and dough and bulk fermentation processes. Such differences will affect both divider weight control and subsequent moulding operations.

## 4.5.1.  Dividing

In order to generate the shape and size of product we require, we must first divide the bulk dough from the mixer into individual portions and then shape them to form the basis of the final product we wish to achieve after proving and baking. Dough is generally divided volumetrically, that is to say, it is cut into portions of a given size either by filling a chamber with dough and cutting off the excess (piston dividing) or by pushing the dough through an orifice at a fixed rate and cutting billets from the end at regular intervals (extrusion dividing). In either case the accuracy of the system depends on the homogeneity of the dough. This is largely decided by the distribution of gas bubbles within the dough. Where the gas structure is comprised of bubbles of uniform size and even distribution, the density of the dough remains constant throughout its volume and dividing is more accurate (for example in CBP-type doughs). Where the bubble structure is comprised of uneven sizes and distribution, dividing is accordingly less accurate (for example, in bulk-fermented and some sponge and dough systems).

### 4.5.1.1.  Dough Damage During Dividing

Compression of the dough during dividing will reduce the effect of weight irregularity due to variations in gas volumes in the dough. Any 'degassing' at this stage will contribute to damage of the dough structure, and so a compromise has to be found between efficiency of dividing and the level of dough damage. This means that different dividers will need to be matched to different dough types in order to give optimum dividing accuracy with minimal compression damage in each instance. For example, typically 'strong' North American bread doughs can withstand high compression loads whereas more delicate French baguette doughs are more readily damaged. To minimize damage to the dough bubble structure, some dividers are available with pressure compensators which permit adjustment for different types of doughs. Some dividers incorporate servo drives and control systems to limit the pressures exerted upon the dough throughout the process of dividing.

Suction damage can be much more serious than compression damage, especially if the rate of suction is not compatible with the rheology of the dough or the size and shape of the hopper and chamber. Once again, some doughs are more sensitive than others to damage. The effects of over-suction on final volume can be considerable, with evidence of individual bubbles in the dough structure bursting during dividing.

Mechanical damage occurs if dough is subjected to aggressive tearing between machine parts during dividing. It can also occur when dough is pumped or transferred to the divider by a screw drive. This should not be compared with mechanical development, the difference being that the mechanical work done during mixing is uniformly distributed throughout the dough structure, whereas mechanical work during such dough transfer systems is not uniformly distributed and confers different changes in dough properties in different areas of the dough mass, which may be manifested as changes in the final product.

### 4.5.1.2.  Two-Stage Oil Suction Divider

The two-stage oil suction divider is probably the most common bread dough divider in use throughout the Western world. Its principles of operation are shown in Figure 4.8. These dividers commonly have a central drive whereby the three different motions of ram, knife and slide are controlled. An essential feature of the divider is an airtight oil seal formed around the main ram and knife. Should this seal leak air back into the primary chamber, then divider weight accuracy is prejudiced and it is wear in this area that most influences long-term operational accuracy. Materials chosen for this important part of the mechanism are generally hard-wearing nickel–iron alloys designed to withstand the wear loads inherent in the system. The main knives are sometimes made from a similar, or slightly softer, alloy to encourage wear in the knife rather than the more expensive ram and main body casting. Some systems use main rams made from hard modern plastics which have self-lubricating properties and so reduce oil consumption in the system. In this case the ram becomes the main wear part. Die materials vary with manufacturer; some using the more dimensionally stable plastics and others opting for food-grade bronze alloys.

Typical cycle speeds range up to 1,800 cycles/h (although purpose-designed single-die dividers for some French doughs have been rated up to 3,000 cycles/h). Where higher outputs are required, multiple dies or pockets (Figure 4.9) are used to achieve outputs of up to 9,000 pieces/h at 1,000 g (2.2 lb) with five pockets and up to 14,400 pieces/h at 120 g (4 oz) with eight pockets.

In suction dividers the dough is pushed into the division box dies under some force, and so upon ejection the release of pressure allows an increase in dough volume. This has no effect on individual weight accuracy at this point, but can cause individual dough pieces to 'balloon' into each other during transfer between division box and belt. These groups of dough pieces should be parted as soon as possible to avoid cross-flow of dough between pieces (this is particularly important with soft or low viscosity doughs). This separation is usually achieved by using a second conveyor running faster than the first to 'snatch' the dough pieces apart. It is important to note that, for the efficient running of down-line equipment, the speed of these conveyors should be linked to the output speed of the divider to provide a continuous flow of evenly separated dough pieces both across and between the batch quantities delivered by the divider.

### 4.5.1.3.  Extrusion Dividers

This type of divider relies upon the ability to pump dough, usually by means of a helical screw, through an orifice at a constant rate and density. As dough emerges

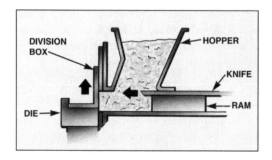

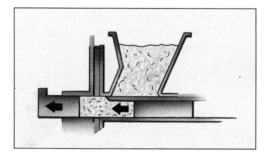

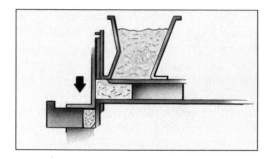

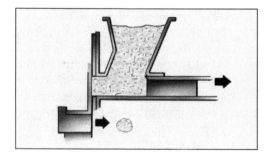

FIGURE 4.8. Two-stage oil suction divider (courtesy APV Baker Ltd).

FIGURE 4.9. Multi-pocket bread dough divider (courtesy Benier).

from the orifice it is cut by a blade or wire at a constant rate to achieve billets of dough of uniform shape and size. The dough is worked considerably during this process and such dividers are best suited to strong doughs which are already highly developed. Typically such dividers are used with doughs formed using a North American bread process, such as sponge and dough. Manufacturers of these divider systems are able to claim high levels of accuracy at high outputs.

#### 4.5.1.4.   Single-Stage Vacuum Dividers

Single-stage dividers extract the dough directly from the hopper into the measuring chamber where the dough volume is set and cut via the action of a rotary chamber or mobile hopper base. The measuring piston ejects the dough piece directly onto a discharge conveyor. As there is no intermediate chamber to pre-pressure the dough, the dough hopper should be maintained at a near constant volume or the dough used should exhibit good fluidity to aid flow into the division chamber. Some manufacturers have assisted dough flow by the use of a semi-porous base to the division chamber to draw excess air from the chamber and provide some compression by suction. Single-stage dividers are commonly used where the final bread product requires a dough which has a low viscosity.

## 4.6.   Rounding and Pre-Moulding

After dividing, the individual dough pieces are almost universally worked in some way before first or intermediate proof. If we look at traditional hand moulding methods we will see the baker kneading the dough with a rotary motion on the make-up table to produce a ball-shaped piece with smooth skin, except one spot on the base (Figure 4.10). The moulding action has forced dough to move from within the body of the piece across the surface of the dough towards the base spot. This is essentially achieved by stretching the surface of the dough piece. The degree to which this can be done without permanent structural damage is a function of the rheology of the initial dough which in turn depends on the ingredients and formulation, and the characteristics of the mixer and divider. Further, one should note that for many hand-moulded products this process would be the one and only time the product is worked after dividing, and so it replaces both rounding and final moulding in more automated bread production systems.

The action of rounding or pre-moulding will add stresses and strains which may lead to damage to the existing dough structure. However, it is clear that some breadmaking processes benefit from limited structural modification at this stage, particularly if followed by a relatively long first proof (e.g. 15 min or more) before extensive moulding takes place. This is seen in some forms of French baguette manufacture and is still the preferred method with many traditional British, Dutch and German bread varieties.

Some breadmaking processes require the rounder to have a degassing effect; however, if the dough has been accurately machine divided, or comes from a breadmaking process which leaves little gas in the dough (e.g. the CBP), then this requirement is unnecessary. In doughmaking processes where the first proof time is short, the rounder adds little or nothing to the structural properties of the final product and for a given first proof time limits the extensibility of the dough piece during final moulding. It does, however, generate a uniform, largely spherical dough piece which makes it suitable for handling in pocket-type provers, rolling down chutes and conveying without concern for orientation. It also plays an

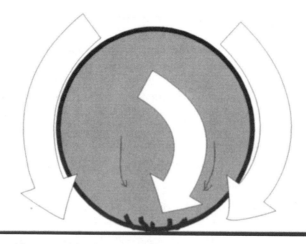

Movement of dough around the dough ball during hand table rounding

FIGURE 4.10. Movement of dough around the ball during hand rounding.

important role in delivering a uniform dough piece to the final moulder. In some processes this may be the sole function of the rounder, but in others the dough piece may be moulded into a cylindrical shape prior to first proof. Ideally, the orientation of this shape is retained throughout the initial first proof, so that it is always presented to the final moulding system with the same side 'leading'.

The action of rounding machines is similar to that of hand rounding described above. Basically the dough piece is rotated on its axis between the two inner surfaces of a 'V', where one side is driven and the other fixed or moving at a lower speed. The dough piece quickly forms the shape of the V (hence many rounders have profiled rounding tracks to encourage a more spherical shape) and moves under the force of the driven side. The difference in speed between the two surfaces is the same but the angular diameter of the dough piece reduces as the two surfaces converge, so that the top of the dough piece is rotated faster than the bottom, effectively attempting to twist it about its axis. However, because the dough piece slips on one of the surfaces (at least one low friction surface is required for most dough types), the action is changed to one of spiralling or rolling. This is deliberately enhanced on moulding systems where the fixed surface pushes the dough piece up or across the driven one. The reader will note that the V-angle, the differential speed, the length of moulding track and the shape of the track all contribute to the moulding effect and the shape of the dough piece.

## 4.7.   Types and Shapes of Rounders

There is a wide variety of rounders available where rotational speed, angle of cone, angle and shape of track, inclination of track and different surface finishes all modify rounder action on the dough.

### 4.7.1.   Conical Rounders

The most common type of rounder is the so-called 'standard' (Figure 4.11) or 'inverted' forms. They consist of a cone which is rotated about a vertical axis, with the track of the fixed moulding surface located in a spiral pattern about the outside of it. An interesting aspect of this design is that the differential speeds are lower at the top, where the axial diameter of the cone is less, hence the rounding effect changes as the dough ball travels up the cone and the forward velocity of the dough piece is reduced, causing the initial gap between dough pieces to become smaller.

In inverted conical rounders the cone is inverted and hollow with the rounding track on the inside. Versions can be found where the cone is fixed and the track rotates within it, although these are less common. With both these types the dough pieces are charged centrally into the bottom of the rounder and driven up the inner wall of the cone. Here the effects described above are reversed.

FIGURE 4.11. Typical conical rounder, note this model has operator adjustable tracks (courtesy Benier).

## 4.7.2.  Cylindrical Rounders

A variation on the conical rounder uses a track around a cylindrical drum (Figure 4.12). The track profile and angle of inclination are important for the final dough shape and consistency of drive on this type of moulder.

## 4.7.3.  Rounding Belts

These can be classified as 'V'-type, vertical and horizontal types. V-types are simply two belts orientated in a V, at least one of which is driven. This system provides the simplest mould with a conical-shaped dough piece coming from it because of the lack of a cross-drive across the moulding surface. Vertical belt

FIGURE 4.12. Typical conical rounder (courtesy Benier).

rounders work in a similar manner to cylindrical moulders, with a track wrapped around a conveyor belt with the end-roller axis in a vertical orientation.

With horizontal belt rounders a track is placed upon or across a conveyor with its axis in the horizontal plane. The track must be shaped to 'trap' the dough piece and cause it to be driven across the belt at an angle to that of the conveyor direction. Such rounders are generally found to be ideal for low-viscosity doughs and can be used with lower speeds and less sharp track angles for light forming applications.

## 4.7.4. *Reciprocating Rounders*

Here neither of the two faces of the rounder may be driven but at least one will reciprocate to present a 'tucking' action to the dough piece and give an action similar to that found with hand moulding. The reciprocating action also pushes the dough piece along the one face and imparts a forward motion to the piece. Such rounders can be either linear or 'drum-like' in operation.

## 4.7.5. *Non-Spherical Pre-Moulding*

This is basically the pre-moulding of a cylindrical shape in which shaping is carried out between horizontal belts or a belt and a board. Given that the dough piece from a divider can often be quite cuboid in shape such pre-moulding can be performed with minimal working of the dough.

## 4.8.    Intermediate or First Proving

In most modern dough make-up processes the intermediate or first proof is used as a period of rest between the work carried out by dividing and pre-moulding, and final sheeting and moulding. The length of time chosen for intermediate proof should be related to the dough rheology after pre-moulding compared with the dough rheology required at final moulding. During first proof the yeast activity begins to generate carbon dioxide gas. The extent of the activity depends on the length of time involved and (mostly) the dough temperature. There is a small effect from the temperature of the intermediate prover but more important is the requirement to prevent skinning. Because of the yeast activity the gas bubbles in the dough begin to increase in size, and first proof time can be used to influence the final bread cell structure. The longer the first proof time, the more open the bread cell structure will be, provided that no degassing occurs in final moulding. A long first proof time is therefore critical in the development of products with an open cell structure, such as French baguette, and even high-speed mixing dough processes (e.g. the CBP) can be used to produce suitable baguette cell structures by lengthening the first proof (Collins, 1983).

The changes which occur in dough properties as it rests are influenced by many factors other than time, and the reader must not assume that the first proof time may be simply extended until the dough has reached a 'suitable' condition for the next-stage moulding process. This is particularly the case where reducing agents (Chapter 3) or proteolytic enzymes are used to improve dough extensibility, since extending the first proof time may eventually adversely affect dough rheology and final bread quality rather than improve it.

In some breadmaking processes the changes in dough rheology which may occur in first proof can have a considerable effect on final bread quality. This is the case in no-time doughmaking processes, such as the CBP, where the elimination of first proof can lead to a reduction of loaf volume and an increase in damage to the bubble structure in the dough when ascorbic acid is the only oxidant in the recipe. Enzymic action may also be enhanced during first proof, the total effect depending on the time and temperature conditions used.

In some dough processes the first proof period is used to enhance the fermentation process, in particular the traditional Dutch green dough process (Chapter 2), where the first proof time may be as long as 50–75 min (sometimes with a second rounding midway through proving). Here it is claimed that the resting time, as well as temperature and humidity within the prover, allow 'natural' dough conditioning to take place, thus requiring lower levels of reducing agents and other dough improvers to be used.

Where proving times are long and the water content of the dough is high, care should be taken to ensure that the prover air is conditioned to prevent skinning (in cool, low-humidity environments the temperature and humidity should be increased) or sticking (in warm, high-humidity environments the temperature and humidity should be reduced). Care should also be taken when trying to enhance the 'resting' process by raising the humidity and temperature in the cabinet of a pocket prover, since condensation occurring on the dough pieces will encourage sticking of the pieces to prover pockets with subsequent transfer problems.

Ideally, the first proof time is a function of recipe and bread type and as such should remain constant. However, within limits first proof times can often vary with actual plant speed in order to ensure that some plant in-feed systems and subsequent moulding machines receive a balanced supply of dough pieces regardless of plant speed.

## 4.8.1.  Pocket-Type Prover

Where first proof times of longer than 1–2 min are required the most common method of achieving this is by means of pocket-type provers where dough balls are transferred into 'pockets' or 'troughs' for the whole of the resting period (Figure 4.13). The pockets are held in 'frames' which are in turn fixed between two chains carrying the 'swings' around the proving cabinet from charging to discharging stations (Figure 4.14). Either due to condensation or capillary action between the dough piece and the pocket surface, the dough piece can stick to the pocket if left in contact with it for too long. Hence, pocket-type first provers with times longer than 5 min often incorporate turnover devices which roll the dough piece from one pocket to another allowing the temporarily empty pocket to dry in some cases and allowing the simple rolling action of the dough ball to alleviate the problem in others. Where the charging method does not fill all of the pockets in a swing, these turnover devices are also used to transfer dough balls across the swings to the discharge side of the prover.

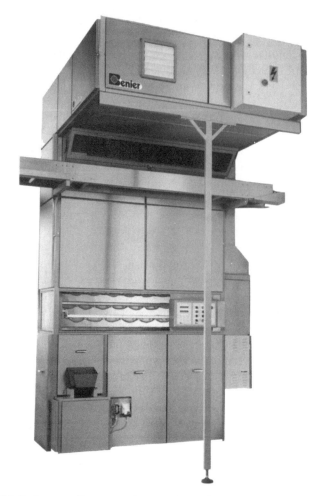

FIGURE 4.13. Pocket type first prover (courtesy Benier).

## 4.8.2.  First Prover Charging Methods

Dough balls from rounders and pre-moulding devices on bread production lines are transferred in a single stream of pieces into the pocket prover, such that at fully rated capacity every pocket of the prover is filled. Because of the slippage in the rounder, the pitch of dough pieces coming from it is not always constant and they must be synchronized to fall properly into the prover pocket. This task becomes more critical at higher throughputs, and so a variety of loading methods are adopted to cater for different dough types, sizes and throughputs.

Single-piece in-feed with intermittent prover drives (sometimes referred to as 'park and ride') receives one dough ball at a time at one charging point. As each dough ball is charged into a pocket, the swing chain is driven until the next swing is at the charging point and waits for the next dough ball. Note that

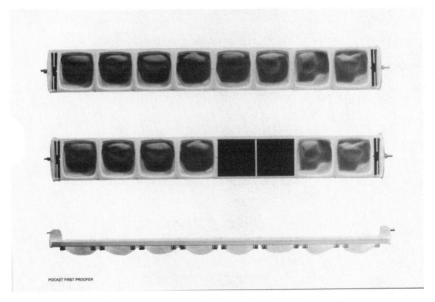

FIGURE 4.14. First prover dough pockets and swings (courtesy Benier).

the actual first proof time achieved is a function of the dough ball supply rate and not preset by the prover drive system. Also, the prover will only discharge whilst dough pieces are arriving, so a discharge switch is required for the end of a production run. This method is very common with throughputs below 1,200 pieces/h.

## 4.8.3.  Indexing Conveyors

These are used when feeding a continuously running prover at up to 2,000 pieces/h (more commonly 1,500 pieces/h) for single-pocket filling or 3,000 pieces/h for twin-pocket filling. Basically the conveyor stops and starts to synchronize the dough ball to the prover operation. Twin-pocket systems feed a valve or 'gate' which diverts the dough piece to side-by-side pockets, while in some systems the conveyor itself swings from one position to another. Note that two pieces fed at a time means two pieces discharged at a time.

## 4.8.4.  Pusher In-Feed Systems

For throughputs of up to 7,000 pieces/h (although the writers know of one recently installed at 8,000 pieces/h) this is the most common form of prover loading system and the type usually found in industrial bakeries with pocket-type first provers. Simply the dough balls are placed on a conveyor travelling across the front of a pocket prover (Figure 4.15). The pitch between the pieces must

FIGURE 4.15. First prover pusher in-feed (courtesy Benier).

be the same as the pocket pitch. When a batch of dough balls are aligned before a swing, a pusher bar rolls them all into the pockets of the swing. Different manufacturers have variations about this theme with different swing widths (6, 8, 10 or 12 pockets per swing) and differences in pusher bar design and action. The most critical aspect is the timing of dough balls onto the transfer belt. Some systems use re-pitching conveyors to correct errors occurring in the rounder and transfer conveyors, while others gear the divider and rounder speed to the prover speed to maintain constant dough ball pitching.

## 4.8.5.  Pallet In-Feed Systems

Pallet loading systems are used at higher speeds but have some limitations with respect to dough type and size at higher outputs. They are commonly used with firm doughs and smaller dough weights at throughputs greater than 7,000 pieces/h. The principle of operation is that of a series of shallow troughs ('pallets') travelling across the front of the prover as the conveyor for the pusher in-feed system. Each of these pallets is filled with a dough ball, and when the pallets are aligned above individual chutes, each feeding a prover pocket, they swing open, dropping the dough balls down chutes into the pockets of the waiting swing. The actions of both pallet conveyor and swings have to be synchronized, and again a critical aspect of the system is the placement of dough balls into the pallet.

## 4.8.6.  Discharging

Where more than one dough piece leaves the first prover at the same time they are often synchronized with valves to the subsequent discharge conveyor. On higher capacity lines, two swings are sometimes discharged simultaneously to feed separate final moulders.

## 4.8.7.  Conveyorized First Provers

In order to eliminate the need for complex pocket prover in-feed systems, some industrial bread producers modify their recipe and process in order to eliminate or considerably reduce the need for first proof times greater than 1–2 min. This then makes simple conveyor transfer of dough pieces from rounder to moulder a practical consideration. If a slightly longer rest time of up to 3 min is required, then spiral conveying systems can be used to provide the residence time required. Some producers eliminate the rounder, so that a simply moulded cylinder of dough is conveyed (appropriately orientated) directly to the moulder.

## 4.9.  Moulding

Throughout this chapter the reader will have noted how dough make-up machinery has been developed to copy or simulate the original manual process and later superseded by subsequent alternative processes. Modern moulding machines still sheet, curl (or roll) and mould the dough in simulation of the traditional manual process, with four-piecing and turning of the dough pieces prior to panning for some bread types.

When moulding dough for single-piece sandwich pan bread the objective is to achieve a cylindrical dough piece with squared ends and a dough piece with a length and diameter equal to those of the bottom of the pan. For single-piece 'farmhouse' bread the objective is to achieve a cylindrical dough piece with hemispherical ends where the length and diameter of the dough piece are equal to those of the farmhouse pan. In the production of four-piece sandwich pan bread the objective of moulding is to achieve a cylindrical dough piece before four-piecing with flat ends where the length is four times the pan width and the diameter is one-quarter of the pan length.

For baguette and bloomer the moulding requirements are such that the dimensions of the moulded dough pieces are close to the size ultimately required when fully proved. Other bread types may require hand finishing to achieve their traditional shapes, with the exception of large cobs or coburgs which may be finished in rounders properly designed for that purpose. All of the dough shapes described above (with the exception of cobs and coburgs) undergo extensive sheeting and curling prior to passing under the final moulding board.

The reader will note from previous sections that the majority of dough make-up processes utilize a rounder for initial handing-up of the dough piece prior to the first proof so that it is usually presented to the moulder as a sphere or more correctly a slightly flattened sphere. By studying Figure 4.16 we can see that when such dough pieces are sheeted they become elliptical in shape which on curling gives an ellipsoid-shaped dough piece. Final moulding of this piece requires additional work to generate the ultimate cylindrical shape with the following effects.

- Because the dough piece is ellipsoidal, the drive between the moulding belt and board is only effective in the centre of the dough piece, where a greater frictional force is exerted. This causes a twisting action between the centre and ends of the dough piece. The reader may test this by drawing a line along the outside of an ellipsoid dough piece and passing it under a flat moulding board. The resultant moulded piece will demonstrate the conversion of this line into two opposite-handed spirals emanating from the centre of the dough piece.
- Because the dough piece reaches its required length before its desired shape the ends begin to press into the side guides of the moulder, which causes additional drag and twisting as the centre pulls the ends of the dough piece along the side guides. When this is done, the internal structure at the centre collapses and dough quickly migrates to the outer edges, forming an 'hour-glass'-type shape. On conventional baguette moulders with no side guides the external surface at the centre is damaged by over-moulding, causing immediate shrinkage of

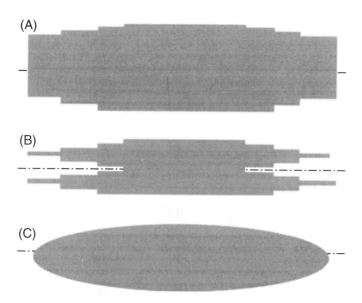

FIGURE 4.16. (A) (B) show the theoretical lamination structure of dough in a sheeted and curled piece ((B) being a cross-section of (A)), and (C) showing the final ellipsoid shape achieved.

the centre of the dough piece as it exits the moulder, and a bulge appears in the centre of the piece.

- Since the initial ellipsoid has concave ends, in some systems large air bubbles are entrained into each end of the final dough piece.

If we now compare the above with a moulding system presented in which a dough piece is already cylindrical in shape (Figure 4.17), we will see a sheeted dough piece of roughly rectangular shape, curled to a cylindrical shape and (particularly if already the correct length) moulded lightly to eliminate moulding seam and finished at the ends.

Given the above, we can observe that in order to achieve the quality of product seen from bread producers today the dough must be fully relaxed (i.e. have a low resistance to deformation) and suitably plastic when entering the final moulding stage. This is particularly true when starting from a rounded dough ball. As discussed above the removal (or significant shortening) of the first proving stage is greatly assisted by the removal of the rounder and the proper presentation of the dough piece to the final moulder, though there can be other adverse effects.

It is essential that the dough piece is presented centrally to the rollers and is maintained centrally throughout the moulding process (Collins, 1993; Cauvain and Collins, 1995). A typical mechanism for achieving such centralization is shown in Figure 4.18.

## 4.9.1. Sheeting Action

Sheeting requires the dough piece to be positioned between pairs of rollers in order to reduce its thickness, either a single set of rollers or consecutive pairs of rollers.

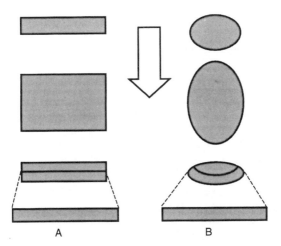

FIGURE 4.17. Sheeting, curling and moulding from (A) a dough cylinder and (B) a dough ball.

FIGURE 4.18. Dough piece being centred before sheeting.

During this process the thickness of the dough piece can be reduced by up to one-tenth and the surface area increased by a factor of more than three. We should note that this represents a considerable reworking of the dough structure. The prime objective is to stretch the cell structure and to close the relaxed open cells delivered by the first prover. The sheeting action cannot degas the dough piece unless the latter has some particularly large gas bubbles (as distinct from bubbles which will become large cells) caused by lengthy first proving, poor distribution of ingredients during mixing or inadequate degassing of fermented doughs during dividing and rounding. Similarly there is little evidence that any gas cells are created during sheeting. However, given the extent of the reworking, some inter-cell walls must be broken whilst others must be stretched and thinned to a considerable extent.

The design of sheeting rollers differs between manufacturers. Some systems favour consecutive sheeting rollers of fixed but progressively narrowing gaps, while others favour larger drum and roller sheeting systems where the dough piece is reduced once between a non-stick roller and a drum (Figure 4.19). Some manufacturers of the latter provide adjustable sheeting pressure using springs or compressed air, claiming that they make the sheeting action more responsive to the rheology of the dough passing through the gap.

Certainly the gap reduction and roller speed ratio reflect the rheology of the dough. If the gap is narrow and the roller speed too high for the dough rheology, then 'scrubbing' will take place either between the roller and the dough piece, or between the inner and the outer structures within the dough piece. When this occurs there is excessive damage which will show as tear marks on the dough surface. Relative roller speeds may also be critical where multiple sheeting sets are combined. Subsequent sheeting sets can be set to run at speed such that the dough sheet may undergo some 'stretching' as well as 'squeezing'.

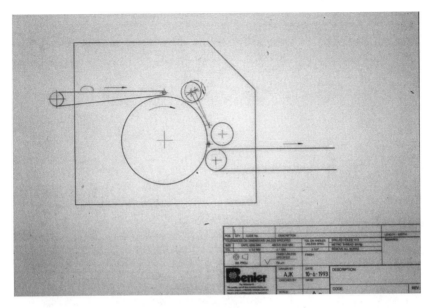

FIGURE 4.19. Drum and roller sheeting action (courtesy Benier).

Some manufacturers supply multiple roller sheeting systems with independent drives capable of being 'tuned' to the rheology of the dough being processed.

### 4.9.2.   Curling

Having achieved a sheet of dough it is then commonly 'curled' to form a 'swiss roll' effect as illustrated in Figure 4.20. Unlike a true swiss roll, there should be no intermediate layers and in particular no air should be trapped between adjacent dough surfaces. Many systems achieve adequate curling by hanging a mesh belt along the moulding belt (Figure 4.21) before the moulding board, so that curling is achieved by the mesh dragging on the leading edge of the dough sheet and causing it to roll back over the following portion of the sheet. Some systems start curling in this manner and finish it under the moulding board.

Other systems curl long sheets at high throughputs and incorporate a driven mesh belt (Figure 4.22) acting against the direction of the moulding belt and so shorten the length of the moulder in this area.

### 4.9.3.   Final Moulding

As discussed above, considerable reworking of the dough can occur during final moulding for only a small dimensional change. It is important that the dough retains a degree of extensibility as it leaves the curling section prior to final moulding. Damage is minimized if the moulding pressure is minimized to encourage 'dough flow' rather than forcing it. It is therefore considered that

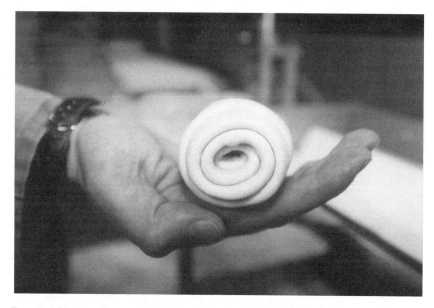

FIGURE 4.20. Dough piece after curling.

FIGURE 4.21. Hanging curling chain.

FIGURE 4.22. A driven curling chain.

a long moulding board with gradually reducing moulding gap is more 'dough friendly' than a short board with a narrow gap. Side guides should not be set in order to reduce the dough piece. This will cause extensive rubbing of the dough piece on the side guide incurring surface damage. In extreme cases the drag offered by the side guide will cause the dough piece to 'ox bow'. Under these conditions the dough centre undergoes a wringing action causing a breakdown in dough structure at the centre of the dough piece. The materials of the belt, board and side guides are chosen to allow some slippage as the dough piece is extended, and will reduce surface damage particularly when moulding doughs for baguettes, sticks or batons.

## 4.9.4.   Four-Piecing

The reasons for four-piecing are further discussed below. The objectives are to cut the dough cylinder into four equal lengths (each equal to just less than the tin width), sometimes each piece is joined by a tail of dough to one or more others and turned through 90° to lie side by side across the tin when panned. In other cases the pieces are completely separated from one another.

To maintain control, the dough pieces are cut under or immediately after the moulding board. Dough drag on the side guides should be minimized or accounted for when setting the gap between cutting blades. Further, the depth and length of the blades should be appropriate if dough tails are required during turning and panning. Side guides should finish before the end of the moulding

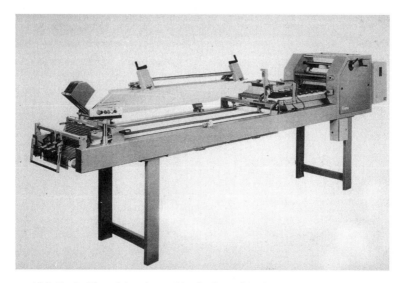

FIGURE 4.23. Typical bread dough moulder for four-piece bread. Dough pieces move from right to left; the pressure board and dough cutting arrangement are raised for illustrative purposes.

board if not during mid-cut, especially if side guide drag is present. A typical bread moulder for four-piece bread is shown in Figure 4.23.

After the moulding board the pieces are 'knocked' into a 'W' (when viewed from the discharge end) as they fall over the end of the moulding belt onto a panning conveyor. The pieces are then pushed parallel between converging belts, rollers or plates and panned into the tin. It is important for subsequent processing that the panned dough pieces lie flat and parallel in the tin.

## 4.9.5.   Cross-Grain Moulding

One method of achieving a wide-sheeted dough piece prior to curling and moulding is to turn the sheet through a right angle so that the wider plane of the elliptical sheet is presented to the curling chain and moulding board. Besides the obvious advantage of reducing the work required during moulding, this method presents the elongated cells achieved during sheeting across the dough piece rather than around it, hence the term 'cross-grain'. Although in practice there are fewer dough layers after curling, equipment manufacturers claim that this presents a cell structure in the finished single-piece loaf which is similar to that found in four-pieced bread.

## 4.9.6.   Other Sheeting and Moulding Systems

Where shapes of particularly long or narrow cross section are required, special equipment is available to enhance the principles of dough moulding previously discussed. Many examples of these are used on automated baguette plants, where different designs of contoured moulding boards are used to encourage dough piece elongation. Similarly, some manufacturers use diverging 'polycord' conveyors above and below the dough piece to pull it out to a greater length. One manufacturer has patented a reciprocating upper moulding belt which reduces the twisting effect seen during final moulding.

## 4.10.  Modification of Gas Bubble Structures during Processing

In the dough processing steps which follow mixing and precede the entry of the dough pieces to the final prover, there are significant changes in the rheological properties of the dough pieces and the gas bubble structures contained within them. In modern no-time doughs the gas bubble structure created in the mixer is essentially the one which will be expanded in the prover and set in the oven. There will be some expansion of gas bubbles during dough processing but it will be relatively small by comparison with that which will be achieved in the prover and the oven. The interaction between dough rheology and moulding operations have to be optimized if damage to gas bubble structures in the dough is to be avoided. The most common manifestations of damage to the dough bubble structure are the formation of large holes and streaks of dull or dark-coloured crumb.

Modern no-time doughs have considerably less gas within them when they reach the divider by comparison with those prepared by bulk fermentation. The upper limit of gas volume with no-time doughs is in the region of 20% depending on how the dough has been prepared, while in the case of bulk fermented doughs then a figure of 70% would be more appropriate. By the time that both doughs reach the final moulder the figures for no-time doughs are 17–18% and for bulk fermented doughs are around 25% (Table 4.2). It is clear from these data that considerable degassing of the bulk fermented doughs occurs but the no-time doughs remain essentially unchanged.

The formation of voids or holes in the bread is commonly associated with dough processing, especially moulding. Often the holes are attributed to pockets

TABLE 4.2. Effects of sheeting on gas volume in dough

| Dough processing stage | Proportion of gas by volume (%) | | |
| --- | --- | --- | --- |
| | Fermented | CBP | Spiral |
| End mixing | 5 | 5 | 7 |
| End fermentation | 70 | – | – |
| End first proof | 27 | 16 | 18 |
| End moulding | 18 | 15 | 17 |

of gas trapped within the dough at various stages. There are a number of opportunities for the occlusion of large gas pockets during dough processing but they are mainly associated with curling in the final moulder. Cauvain (1996) used Computerized Tomography (CT) to show that while voids may be occluded in dough pieces leaving the divider they did not survive the sheeting rolls, or if they did they had to be smaller in size than the gap of the last pair of rolls. This showed that the origins of many of the larger holes were most likely to come during the curling process since they were mostly situated towards the ends of the dough piece.

Other holes which may form in the dough are likely to do so later during proof and the early stages of baking. Such holes are most likely to arise because the delicate bubble structure in the dough has been damaged, often because of the application of high pressures when the dough piece passes underneath the pressure board. The high pressures are often used to 'mould out' trapped pockets of gas but in many cases create the very problem concerned. The mechanical breakdown of the gluten network between gas bubbles allows them to expand more readily and coalesce when they touch. The increase in bubble size which occurs creates localities of relatively lower pressure and the carbon dioxide gas from yeast fermentation preferentially diffuses into them. Thus the larger gas bubbles expand while the smaller ones remain relatively unexpanded. If the expansion is sufficient then a hole may remain in the final product.

## 4.11.   Sheet and Cut Dough Processing Systems

It has long been recognized that the gluten network and the gas bubble populations in dough should be subjected to as little mechanical pressure as possible. This is especially true of doughs required for the manufacture of baguette and ciabatta which also require high water additions to create the characteristic open and random cell structure. In recent years the concept of 'stress-free' dough processing equipment has developed in order to handle high water doughs and manufacture open cell structure products. In one sense the term 'stress-free' is misleading since any handling of the dough to change its shape subjects it to stress. A more appropriate term would be 'reduced-stress'.

In reduced-stress dough processing systems the bulk dough is fed as a sheet onto a conveyor. The width of the dough sheet is adjusted to be constant before it is split into narrow strips. A guillotine knife arrangement divides the narrow strip into a series of units of the required weight and individual dough pieces move on for moulding into the appropriate shape. The avoidance of sheeting rolls ensures that the dough is not de-gassed and the bubble structure within is largely preserved. Such sheet and cut systems are suited to the manufacture of artisan-type products which do not require the regularity of shape associated with products such as sandwich breads though they can be used to manufacture the latter.

## 4.12.   Panning and Traying Methods

As bakery plant speeds and moulder speeds have increased so panning processes have to cope with faster throughput capacities. Most manufactures offer simple drop systems whereby the tin or tray is indexed under the discharge of the moulder as each dough piece exits the machine. Faster systems use retraction belts in order to fill a complete stationary strap or tray of sufficient dimension to allow it to be indexed between fillings. Other systems will synchronize the flow of tin strap/tray and dough product such that dough pieces are panned 'on the move'.

In plants where the dough piece has been cut in order to make several products from one dough piece (e.g. baguettes strings cut to form petit pains) the panning unit may be combined with a separating unit in order to space products on the receiving tray or fillet.

## 4.13.   Equipment for Small Bread and Rolls

Previous sections have referred to dough handling equipment after the mixing process as separate dividing, rounding, proving and moulding machines. In the production of small breads and rolls it is common for these functions to be brought together in one piece of equipment or plant. A roll plant, however, still contains within it the various stages of dough make-up used for bread processing, and the same dough-handling constraints will be relevant.

### *4.13.1.   Small Bun Divider Moulders*

The divider principle used here is the same as that found with hydraulic dividers previously mentioned. Manual versions are available where pressure (to distribute the dough evenly under the knives) is applied by a lever. The dough is cut by the downward movement of the knives from the head mechanism. After dividing, the cut dough pieces move in a rotary motion to round the pieces between the plate, the knife walls and the top platen. Dough is manually loaded into the machine on the plate, which is later removed with the rounded dough balls on it.

Typical small bun divider moulders will produce between 15 and 36 dough balls per cycle with weights from 18 to 160 g depending on the model and manufacturer. Automatic versions which pressurize, cut and mould the dough are available (Figure 4.24). Some have adjustable timers so that the length of the moulding period can be adjusted. Typical production rates are operator dependent, but capacities of up to 5,000 pieces/h are possible.

FIGURE 4.24. An automatic bun divider–moulder (courtesy Daub Verhoeven).

## 4.13.2.   Integrated, Multi-Lane Roll Plants

Such plants are the most common type of equipment used for roll production. Output capacities vary from 4,000 to 36,000 pieces/h depending on the number of 'rows' of product being processed and the speed of the plant per row. Depending on the type of product to be made, processing modules are added to the basic specification in order to achieve the different moulding, cutting, seeding or stamping effects.

All such lines usually incorporate a divider–moulder which can use a combination of dividing and rounding, as already discussed. Two-stage dividing is still the most common, but without the use of oil or a knife, as found in the oil suction divider. Extrusion dividing is more commonly found, particularly in the production of burger buns and hot dog rolls, where the dough consistency is usually based on highly developed North American doughs with low viscosity and good flow characteristics (Chapter 9).

Rounding is largely based on the oscillating system described previously for the bun divider–moulder and the linear rounder. This is applied differently by different manufacturers, some moulding the dough piece between an

FIGURE 4.25. Combination bread and roll plant (courtesy J. Sainsbury).

oscillating cup and a processing belt, some incorporating the oscillating motion into a honeycomb frame mounted around a drum and others combining the dividing chamber with the moulding system to mould against an oscillating plate. Extrusion dividers commonly round by means of a linear rounding track mounted over a processing belt.

Spreading belts are usually incorporated into the divider–rounder to alter the pitch dimension between rows before the next processing module or unit. This allows the individual rolls to be pinned out later to longer finger or hot dog rolls whilst maintaining clearance for later expansion during proving.

Swing-type first proving modules are incorporated when higher levels of dough reworking are required to develop the final product shape and always when the dividing and rounding action is aggressive or the rheology of the initial dough is particularly poor.

Moulding can be achieved with a board or contra-rotating belt as described earlier. Some products may be simply rolled to shape while others may be sheeted and rolled. Dough pieces for burger buns can be pinned simply to flatten the rounded dough piece to a disc.

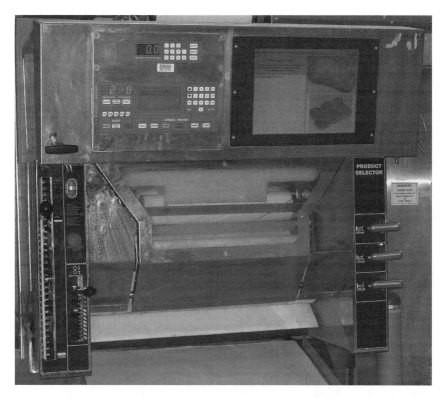

FIGURE 4.26. Computer panel from combination bread and roll plant (courtesy J.Sainsbury).

The seeding and topping of fermented products are more commonly found prior to the final proof on roll lines than on bread plants. Hence it is common to find seeding systems incorporated into roll plants in which the roll dough piece is first wetted and then sprinkled with seed. Excess seed is returned to a collection bin for later use.

Panning is usually onto flat or indented trays. Panning methods can consist of row-by-row panning similar to that seen with bread lines where the tray indexes under the panning point are in line with the flow of product or more commonly by a retracting belt which rapidly withdraws under the dough pieces to drop up to a tray full of dough pieces every cycle.

## 4.14.  Combination Bread and Roll Plants

Some smaller plants combine the requirements of bread and roll production into one largely automatic operation. An example suitable for the production of bread and rolls in smaller and in-store bakeries is illustrated in Figure 4.25. The dough is mixed in the spiral mixer located to the left of the illustration. After mixing is completed the dough is automatically extracted from the bowl and fed directly into the divider. The divided dough pieces are rested and transported to the appropriate final moulder and afterwards panned or trayed up by the one operator needed to run the plant. The settings required for the various product types are stored in a computer. The program chosen by the operator (Figure 4.26) automatically controls the various process stages and choice of equipment so that the skill and manual input required of the operator are limited.

## *References*

Alava, J.M., Navarro, E., Nieto, A. and Scauble, O.W. (2005) COVAD – the continuous vacuum dough process. In (eds S.P. Cauavin, S.E. Salmon and L.S. Young) *Using Cereal Science and Technology for the Benefit of Consumers*, Woodead Publishing Ltd., Cambridge, pp. 169–173.

APV Corporation Ltd (1992) *Dough Mixing*, UK Patent GB 2 264 623A, HMSO, London, UK.

Cauvain, S.P. (1994) New mixer for variety bread production. *European Food and Drink Review*, Autumn, 51–3.

Cauvain, S.P. (1995) Controlling the structure is the key to quality. *South African Food Review*, **21**, April/May, 33, 35, 37.

Cauvain, S.P. (1996) Controlling the structure is the key to bread quality. *Proceedings of the Fiftieth Anniversary Meeting of the Australian Society of Baking*, Sydney, Australia, pp. 6–11.

Cauvain, S.P. (1999) The evolution of bubble structure in bread doughs and its effect on bread structure. In (eds G.M. Campbell, C. Webb, S.S. Pandiella and K. Niranjan) *Bubbles in Food*, Eagan Press, St. Paul, Minnesota, USA, pp. 85–88.

Cauvain, S.P. (2004) Improving the texture of bread. In (ed. D. Kilcast) *Texture in Food, Volume 2: Solid Foods*, Woodhead Publishing Ltd., Cambridge, UK.

Cauvain, S.P. and Collins, T.H. (1995) Mixing, moulding and processing bread dough. *Baking Industry Europe*, London, UK, 41–3.

Cauvain, S.P. and Young, L.S. (2006) *The Chorleywood Bread Process*, Woodhead Publishing Ltd., Cambridge, UK.

Collins, T.H. (1983) The creation and control of crumb cell structure. *FMBRA Report No. 104*, July, CCFRA, Chipping Campden, UK.

Collins, T.H. (1993) Mixing, moulding and processing doughs in the UK, in *Proceedings on an International Conference on 'Bread – Breeding to Baking'*, 15–16 June, FMBRA, Chorleywood, CCFRA, Chipping Campden, UK.

Chamberlain, N. and Collins, T.H. (1979) The Chorleywood Bread Process – the roles of oxygen and nitrogen. *Bakers' Digest*, **53**, 18–24.

French, F.D. and Fisher, A.R. (1981) High speed mechanical dough development. *Bakers' Digest*, **55**, October, 80–2.

Tweedy of Burnley Ltd (1982) *Dough Mixing for Farinaceous Foodstuffs*, UK Patent GB 2 030 883B, HMSO, London, UK.

# 5
# Proving, Baking and Cooling

Chris Wiggins and Stanley P. Cauvain

## 5.1. Introduction

Proving, baking and cooling are the stages of breadmaking that convert a fermenting dough into a stable product ready for consumption. There is evidence that leavened products were made in Egypt around 2000 BC and unleavened bread has been made since prehistoric times. The three operations – proving, baking and cooling – have been essentially the same ever since, relying on the properties of the raw materials and the way they behave when heated to produce a staple product that is both nutritious and good to eat. How the process was discovered we shall never know in detail but the huge variety of leavened bread products eaten across the world today all rely on the same basic principles. Proving, or proofing, allows time under favourable conditions for the yeast and enzymes in the flour to be active. Then, during baking, the rate of heat transfer is increased so that the outside of the loaf dries to a crust, and inside, the starch swells and the protein coagulates. Cooling reverses the direction of heat transfer and aims to produce loaves that are ready for wrapping, often with slicing as an intermediate operation. A typical timescale, with process conditions and their effect on loaf core temperature, is shown in Figure 5.1.

Like many other processes in the food and drink industries, the details of the physical mechanisms and chemical changes occurring inside the dough during its various processing stages are extremely complex and only comparatively recently has it been possible to argue that breadmaking has become more of a science than an art.

Since the mid-1950s, as more sophisticated measurement and microscopic techniques have been developed, understanding of the changes that occur during baking has advanced dramatically, increasing many-fold the resources available to the baker, particularly in ingredients, where flours, enzymes and special yeasts can be obtained, all tailored for specific purposes (Chapter 3).

The ready availability of significant computing power means that dynamic models can now be constructed for the heat and mass transfer during baking, and computational fluid dynamics (CFD) can be used to visualize the air flows and heat transfer both in the processing chamber and also around and within

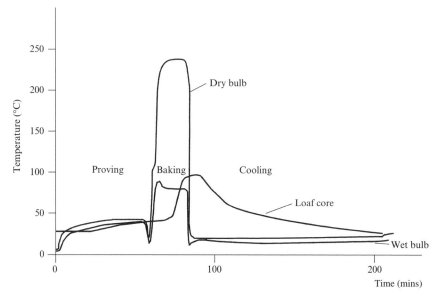

FIGURE 5.1. Proving, baking and cooling.

the product. A plot of air velocity vectors in an oven chamber with a blowing nozzle, strip burners and an extraction point is shown in Figure 5.2. The application of imaging techniques has now extended to the use of X-ray tomography which has enabled the dynamic development of the internal processes which occur during proving and baking to be visualized (Whitworth and Alava, 1999; Cauvain, 2004).

A better understanding of these heat transfer mechanisms has not led to any breakthrough in the design of the equipment to prove, bake and cool bread. The basic designs are well established and have developed in an evolutionary fashion, mainly to support increases in output and improvements in consistency and reliability.

Every country has traditional breads, for some of which quite special baking conditions are required, but as a popular product spreads around the world, there is a tendency for its characteristics to change slightly to conform to the requirements of high-volume production on modern equipment. This tendency towards standardization makes it legitimate for this chapter to assume that the bread is being baked in tins or pans, as the processes used are substantially the same as for other specialty breads that may be proved and baked on flat plates.

Figure 5.3 illustrates a variety of popular bread products and the choice of the range of products to be made, and their characteristics will determine how the plant is configured and adjusted. Some important considerations are as follows:

- What are the required crust and crumb properties?
- Is the surface to be glossy?

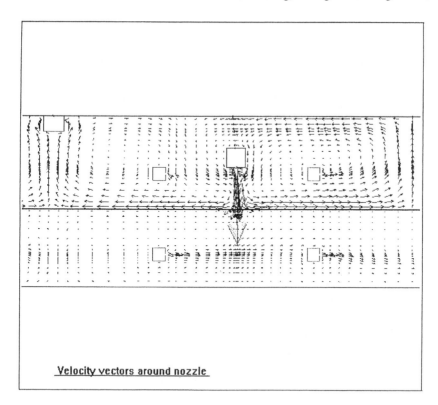

FIGURE 5.2. CFD representation of a baking chamber.

- Does the loaf have to be sliced?
- What must the final moisture content be to meet the yield target or legislative requirements?

All these properties can be manipulated by changing the heat transfer programme, although they are not, of course, independent of the recipe formulation and mixing regime.

This chapter does not contain any formulae, but attempts to explain qualitatively the mechanisms at work, in the belief that an understanding of what is happening during the later stages of the breadmaking process will be of more use in achieving the required product characteristics and in overcoming quality problems than will a list of specific instructions or pages of photographs of defective loaves.

The properties of air and water mixtures are fundamental to the understanding of proving, baking and cooling, so a short refresher course on psychrometry has been included after this introduction.

Production bakers will also be interested in production costs and in maintaining the performance of their heat transfer equipment, and some routine procedures are suggested to help them to monitor the state of the plant.

FIGURE 5.3. Bread varieties.

Trends in equipment design are also indicated in each section and the promise of emerging technologies is discussed.

## 5.2.  Psychrometry

### 5.2.1.  Definitions

**Dry bulb temperature** is the actual temperature of a gas. **Dew point temperature** is the temperature of an air and water vapour mixture below which condensation of vapour will begin. **Saturation** is the condition of the mixture at the dew point temperature.

**Relative humidity** is the ratio of the partial pressure of the water vapour in a mixture to the saturation pressure at the same temperature. It is a concept that is very useful for air conditioning work, but for process work at higher temperatures, dew point is often more useful when trying to visualize whether the dough surface will be experiencing condensation or evaporation.

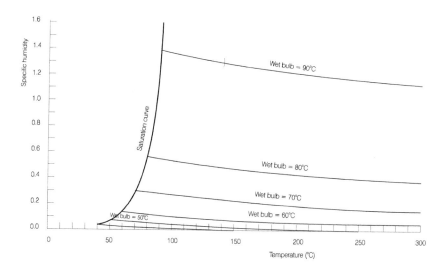

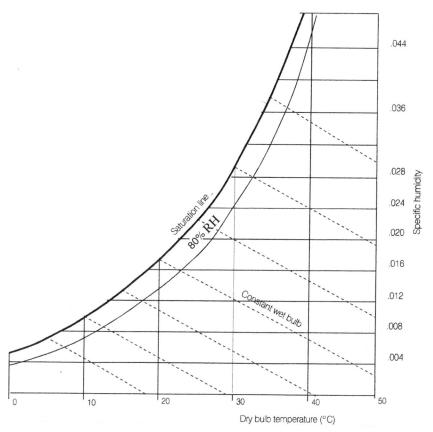

FIGURE 5.4. Psychrometric chart.

**Wet bulb temperature** is the temperature indicated by a thermometer in an airstream when the bulb is kept wet. It is not an absolute physical property like dry bulb and dew point temperatures, but it is easier to measure and can be used to estimate the relative humidity, from tables.

**Specific humidity** is the mass of water per unit mass of dry air in a mixture of air and water vapour.

A psychrometric chart (see example in Figure 5.4) is used to compute the condition of air and water vapour mixtures during heat and mass transfer processes, of which proving, baking and cooling are typical examples.

## 5.3.   The Proving Process

Proving, or proofing, is the name given to the dough resting period, after the moulded pieces have been put into tins, during which fermentation continues in a controlled atmosphere. Fermentation is a blanket description for a series of complex, interlinked reactions, which are dealt with in more detail in other chapters. To understand proving, a simpler model is sufficient. Starch is converted into sugars by enzyme action. The sugars feed the yeast and the breakdown products are carbon dioxide and alcohol. As carbon dioxide is produced it is retained in the tiny cells formed in the protein matrix during mixing, causing the cells to grow and the dough to expand. The number of cells cannot be increased during proving, but the structure can be coarsened, and if the dough is over-proved then the cell walls will start to collapse. Other products of yeast activity, mainly acids, are also formed during proving, and they can contribute significantly to flavour development.

The dough is confined by the walls of the tin, and this helps to determine the shape and orientation of the cells in the final product. For example, a dough piece that has been divided into four or six pieces and laid to fill the bottom of the tin will finish with uniform, vertically elongated cells, whereas a single piece, too short to fill the tin, can expand in length as well as height and the final loaf crumb will not look as white or bright as its multi-piece competitor. Figure 5.5 illustrates diagrammatically how the cells grow and elongate as the dough piece volume increases and the outside skin of the piece slides up the oiled surface of the tin. In reality the cells in the dough will not be round to start with and they will have been aligned by whatever moulding has taken place into a starting pattern, which will then be subjected to expansion.

Examination of CT images of dough pieces proving in a pan (Whitworth and Alava, 1999) show that there is greater expansion in the lower half of the piece than the upper portion. In part this occurs because conducted heat readily reaches the dough piece from the walls of the pan in the early stages of proving. As the dough piece begins to fill the pan, heat transfers to the centre of the dough is slow but the gas pressures generated by the rapidly proving portions of the dough push the centre of the dough upwards. By the end of proof the original centre of the dough piece is about two-thirds of the way up the pan. This portion of the

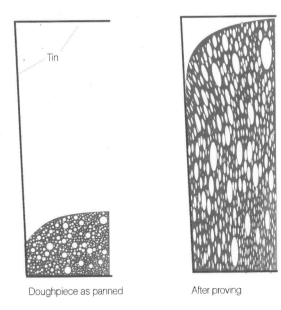

Doughpiece as panned          After proving

FIGURE 5.5. Effect of dough piece and tin shape on cell structure.

dough piece is somewhat cooler than the rest of the dough and will make a major contribution to oven spring during baking. In oven-bottom (hearth) breads and those supported during proof and baking in slings, the expansion of the dough during proof is more uniform and the center portion of the piece remains more or less central during proof.

As the gas bubbles in the dough begin to expand during proof, the gluten network which surrounds them begins to stretch. As the dough continues to warm, a point will be reached when gas bubbles begin to touch. If the gluten network is not able to stretch sufficiently to cope with the bubble expansion, the gluten films may rupture and touching gas bubbles may coalesce to form a single larger gas bubble. The coalescence of gas bubbles may occur throughout the dough piece during the proof phase, especially if the gluten network does not have the necessary properties to accommodate bubble expansion (Campbell, 2003). It is now evident from X-ray tomography (Cauvain, 2004) that significant coalescence of gas bubbles does occur in doughs made with low-protein (weak) flours in the prover, as well as in the oven.

Providing it can be done without compromising product quality, there is an obvious requirement to minimize proof time – a longer time represents a bigger proving chamber, which means higher capital cost and more space, more work in progress and longer product changeover times. This means optimizing the conditions for fermentation so that the required biochemical changes will happen as quickly as possible.

When the dough enters the prover, it will be at a temperature of 28–32 °C (82–90 °F), which is the maximum at which most modern moulding equipment will work efficiently – any hotter and the dough will be so sticky that problems of product transfer will outweigh the advantage of any possible reductions in proof time.

The dough expands by a factor of three or four during proving, to almost its final volume, and it is important that the skin remains flexible so that it does not tear as it expands. A flexible skin is one which has not been allowed to dry out, so that controlled humidity is essential. A humid atmosphere is also required to minimize weight loss during proving.

Yeast is at its most active at 35–40 °C (95–104 °F), so to minimize proof time, heat transfer to the dough is necessary, to raise its temperature typically by 10–15 °C (18–27 °F). Part of the heat is supplied as moisture condenses on the loaf surface and on the baking tin, and gives up its latent heat. This quickly establishes a temperature gradient, which is the driving force for heat transfer, by conduction, to the centre of the dough piece. At this stage of the process the dough is dense and is a relatively good conductor, so the temperature of the dough piece will stabilize quite quickly.

A prover air inlet temperature of 42 °C (108 °F), with a dew point of 35 °C (95 °F), is typical. This warms the dough gently, and after about 20 min the dough will be at a uniform temperature of about 35 °C, the peak gassing temperature. The temperature of the dough stabilizes at the dew point temperature and further heat transfer evaporates moisture from the surface instead of raising the temperature. The evaporated moisture raises the specific humidity as the heat transfer reduces the dry bulb temperature, so that at the outlet the circulating air may be almost saturated. Typical air conditions and dough temperatures are shown in Figure 5.6.

The dough piece gains weight at the start of proving and then loses it again – the nett loss on a dough piece weighing 880 g (31 oz) will be typically about 3 g.

## 5.3.1.  Practical Proving

The prover air conditioning system has to be able to handle the full range of operational conditions. At one extreme will be the start-up on a cold morning, when the bakery and all the mixing and forming equipment are cold and, even if the prover inlet conditions are right, the average condition will be colder than standard because the heat load on the system is greater. It is normal practice under these conditions not only to raise the prover temperature for start-up, but also to increase the yeast level in the dough. This is a typical example of the pragmatic measures that are taken in production to maintain acceptable quality when a process solution, in this case providing a temperature-controlled dough-forming area, is available but could not be economically justified. At the other extreme is the hottest, most humid day of summer, when flour temperatures are high and tins returning to the moulder are hot. The process is always the most demanding when the external circumstances make it most difficult for the prover air conditioning to respond. However, it will help to reduce yeast level and run the prover at a higher speed than normal.

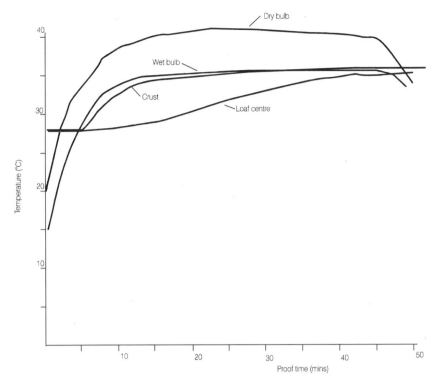

FIGURE 5.6. Changes during proving.

## 5.3.2.  Prover Checklist

It is important that bakers understand how the air is supposed to circulate in their provers and how the air conditioning systems cope under different ambient situations. Then they are in a position to take regular readings and detect any deterioration in air quality before process performance starts to get worse. The minimum checklist should include air temperatures and humidities at entry and exit of the prover, as well as a velocity measurement to check that airflow is as it should be. To check uniformity it is best to compare proof height and centre temperature for loaves that have taken different paths through the prover. Many large plants now convey the products through the prover and oven in single file, often in a figure of eight, and this concept largely removes the possibility of differences in air flow, and therefore heat transfer, from one tin strap to the next, although badly designed tin straps can still cause loaf-to-loaf differences within a strap, especially if the design of the strapping is such that air cannot circulate between the pans.

Finally, for each of their products, bakers should record how the product looks, in terms of temperature and height in the tin, at convenient points in the prover. This knowledge will often allow them to take corrective action when something changes in the mixing room and they may be able to save loaves that would otherwise be out of specification.

## 5.4.  Modern Prover Design

### 5.4.1.  Airflow

The air condition that the dough pieces encounter when they enter the prover must encourage condensation as an essential part of the process, but excessive condensation can lead to unsightly streaks or spots on the crust. Also, saturation of the air should be avoided since condensation is then likely to occur on the structural surfaces of the prover, which may lead to hygiene problems, or could cause corrosion in older equipment which is not satisfactorily protected.

These potential problems are avoided by minimizing the temperature drop between the air inlet and return, with the air approaching saturation as it leaves the prover. The temperature step is proportional to the heat transferred so a smaller step means that more air must be circulated in order to supply the required heat transfer to the dough. Higher air flows give the potential added advantage of more even side-to-side conditions and less danger of stagnation of air.

### 5.4.2.  Ambient Conditions

There are other practical issues of prover design which need to be considered, particularly where ambient conditions can be hot and humid and air of the right quality is not available to be mixed or conditioned to produce air for the prover at the specified temperature and humidity. In the UK, air at 42 °C (108 °F) and 80% relative humidity can always be provided by taking fresh air, passing it through heating coils and then adding steam to raise the humidity. Often, to improve thermodynamic efficiency, most of the air is re-circulated and only a proportion of fresh air is added, to return the inlet stream to its design condition. However, in hot climates the fresh air may be too hot to provide this function and extra air conditioning equipment must be used. If conditions are dry, as is often the case in the tropics at high altitude, water sprays can be supplied in the fresh air stream to cool the air as the water evaporates, as well as to provide the planned humidity level. If the ambient conditions are humid as well as hot, then refrigeration may have to be supplied so that the fresh, or the return, air can be cooled, in order to condense some of the moisture, and then reheated to the operating point. The process path on the psychrometric chart is shown in Figure 5.7.

The potential problem of an excessive heat load on the prover air conditioning system caused by hot tins returning from depanning can be economically averted by adding forced convection cooling on the return tin conveyor.

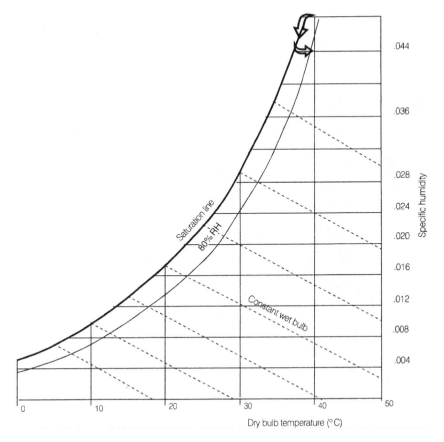

FIGURE 5.7. Refrigerated proving in tropical climates.

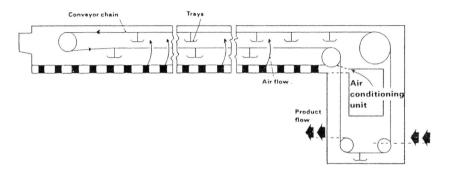

FIGURE 5.8. Prover with carriers.

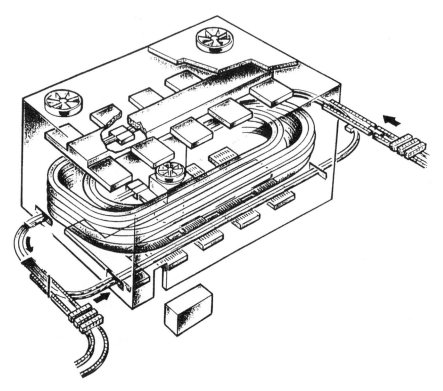

FIGURE 5.9. Spiral prover.

## 5.4.3. Mechanical Handling

As well as providing the right process conditions for the proving dough pieces, the prover must be capable of accepting tins at the specified plant rates, holding them for the proof time and delivering them to the oven. Two configurations are common, one where the tin straps are loaded on shelves in a carrier moving on a chain circuit, and the other where the straps are fed onto a continuous conveyor that carries them through the proving chamber in a spiral path. Both principles are shown in Figures 5.8 and 5.9.

## 5.5.  Developments in Proving

### 5.5.1.  Higher Proof Temperatures

Attempts have been made to run proving commercially at higher temperatures. Higher proof temperatures will warm the dough more quickly, but a steeper temperature gradient through the dough will certainly have an effect on product quality, because it will result in uneven gas production rates and, ultimately, uneven cell structure and texture in the finished loaf. A further disadvantage of proving at higher temperatures is that it makes the effect of a plant stoppage much more dramatic. Although most provers have some extraction facility that can be used to bring down the temperature quickly when there is a plant stoppage, to give more time for the fault to be found and corrected without over-proving the dough that is already in the system, it is still a fact that the higher the operating temperature, the shorter this critical time will be.

As so often in real processes, the requirements of economics and quality conflict and it is the job of the engineer and the baker to arrive at a sensible compromise.

## 5.5.2.  Shorter Proof Time

The other way to reduce proof time is to start proving with the dough already at its optimum temperature of 35–40 °C (95–104 °F). This has been done on an experimental basis both by using conventional make-up plant followed by a microwave pre-heater and by using special make-up plant which will handle dough that has been mixed to the required temperature. Neither method has yet proved sufficiently attractive to progress to a full production situation.

Microwave heating has also been applied to reduce the proof time of dough pieces. The most successful application was for proving doughnuts and similar fermented goods which do not require a pan. The claim (Schiffmann et al., 1971) was for a reduction in final proof time from 45 to 4 min without loss of quality. However, while there have been several attempts to introduce microwave proving, the adoption of the technology has been limited. In part this is associated with capital and operating costs.

## 5.6.   Prover to Oven

The transition from prover to oven is a critical phase in the breadmaking process. The loaf is approaching its final volume but is still a completely flexible structure, maintained only by the continuing production of gas within the semi-porous cells formed by the membranes of hydrated protein.

The first essential is to protect the young loaf from physical damage because if gas bubble stability is poor, even quite small impacts can destroy the fragile structure. The structure can be particularly delicate in some circumstances, for instance if there are flour quality problems or if the dough has been over-mixed, and it is in these circumstances that the product handling system is really tested.

## 5.7.   The Baking Process

### 5.7.1.  Crumb Structure

For the centre of the dough piece, the move into the oven is undramatic. It is so well insulated by surrounding dough that it is completely insensitive to any change for the first several minutes of baking and continues at peak gas production. Effectively, the centre of the loaf gets additional proof time which compensates for its slower start at the beginning of proof. Eventually the centre of the loaf does start to warm up, and as the temperature rises, it goes through a complex progression of physical, chemical and biochemical changes which are independent of the precise conditions in the oven and, therefore, outside the control of the oven operator.

Once the crust has formed, it creates a physical barrier to further expansion of the dough piece. Where the dough piece is being baked in the pan the sides of the pan create a further barrier to expansion. The central portions of the dough pieces continue to expand in the early stages of baking, and the crumb is forced outwards where it meets the crust and the pans. As a result of the expansion, thin layers of crumb become compressed against the inner surfaces of the crust, and there is a loss of cellular structure in these regions. Once the central crumb expansion has ceased, the main features of the crumb cellular structure are in place.

Coalescence of gas bubbles occurs while the dough is expanding and the crumb cell structure begins to form. Gas bubbles that started with sizes in the range 20–250 μm now coalesce to yield crumb cells of 1–4 mm.

Thermodynamically, the situation in the centre is fairly simple. The driving force for heat transfer is the temperature gradient from the region near the crust, where the temperature is limited to the boiling point of water, to the centre. The heat transfer mechanism is conduction along the cell walls, and the centre temperature will rise independently of the oven temperature and approach boiling point asymptotically. There is no significant movement of moisture, and the moisture content at the loaf centre will be more or less the same at the end of baking as it was at the beginning. Figure 5.10 shows the loaf core temperature (LCT) during baking for three different tin sizes – loaves with a smaller cross-sectional area will bake more quickly.

## 5.7.2.   Yeast Activity and the Foam-to-Sponge Conversion

Yeast activity decreases as the dough warms and the yeast is inactivated by the time the temperature has reached 55 °C (131 °F). Stability of the structure is maintained because the trapped gases expand as they warm and maintain the positive internal cell pressure.

As long as the gas cells in the dough remain intact, the dough piece can expand, but as the foam in the dough makes the transition to sponge in the baked product, gas pressure becomes equalized and expansion ceases. Yeast activity is a significant contributor to the gas cell expansion along with thermal expansion of the gases in the closed cells and the evolution of steam. The transition from foam to sponge takes place after the yeast has been inactivated, the precise point depending on a number of factors as discussed below.

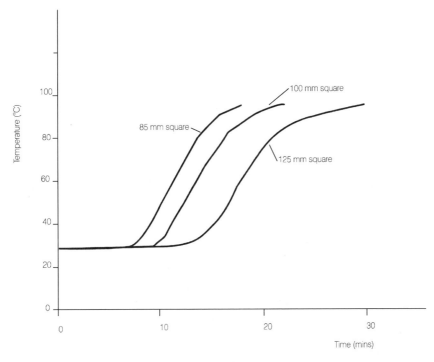

FIGURE 5.10. Variation of loaf core temperature with time.

### 5.7.3.   Starch Gelatinization

Gelatinization of wheat starch starts at about 60 °C (140 °F), and initially the starch granules absorb any free water in the dough. The gelatinized starch will eventually be self-supporting and will take over from the protein membranes which, by their gas-retaining properties, have established the form of the loaf but have little intrinsic strength. There is insufficient water in the dough to gelatinize the starch fully, and water will be transfered from the protein membranes to the starch as baking proceeds. Starch has remarkable water-retaining capacity, and the ease with which gelatinization continues is affected both by starch damage, deliberately imparted during milling (Chapter 12) and by enzyme activity during baking.

The precise temperature of gelatinization will be affected by the dough recipe. This is particularly important when sugar is present in the recipe since this has the effect of delaying the gelatinization of the starch by many degrees, and in some cases may prevent a complete foam-to-sponge conversion taking place. A late or incomplete foam-to-sponge conversion may be seen as sidewalls collapse in pan loaves and in the sinking back and wrinkling sometimes seen with sugar- containing fermented goods, for example soft rolls, fruit buns.

## 5.7.4.  Enzyme Activity

Cereal *alpha*-amylase activity, which is responsible for converting starch into sugars, is at its maximum between 60 and 70 °C (140–158 °F), and the amylase is not completely destroyed until the temperature reaches 85 °C (185 °F). Insufficient amylase activity can restrict loaf volume, because the starch structure becomes rigid too soon, while too much activity, which may be the case after wet harvests, can cause the dough structure to become so fluid that the loaf collapses completely (Chapter 3).

It is the timing and volatility of the complex interactions of gelatinization and enzyme activity that determine the quality of crumb in the finished loaf. It is worth reiterating that poor crumb quality is not a baking fault, it is caused by loss of control at an earlier stage in ingredient or dough preparation.

## 5.7.5.  Baked Temperature

A key parameter of loaf quality that the oven operator must monitor is the final core temperature. The definition of 'baked' is an arbitrary one, often determined by the loaf's ability to withstand slicing, but a temperature in the order of 92–95 °C (198–203 °F) at the centre of the loaf at the end of baking is generally accepted as being necessary for the structure to be adequately rigid throughout the loaf, in part because of the loss of water.

In most bread and fermented products the foam-to-sponge conversion will have taken place before the product reaches a core temperature of 92–95 °C (198–203 °F). This transition can be seen in the baking product when the height of the product falls slightly just before the end of baking. If the loaf is taken from the oven immediately when this point is reached, it is usually too weak to support its own weight, and so it appears that the continued loss of water contributes to the structural integrity of the product.

## 5.7.6.  Crust Formation

Formation of a satisfactory crust is one of the most important aspects of baking – the crust provides most of the strength of the finished loaf and the greater part of the flavour. The thickness and characteristics of the crust to a large extent define the product (Chapter 1).

In contrast to the crumb, where changes during baking are largely chemical and biochemical, though initiated by the rising temperature, in the region of the crust very complicated physical mechanisms are at work. Condensation on the surface of the loaf at the start of baking is essential for the formation of gloss, but quite soon the temperature of the surface rises above the local dew point temperature and evaporation starts. Very soon after that, the surface reaches the boiling point of the free liquid, which will be close to 100 °C (212 °F), and the rate of moisture loss accelerates. As the loaf surface dries, so the evaporation front moves below the surface and the crust of the loaf starts to form. It is important to understand that moisture loss, and therefore weight loss, is an essential part of crust formation, and although it is possible to manipulate oven conditions to reduce weight loss, the characteristics of the loaf, specifically the qualities of the crust, will also change. For a typical plant loaf, with a finished weight of 800 g (28 oz), the weight lost during baking will be 50–55 g (1.8 oz), and anything produced with more crust will inevitably have a higher weight loss.

The heat transfer mechanisms at the evaporation front are complex. As well as conduction within the cell walls, each cell acts rather like a heat pipe – water is evaporated at the hot end of the cell, some is lost to the outside but the rest moves across the cell, in the direction of the centre, and condenses at the cold end of the cell, transferring its latent heat before diffusing along the cell wall to evaporate again at the hot end. The evaporation front may develop more quickly at the top surface of an open-topped loaf, but it is present also on all the faces of the loaf. The tin, the supporting grid and any surrounding tins may shield the loaf from convection and radiation to some extent, and so slow down the formation of crust, but the presence of the impermeable tin surface does not prevent evaporation, the steam escaping freely from between the inside walls of the tin and the dough surface. The operation of the evaporation front is shown schematically in Figure 5.11.

If the steam is not able to readily escape from between the dough piece and the pan, then pockets of high pressure can build up and these may lead to 'indents' in the bottom and occasionally the side-wall crusts. Commonly the indented areas are pale in colour showing that these portions of the crust have not been in contact with the pan surfaces for long enough for them to assume the normal colour. This phenomenon has been referred to as 'pan-lock' by bakers (Cauvain and Young, 2001) and its presence is exacerbated by strapping pans too close together or by placing individual pans too close together in the oven. Small

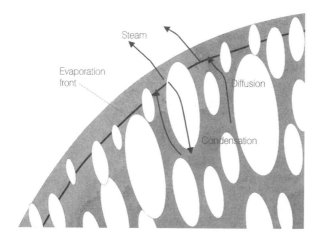

Figure 5.11. Heat transfer in the evaporation zone.

venting holes placed at the angle of the side wall and base of the pan readily prevent this problem. Product baked in pans with slightly ridged sides do not normally exhibit this problem. Oven bottom loaves and other fermented goods baked on trays do not normally exhibit this problem because the steam pressure is more readily dissipated.

Outside the evaporation front is the region that is called the crust, and here the temperature will continue to rise towards the air temperature in the oven. As it heats further, bound water will be driven off and the crust will acquire its characteristic crispness. Colour comes mainly from Maillard reactions, which start at temperatures above 115 °C (240 °F). These reactions also produce bread flavour and the aromas of baking, which are so important for in-store bakeries where they act as an advertisement for fresh bread. The rate of formation of crust is approximately linear, so that a loaf baked in 30 min may have a 3-mm (0.12-in) crust thickness, whereas a larger loaf that takes 50 min to bake will have a 5-mm thick crust. Cauvain and Young (2000) showed that an increase in crust thickness from 1 to 2 mm reduced the overall moisture content of an 800-g pan loaf by about 1%.

The other contributor to crust formation is the continuing expansion of the inside of the loaf, caused first by the final burst of carbon dioxide production from yeast fermentation and then by the thermal expansion of the gases trapped in the cellular structure of the dough. The loaf, if it is in a tin, can only expand upwards. But the sides of the loaf have formed a thin crust and will not move, so cells just inside the crust, in fact those at the evaporation front, are subjected to shear and compression and become flattened and elongated. This effect is most obvious at the top edges of the loaf, where the displacement is greatest and where a split develops as the top crust lifts, exposing a band of elongated inner crust cells, called the 'oven break'. Figure 5.5, which showed how the cells in the dough expand during proof, is magnified further in Figure 5.12, to show how cells near the crust shear to give a compacted layer and expose the shredded oven break. Specific characteristics of the crust can be developed independently.

## 5.7.7.   Gloss Formation

The first seconds of baking are vital for the formation of a glossy crust on the loaf. Again it is a psychrometric problem. As it enters the oven, the surface of the loaf is exposed to high levels of radiation, and sometimes convection, and will increase in temperature very rapidly. To obtain gloss, it is essential that vapour condenses on the surface to form a starch paste that will gelatinize, form dextrins and eventually caramelize to give both colour and shine. The pioneering experiments on gloss formation were conducted at the University of Michigan, Ann Arbor, by Brown and Brownell (1941), and their results are still valuable to the oven designer today. They identified a difference in the way starch can gelatinize on the loaf crust. If there is excess water, paste-type gelation takes place; if the water availability is insufficient, then crumb-type gelation occurs.

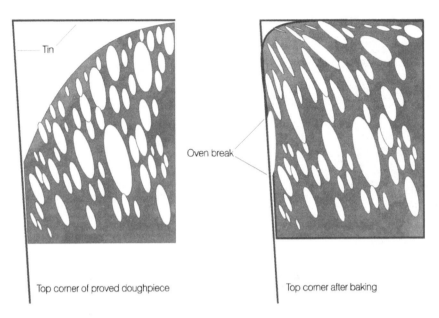

Tin

Oven break

Top corner of proved doughpiece

Top corner after baking

FIGURE 5.12. Crust development and oven break.

The conditions necessary for a glossy crust are as follows:

- The dough must not be over-proved. If the dough has reached its maximum volume before leaving the prover, it will not develop satisfactory gloss.
- Paste-type gelation must take place before there is any chance of crumb-type gelation.
- A minimum oven temperature of 74 °C (165 °F) must be maintained for sufficient time (at this minimum temperature, the time required would be too long to be practically useful). In fact, the required time varies from 10 min at an oven temperature of 77 °C (171 °F) to only 15 s at 99 °C (210 °F).

The important condition of excess water on the crust will only happen when the crust temperature is below the dew point temperature of the oven atmosphere at the oven inlet. In practice, this means that if the oven is at a typical temperature of, say, 225 °C (437 °F), the dew point temperature needs to be above 93 °C (199 °F) to ensure that the excess water condition is maintained for long enough. To guarantee suitable conditions usually requires a special inlet section on the oven, called a steam tunnel. The way in which crust temperature depends on dew point temperature is shown in Figure 5.13, with only the top curve satisfying the conditions for high gloss. Notice how the rate of increase of the crust temperature levels out as it approaches the dew point temperature. This is the point where condensation changes to evaporation, and exactly how this happens will depend on the air velocity in the region. To make sure that all loaves are

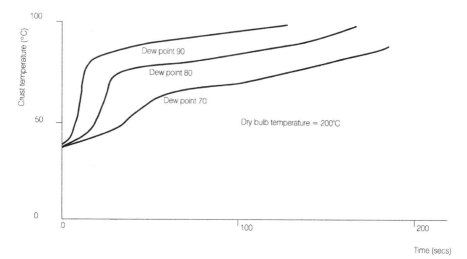

FIGURE 5.13. Crust temperature against time for different dew point temperatures.

treated equally, steam tunnels usually have low air velocities, but this is not a necessary requirement for gloss, it just makes it easier to maintain a high dew point temperature.

## 5.7.8. Crust Crispness

The pattern of humidity during baking affects another important crust quality besides gloss. Sometimes a smooth, elastic crust is required that can be sliced without disintegrating, but at other times a brittle crust, which will form a mosaic of cracks when cooled, is the marketing specification. The behavior of the crust will depend mainly on the thickness of the starch paste layer – the thicker the layer, the more likely it is to craze as it cools. This means that when a pliable and glossy crust is required, the conditions for gloss formation must be maintained for the minimum time necessary and thereafter the baking conditions must be comparatively dry.

## 5.7.9. Oven Break

The shredded band of exposed inner crust that develops during baking at the edges of the loaf, on the join between the top crust and the side crusts, is called the 'oven break'. Most products look best with a regular break some 20 mm (0.75 in) high, but sometimes less is required (e.g. in the manufacture of lidded sandwich breads), and it is important to understand which parameters can be used to control the break.

Break will increase with volume increase in the oven, the so-called 'oven spring', and decrease with crust toughness. Oven spring depends to a large extent on the state of fermentation in the dough when it enters the oven. If fermentation is almost complete, in other words if all the sugars have been converted, then oven spring will be limited. If this point has not been reached, which could happen if not enough yeast has been used, or if the proof temperature is too low or the time too short, then oven spring will be greater and there is more chance of an excessive break, which can become what is colourfully described as a 'flying top'.

One of the mysteries of baking is why the oven break forms in the part of the loaf that it does. There is no clear pattern of oven break from loaf to loaf from a given batch of dough. There is no doubt that the oven break is exploiting small (perhaps microscopic) variations in the dough structure of individual dough pieces. One factor which has been overlooked in understanding the physical manifestation of oven break has been the placing of the dough in the pan. In particular the positioning of the 'swiss roll' or curl of the dough piece. Craft wisdom dictates that the 'seam' or trailing edge of the dough piece should be placed on the bottom of the pan or tray in order to obtain uniform expansion. In modern plant such precise positioning is seldom achieved. There is no doubt that the random positioning of dough pieces in modern production confounds the location of oven spring. There is considerable tension in the curled dough piece, and there will be a tendency for the swiss roll to uncurl in the early stages of baking. It is possible that it is the spatial positioning of the leading edge, the first part of the curl, which is responsible for the way in which oven break is manifest.

With many bread products oven break is controlled by cutting or marking the surface. The act of cutting allows for the controlled release of tensions in the dough, and the increased surface area of dough so created increases the rate of transfer of heat transfer to the centre of the dough and adds to the development of crust.

## 5.7.10.  Practical Baking

It is important to check regularly that the heat transfer equipment is in good condition, but for an oven the measurements are more difficult to make than in a prover and may require specialized help, perhaps from the oven manufacturer. A regular flue gas analysis is the simplest way of monitoring the condition of the combustion equipment, while measurements of air temperature and humidity in each oven zone for each product should also be recorded and checked regularly.

The best confirmation that all is well is provided by quality control checks on the products themselves, and the most important reading is core temperature at the end of baking. When establishing a quality control procedure to monitor centre loaf temperature, it is important to remember that, just as there is a lag at the start of baking, so the temperature at the loaf centre will continue to rise for several minutes after the loaf has come out of the oven. No oven gives perfectly uniform baking, and in some ovens there will be significant differences in temperature depending on the position in the oven. The control procedure reflects changes with time and must not be confused with information reflecting unevenness in the oven. Thus it is important to understand the particular oven and always to take samples from the coolest position. The position of the loaf within a tin strap can also affect the heat transfer rates. To be on the safe side, the sample should always be taken from the centre of the strap.

## 5.7.11.  Pan Strapping

As with all baking systems, in order to obtain optimum results it is important to design the tins and tin straps to suit the heat transfer capabilities of the oven. Compact straps may increase oven loading, but the overall profitability of the line also depends on loaf-to-loaf variations in centre temperature (and therefore weight loss), and a strap with narrow air passages between tins may cause significant weight loss variations between the end loaves in the strap and those at the centre. Most tins are made from steel, usually 0.8 mm thick with a non-stick coating, but baking time can be reduced by using black anodized aluminium, with the penalty that it is less robust and will not last as long in an automatic handling system.

Narrow straps between pans reduce the air flow and the rate of heat transfer in all ovens. The slower transfer of heat allows yeast to continue working longer in the early stages of baking. In effect this means that the some of the loaves, usually the ones in the centre of the strap, get extra proof. In such circumstances, loaves may touch one another during baking and prevent the formation of a crust on the sides of the loaf (Figure 5.14). Similar problems may be encountered with crusty baguette baked in unsuitable slings and with bloomers and crusty rolls not spaced far enough apart on trays. The areas of the product without adequate crust formation are weak points in the product structure and contribute to more rapid loss of the crisp crust which characterizes oven-bottom and hearth breads.

## 5.7.12.  Oven Design

As with provers, the biggest change in plant bakeries in recent years has been away from tunnel ovens towards the single-file conveyorized spiral oven. The simplicity of the concept improves reliability and uniformity of bake while the small inlet and outlet ports, together with the low thermal mass of the conveyor, lead to useful increases in thermal efficiency. An indirect-fired conveyorized oven is shown schematically in Figure 5.15.

There has been a trend to design for radiating surfaces in the oven to be cooler than used to be the case and to rely on forced convection for a higher percentage of the required heat transfer. This tends to reduce the thermal mass of the oven so that it warms up more quickly and can handle product changes and gaps in production more easily. Baking by convection also results in a more even surface colour than by radiation because the latter possesses the destabilizing

FIGURE 5.14. Closely strapped bread pans.

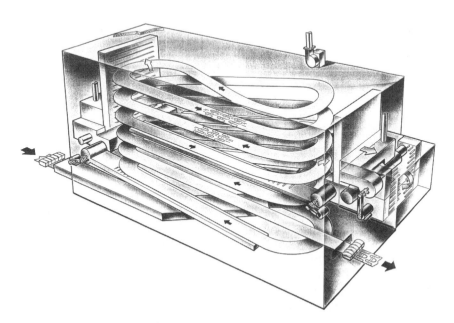

FIGURE 5.15. Conveyorized oven.

characteristic of being absorbed more readily by darker surfaces. In other words, radiation will accentuate contrasts. It is interesting that there has also been a trend towards convection in biscuit ovens since the 1970s, but there are now signs of the pendulum swinging back, as the required textural and colour-related properties for some products can only be obtained by increasing the radiative component of the heat transfer.

## 5.7.13.  Developments in Baking

As with proving, most of the developments in baking technology have been incremental in nature and have been related to increasing plant output, reducing product variability and improving mechanical, reliability. Three technologies that are proven experimentally but not fully implemented commercially offer the potential for markedly reducing baking times. One is dielectric heating, using either radio frequency or microwave power, where the energy is supplied through the bulk of the dough piece, and it is therefore not surprising that baking times can be halved. The volumetric transfer of energy will not give rise to an evaporation front in the same way as conventional baking, so any crust formed will be thin and delicate, nor will the skin temperatures reached be high enough to generate colour and flavour in the crust. Possibly these characteristics of dielectric heating will become advantages in the search for the perfect part-baked process, a subject discussed later in this chapter. (For a detailed history of dielectric baking of bread, see Ovadia, 1994.)

The second technique uses infrared radiators to increase heat transfer to the loaves. Again, marked reductions in baking time are recorded, but for this technology, where penetration of energy into the loaf surface is shallow, the result is less easy to explain. It seems possible that the crust region, outside the evaporation front, is the heat transfer bottleneck during conventional baking and that the limited penetration ($< 1$ mm) of infrared radiation can overcome this barrier and increase the vigour of the heat pipe mechanism proposed for the evaporation front. Thus although the temperature driving force to heat the crumb is unchanged, this second mechanism increases the heat transfer and rate of temperature rise of the loaf core.

The third innovation uses high-velocity convection to improve the rate of heat transfer to the products. The main effect is in the warming-up phase, where baking time is saved by warming up the tin and dough surface more quickly than in a conventional oven and as much heat as possible is transferred while the conductivity of the dough is still relatively high. Such a process is described in an RHM patent (1989), where one embodiment, by way of example, specifies, for the first 10% of the oven length, air nozzle velocities of 25 m/s (82 ft/s) and a temperature of 375 °C (707 °F).

All these developments are designed to reduce baking time, but they will also have an effect on the interaction of the mechanisms at work during baking and will inevitably change the characteristics of the final loaf.

Relatively small reductions in baking time may be very valuable if they allow existing plant to be up-rated, but when it is a question of specifying new equipment for a particular output, it is usually more cost effective to use conventional technology in an oven of the appropriate size than to save space by incorporating components to enhance the heat transfer.

## 5.8.  Oven to Cooler

As noted earlier, baking and cooling overlap; cooling starts before baking has finished. As soon as the bread reaches the exit of the oven the air temperature will fall below the crust temperature. But the evaporation zone is still operating and the loaf centre is temporarily unaware of the changing outside environment. There is still a temperature gradient and it will continue to get hotter in the centre of the product for some minutes after it has left the oven.

De-panning is usually the next step and this is an operation that can become very difficult if the mixing, forming, proving and baking have not been properly controlled.

Faults that cause problems include the following:

- under-proofed bread that has too much oven spring and such a dislocation around the top crust that the whole top crust comes off under suction;
- over-proved bread that may have a very weak structure or may be too tall to fit under the suction head;
- under-baked bread that may collapse or tear;
- sticking of some of the loaves in the tins, possibly caused by erratic tin greasing.

Cooling continues as the bread is de-panned and transferred to the cooler. During all this time it is in an uncontrolled atmosphere, and as well as losing heat, it is rapidly losing moisture. Whilst in the oven, moisture loss is an essential part of crust formation, but once out, any loss is wasted profit where bread is sold by weight and the loaves should be transferred as quickly as possible to the controlled conditions in the cooler. It can easily take 3 min to move a loaf from oven to cooler, and in that time it can lose 12 g (about 0.5 oz). Thinking negatively, this represents 12 g given away unnecessarily, but from a positive viewpoint, evaporating 12 g of water has cooled the loaf significantly and reduced the heat load on the cooler by about one-fifth.

## 5.9.  The Cooling Process

There are two distinct mechanisms involved in cooling a loaf. The first is by heat transfer, mainly by convection to the surrounding air, but also by radiation and conduction to the structure of the cooler. The second is by evaporation – moisture

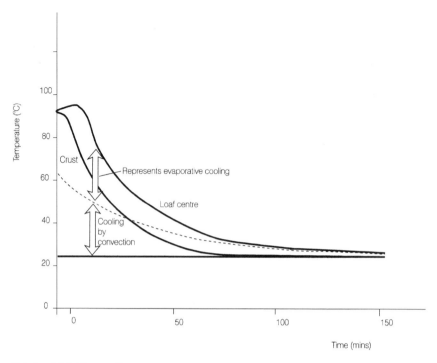

FIGURE 5.16. The cooling process.

evaporates from the crust, and the energy to evaporate it is drawn from the crust. Typically, an 800-g loaf will lose 20–25 g (about 1 oz) during cooling, which, with the weight it has lost in the transfer from oven to cooler, represents about half the load required to cool the loaf to 25 °C (77 °F). This is illustrated schematically in Figure 5.16.

It would be quite possible to design a cooler where the evaporation loss was much lower, for instance by having many small, independent cooling zones fed with cold saturated air, but the cooling times would be longer, because most of the heat would have to be removed by convection, and the capital and maintenance costs would be much higher. In practice the most common solution, combining an economic capital installation with a reduced evaporation loss is to have the section where the bread enters the cooler as cold as possible without refrigeration (with air velocities in the region of 1 m/s (about 3 ft/s), an air temperature of 20 °C (68 °F) and a relative humidity of 80%), causing the crust temperature to drop so quickly that the crust will act as a barrier to further moisture loss.

In countries where summer ambient conditions are hot, refrigeration will be necessary to obtain reasonable cooling times combined with low weight loss. Once refrigeration has been decided upon, however, it is worth installing sufficient capacity to run the inlet section of the cooler at 5 °C (41 °F). The initial rapid cooling rate will save another 3–4 g of moisture loss compared with typical, non-refrigerated cooling conditions.

## 5.9.1.  Practical Cooling

Most coolers have no refrigeration and rely on using more fresh air in summer than in winter to maintain constant cooling conditions. The dampers that control the air flow are usually fairly simple, but there may be manual balancing adjustments, and it is important to check that airflow as well as temperature and humidity are maintained over the whole operating range. Understanding the cooler and keeping it in good condition is every bit as important as it is for the oven and prover, and a continuous record of temperature and humidity near the beginning and near the end of the product path provides a valuable check, providing the humidity instrumentation is regularly calibrated. Regular cleaning is vital, not only for hygienic reasons, but also to maintain performance, which could be seriously compromised by a furred spray nozzle or blocked filter.

Uniformity of cooling and weight loss is just as important as absolute performance. If the loaves are at different temperatures, the warmer ones will tend to get squashed in the slicers, and weight loss variability directly affects the target dough piece scaling weight. The target temperature for the centre of the cooled loaf will be about 25 °C (77 °F) for a sandwich loaf that is sliced on a reciprocating slicer, but if the recipe contains more fat, as in America for example, or if more sophisticated band slicers are installed, then satisfactory slicing will be practicable at higher temperatures and cooling capacity can be saved. Temperatures should be recorded regularly, using the principles described in the baking section, but two techniques are necessary to check the weight loss performance. Regular, planned sampling, which is non-invasive and therefore does not create scrap, will allow the weight variance for the plant to be calculated, and by subtracting the variance obtained at the end of the oven, a figure for cooler variance can be found. The variance is found by summing the squares of the weight differences from the average and dividing the total by one less than the number of readings. Variances can be subtracted – standard deviations cannot! At this point I shall give up on my stated intention not to include any formulae and give the definition of variance mathematically as:

$$\text{variance} = \Sigma(x - \bar{x})^2/(n-1)$$

where $x$ is the measured variable, $\bar{x}$ is the mean and $n$ is the number of readings.

If there is any sign that cooler variance is increasing, it will be necessary to conduct detailed tests by weighing loaves into the cooler, marking them and weighing them again when they come out. Remember that the loaves are losing weight at their maximum rate on the cooler in-feed, so the initial weighing and marking must be done without upsetting the normal loading progression.

## 5.9.2.  Cooler Design

Just as tunnel ovens and tray provers in plant bakeries are being superseded by systems that use a conveyor that is only one tin strap wide, so rack coolers, where the bread is loaded onto shelves within a rack which is indexed along the top of the cooler then lowered and returned along the bottom, are being replaced with spiral designs. In fact the technology was born in the refrigeration sector, where the first spiral coolers were built in the 1960s and, because the concept is mechanically simple and inherently subjects all the products to the same processing cycle, it has migrated to proving and baking equipment also. The conveyorized designs have the added advantage that they are generally much easier to clean and access to the air conditioning equipment is better.

## 5.9.3. Developments in Cooling

As already discussed, conventional bread cooling is by convective heat transfer and evaporative cooling, with the balance slightly in favour of convection. Process development is active at the two extremes, tilting the balance one way or the other. On the one hand, improvements to the distribution of the air and the control of its condition can further reduce weight loss by minimizing evaporation, and, at the other end of the spectrum, most of the heat can deliberately be removed by evaporation, by using a vacuum cooler.

Two patents describing ways of improving the condition of the air have been granted to RHM (1988) and Allied Bakeries Ltd. (1986). The first describes a spiral cooler with multiple zones with successive zones fed from alternate sides of the conveyor, a scheme that allows the contacting air to be maintained at low temperature and high humidity. The second patent describes another method of maintaining the circulating air in peak condition by spraying water onto the surface of the loaves in measured quantities so that it all evaporates before impregnating the crust. In other words, the air is continuously conditioned in the cooling chamber itself. The limitations of this method can be calculated directly from the psychrometric chart.

Vacuum cooling is invariably done as a batch process; bread is loaded into a vacuum chamber, which is sealed and partially evacuated. As the pressure falls, so there is a fall in the temperature at which the free water in the loaf boils, and the latent heat for the evaporation is extracted from the loaf, cooling it to slicing temperature in as little as 3 min, as shown in Figure 5.17.

In practice, the operation is quite sophisticated and the vacuum curve is modified to produce the particular product characteristics that are required. The process has been particularly successful on products where the crust is an important part of the structure, crusty rolls and croissants, for example, and on products which are very delicate and difficult to handle in the warm state, such as malt bread or panettoni. Two of the less obvious advantages of vacuum cooling are that it is relatively easy to maintain sterile conditions and thus gain a little extra product mould-free shelf-life, which can be critical in large countries with extended distribution systems, and it is also possible to reduce baking times, often increasing output by as much as 10%. The reduction in baking time is possible because the loaf centre temperature

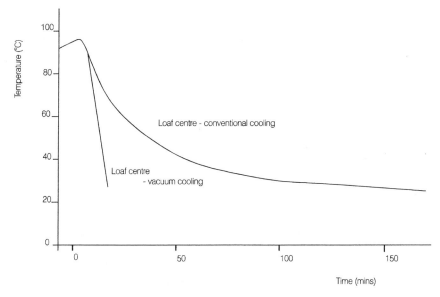

FIGURE 5.17. Loaf cooling in a vacuum.

which satisfies our condition for structure stability is less when the product is vacuum cooled – the speed of cooling, combined with the rapid expansion of the gases within the loaf throughout the cooling period, ensures that the loaf sets at its maximum volume and does not shrink or collapse. The disadvantages of vacuum cooling are the high capital cost of the equipment and the high losses of moisture and the volatile fraction, although the latter can be partially compensated for by reducing the baking time and by introducing a pre-cooling stage to reduce the crust temperature rapidly before applying vacuum. The ideal system of the future might use convection to keep the crust cool and vacuum to accelerate internal heat transfer, reversing the heat pipe mechanism postulated for heat transfer at the evaporation front during baking.

## 5.10.   Part-Baked Processes

Part-baked bread is a growing market sector for the baking industry, not so much for products for finishing at home, but for products that can be distributed for final baking at the point of sale. The system beautifully combines the economies of scale available where the part-baked loaves are manufactured with the flexibility and freshness obtained by finally baking to satisfy current demand. As the technology develops and the final product quality improves, this process could well come to dominate the industry.

### 5.10.1.  Frozen Processes

In mainland Europe most of the intermediate products are frozen for distribution, often using conventional spiral freezers that have not been specially developed for the bakery industry. This can mean that weight loss is not minimized, and this directly results in problems of icing on the refrigeration coils. The technology will improve, as will the product quality, but it does have the fundamental cost burden that energy has to be expended to freeze each loaf and then again to thaw it.

### 5.10.2.  Ambient Processing

A leading contender for the ambient part-baked market was the Milton Keynes Process, where vacuum cooling was used after a slow bake to stabilize the intermediate product. The loaves could be distributed and stored in ambient conditions and then be finish baked at the point of sale (Chapter 9). The Milton Keynes Process as such is no longer but several similar approaches have been used by specialist manufacturers to supply ambient stored, part-baked products for bake-off. Typically such products are described as '80 or 90% baked' before delivery to the end user.

## 5.11.  Process Economics

### 5.11.1.  Weight Loss

Weight loss at all stages – proving, baking and cooling – is an important parameter affecting the economics of plant operation. This is particularly true in countries like Britain, where bread is sold by finished weight, in countries like the USA, where there is a specified maximum water content, or in other countries where there is minimum solids legislation. Under this sort of legislation, if giveaway is to be minimized, weight loss must be tightly controlled. Minimization of weight loss is less critical in those countries which sell on a dry weight basis, but uniformity is still important as a quality parameter, as moisture content will affect both eating qualities and shelf-life. Typical commercial figures for each stage of the process are given in Figure 5.18.

Selling bread on a dry weight basis is, on the face of it, a sensible arrangement since it removes the competitive temptation to maximize the water content, but it seems unlikely that legislation in this area will be harmonized throughout Europe in the near future.

### 5.11.2.  Life Cycle Costs

Life cycle costing (LCC) is an industrial management technique, aiming to:

- provide a comprehensive understanding of the total commitment of asset ownership;

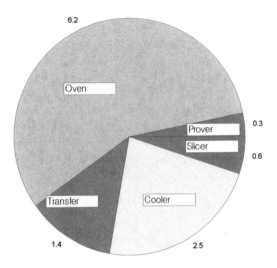

FIGURE 5.18. Weight loss during processing.

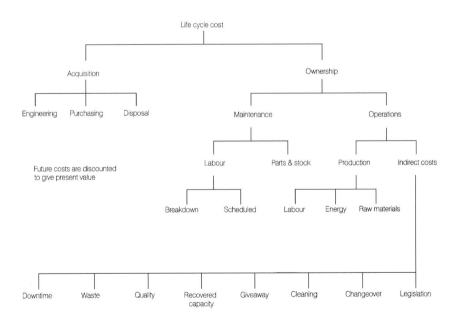

FIGURE 5.19. Breakdown of life cycle costs.

- identify areas where improvement can be achieved through redesign or reallocation of resources;
- improve profitability and increase industrial efficiency.

The practice is not widespread in the bakery world, but we often lag behind the chemical and pharmaceutical industries. It will come and when it does the leading companies will enjoy a competitive advantage during the transition period.

The methodology for LCC was recently standardized in IEC (TC56) (1996), which puts costs into perspective by modelling the total cost of operating a process plant for the whole of its planned life. It is particularly valuable when new equipment is being considered, when quite a small change in specification can yield huge savings during the life of the plant, but unless data are collected in a useful format during normal bakery operation, the costing model will be based on conjecture rather than fact. For instance, some bakeries regard maintenance and cleaning as overhead charges and this makes it impossible to justify extra capital investment that might reduce maintenance requirements or make cleaning easier.

The life cycle cost approach also forces the equipment supplier and the user to work closely together, ensuring that the supplier fully understands all the operational implications of design decisions. By specifying performance guarantees in terms of the model for life cycle costs, the supplier is committed to a realistic initial model and the user is much better placed to make the investment decisions that will maximize profitability. Suitable categories for structuring life cycle costs for a bakery are given in Figure 5.19.

## References

Allied Bakeries Ltd. (1986) *Method of cooling baked goods*, GB Patent 2 134 636B, 20 August, HMSO, London.

Brown, G.G. and Brownell, L.E. (1941) University of Michigan, Ann Arbor, sponsored project for APV Baker Inc.

Campbell, G.M. (2003) Bread aeration. In (ed. S.P. Cauvain) *Bread Making: Improving Quality*, Woodhead Publishing Ltd., Cambridge, UK, pp. 352–374.

Cauvain, S.P. (2004) Filling in the holes. *British Baker*, February, 6, 12–13.

Cauvain, S.P. and Young, L.S. (2000) *Bakery Food Manufacture & Quality: Water Control & Effects*, Blackwell Science, Oxford, UK.

Cauvain, S.P. and Young, L.S. (2001) *Baking Problems Solved*, Woodhead Publishing Ltd., Cambridge, UK, pp. 81–82.

Ovadia, D. (1994) Dielectric baking of bread – past and future. *Microwave World*, 15(2), 16–22.

Rank Hovis McDougall (1988) *A Method and Apparatus for Cooling Foodstuffs*, GB Patent 2 206 190A, 29 December, HMSO, London.

Rank Hovis McDougall (1989) *Travelling Bread Cooler*, GB Patent 2 217 969A, 8 November, HMSO, London.

Schiffmann, R.F., Roth, H., Stein, E.W, Kaufman, H.B., Hochhauser, A. and Clark, F. (1971) Appications of microwave energy to doughnut production. *Food Technology*, **25**, 718–722.

Whitworth, M.B. and Alava, J.M. (1999) The imaging and measurement of bubbles in bread doughs. In (eds G.M. Campbell, C. Webb, S.S. Pandiella and K. Niranjan) *Bubbles in Food*, Eagan Press, St. Paul, Minnesota, USA, pp. 221–232.

# 6
# Dough Retarding and Freezing

Stanley P. Cauvain

## 6.1. Introduction

Over the years, bakers have sought ways to extend the working life of dough, which is otherwise largely limited by the natural processes of yeast fermentation, enzymic activity and structural relaxation of the gluten. Improvements in production efficiency have been the main driving force for the baker to seek such extensions in dough life. For example, some products are required in small numbers, which necessitates mixing small batches of dough, but maximum production efficiency is more often achieved by producing larger quantities of dough than those needed for a single day's product sales. Other driving forces for extending dough shelf-life have included the avoidance of unsocial working hours and local restrictions on night baking.

Since yeast fermentation is significantly slowed down by lowering the dough temperature, it was only natural that attention would be focused on the application of refrigerated and deep-freeze temperatures to yeasted doughs. If the dough temperature is reduced low enough, yeast fermentation will cease altogether and the dough can be held in what approximates to a state of suspension. Early experiments with the refrigeration of yeasted doughs was to lead eventually to the development of the process which has become known as 'dough retarding' and which utilizes specialist refrigeration equipment. There are both similarities and significant differences between retarded and frozen doughs. They share the common problems associated with the poor conductivity of dough but require quite different equipment if both processes are to run under optimized conditions.

Retarding and deep freezing of yeasted doughs are not necessarily the 'convenience' products that they are sometimes claimed to be. For example, the early introduction of retarding and deep freezing techniques into UK craft bakeries during the 1950s met with some success but gradually the limitations of the technology and the equipment reduced their impact on the baking industry. The evolution of the 'in-store' bakery resulted in a revival of interest in retarded doughs in the UK. In other European countries the use of retarders has remained more common since their introduction, perhaps in part because of the continued survival of large numbers of small, craft bakers where there are limitations on working practices, or where the retarding process is seen to be making a positive contribution to product quality.

The production of frozen doughs had less appeal to the craft baker and was more suited to centralized production of the frozen units, with distribution to satellite storage and bake-off. The popularity of frozen dough products has always been less in Europe than that in the USA, where there are significant numbers of bake-off units using frozen dough technology. With improvements in technical understanding and process control, the use of frozen dough in Europe has increased.

## 6.2.    Retarding Fermented Doughs*

Most fermented goods can be produced successfully via retarded dough, provided that retarding and proving conditions are chosen to suit the product type. As discussed in more detail in the section on Freezing Fermented Doughs, dough is a relatively poor conductor of heat and in consequence products having a large radius, such as pan bread dough pieces, will take longer to cool and warm than those with a small radius, such as baguettes and rolls. Further, dough pieces with large radii are much more likely to cool, or warm, with larger temperature differentials between the centre and the surface of a piece of dough, which are likely to have an adverse effect on final product quality. Therefore the range of conditions that can be used to produce acceptable bread is less than that for smaller fermented goods so that if retarders or retarder–provers are to be used for mixed large and small goods production, the conditions in the units should be set to achieve optimum large-bread quality.

### 6.2.1.    Suitability of Breadmaking Processes

All the major breadmaking processes described in Chapter 2 may be used for producing dough intended for retarding, though the most suitable are those which are based on no-time breadmaking processes. The process of retarding dough will not improve the overall product quality, so the recipe used, the choice of improver and processing conditions must be compatible with the breadmaking process chosen.

Breadmaking processes which use periods of bulk fermentation as a major part of the dough development mechanism are least suited to retarding for two main reasons. First, the level of yeast in the recipe is an integral part of dough development so that any changes in yeast level for purposes described later would

---

* To avoid unnecessary repetition the term 'retarding' will refer to the cooling process which takes place in both 'retarders' and 'retarder–provers' so that references to retarders should be taken to include retarder–provers, and vice versa, unless specified otherwise.

require compensatory changes in bulk fermentation time or dough temperature, or both. Second, the greater quantity of gas present in bulk-fermented doughs as they enter the retarder may have an adverse effect on final product quality. Sponge and dough breadmaking processes may be used to produce doughs for retarding, the best results being obtained if no bulk fermentation time is given after the sponge has been mixed into the final dough. Under most retarding temperatures fermentation continues, albeit at a low rate. Using sponge and dough processes in such circumstances may raise the 'acid' flavour of the baked product to an unacceptably high level and will require adjustment of the sponge conditions.

No significant changes are necessary to the mixing or processing conditions used with short or no-time breadmaking processes. It is unnecessary to lower the final dough temperatures, although this practice is often used as a means of reducing yeast fermentation before and during the early stages of retarding. However, it is not recommended, especially in breadmaking processes where improvers are a necessary part of dough development since lowering of the dough temperature will reduce the contribution of any oxidizing agents and enzymes present and may lead to problems associated with lack of gas retention in the dough, such as loss of product volume and white spot formation (see below).

As a general principle doughs should be processed and loaded into the retarding unit as quickly as possible in order to avoid excessive gas production and any losses of quality which may arise from undue delays. This requirement for minimizing delays before the dough enters the retarder must be balanced against the practical requirements of the bakery. For instance, it may take some time for the bulk of the dough from the mixer to be divided and processed. The length of this operation will depend, in part, on the size of mixing that has been produced. As a general rule delays greater than 20 and certainly 30 min should be avoided. It may be necessary to mix smaller quantities of dough than normal in order to achieve more rapid processing.

The frequency with which processed dough batches are transferred to the retarder also requires some control, otherwise frequent opening of the retarder doors will continually raise the temperature in the retarder and reduce its effectiveness in cooling the dough pieces. Rapid chilling of the dough pieces is necessary if gas production in the retarding phase is to be minimized and is especially important with dough pieces of larger radius. Each time the retarder door is opened, there will be an exchange of air with the bakery atmosphere. The air in the retarder will have a higher relative humidity (RH) than that of the bakery and so there is a tendency for the RH to fall. This exchange will increase the risk of dough pieces 'skinning'. Similar problems will be encountered if the retarder is only partly loaded. The relationship between products and retarder RH is discussed further below.

## 6.2.2.  Recipe and Yeast Level

When deciding on a recipe to use for retarded doughs it is essential to choose one which will give products of acceptable quality in fresh production. Usually the only change in dough recipe required for retarding is to adjust the yeast level, the degree of adjustment depending on the process and equipment being used. With no-time doughmaking processes the adjustment of yeast level is not a problem since it is relatively easy to adjust subsequent proof temperatures, times or both. However, with bulk fermentation processes, changes in yeast level will affect the balance between yeast quantity, dough temperature and bulk fermentation time which is required in order to achieve optimum dough development. For example, if yeast levels were to decrease, it would be necessary to combat the under-fermentation which would arise by a simultaneous lengthening of the bulk fermentation period or an increase in bulk dough temperature, or both. As a general rule, raising the dough temperature by 2 °C will require a compensatory decrease of about 25% of the current yeast level; this alteration will slow down the rate at which the dough proves after retarding.

The choice of yeast level in retarded doughs is intimately linked with dough piece size, storage temperature and time, proving conditions and, of course, the desired product quality. There are such a large number of potential combinations of these factors that it is not possible to set down precise rules for adjusting the yeast level in retarded doughs. The choice of level also depends on whether the baker is using a separate retarder and prover or a combined retarder–prover. In the former case the practice is often to keep yeast at the same levels as for scratch production, or even to increase yeast levels so that when the colder than normal dough enters the prover, final proof times remain conveniently short. Such practices limit the types of products which can be successfully retarded to those of small radius or with higher levels of enrichment. The automatic control on retarder–provers (see below) allows for greater flexibility and choice of retarding and proving conditions. In such cases a wider range of products can be made and product quality can be more closely controlled. When using retarder–provers it has become common practice to reduce yeast levels, especially with larger dough pieces, to avoid losses in quality associated with 'skinning' and 'white spots'.

The RH of the dough when it first enters the retarder is much higher than that of the air in the retarder. Typically the equilibrium RH of the dough is between 90 and 95% (equivalent to water activities of between 0.90 and 0.95) and the RH of the air around 60%. During the early stages of retarding, the dough loses moisture from its surface as it tries to achieve equilibrium with the air. The cooling coil in the retarder is designed to maintain as high an RH as possible, but the dough pieces will progressively lose moisture during storage as shown by the example for 60 g (2 oz) rolls illustrated in Figure 6.1. The overall weight loss from a dough piece during retarded storage is directly affected by the yeast level and is greater with increasing yeast level. The more moisture that is lost from the dough piece, the greater is the risk of the formation of the dry surface which bakers call skinning. The mechanism of the formation of this skin is not fully understood but the phenomenon is largely irreversible.

The potential for greater weight losses from retarded doughs with higher yeast levels is due in part to the greater expansion and accompanying increase in

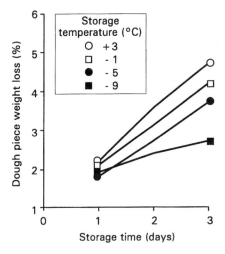

FIGURE 6.1. Effect of storage time and temperature on weight loss from 60 g dough pieces.

the surface area of the dough piece. An example of this effect is illustrated in Figure 6.2 where the volume of 60 g dough pieces made with four different yeast levels and stored for 3 days at 3 °C (38 °F) are compared. For the highest yeast level illustrated, 3% flour weight, the increase in volume approaches threefold. Such an increase is very similar to that observed under normal proof at much higher temperatures, though with shorter times.

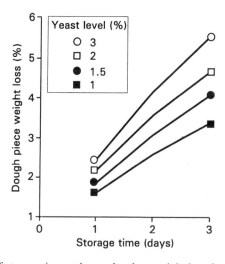

FIGURE 6.2. Effect of storage time and yeast level on weight loss from 60 g dough pieces.

## 6.2.3.  Retarding Temperature

The initial cooling rate of the dough and the temperature at which the storage phase is carried out have a considerable influence on final product quality. The effect of retarding temperature, whatever the equipment used, can be considered in three broad temperature ranges:

- above the freezing point of water, 0 up to 5 °C (32–41 °F);
- between 0 and −5 °C (32–23 °F), the latter being the approximate freezing point of bread dough because of the dissolved recipe salt (lower if sugar is also present);
- below −5 °C (23 °F).

The rate at which dough pieces cool when placed in a retarder depends to a large extent on the temperature within the unit; in general, the lower the retarding temperature the faster the cooling rate. An example of how retarding temperature affects the rate of cooling is given in Figure 6.3, where the effects of four different retarding temperatures on the temperature at the centre of dough pieces for 60 g rolls are compared. The sharp rise in temperature at the centre of the dough pieces cooled at −9 °C (16 °F) coincides approximately with the temperature at which the roll dough pieces froze. The freezing point of the roll dough is somewhat lower than that of a standard UK bread dough because of the presence of sugar in the formulation.

The actual rate at which dough pieces cool is more complicated than the example given in Figure 6.3. Fermented doughs are poor conductors of heat, and because of this the rate at which the centre cools is much lower than that of the surface. The actual differences in cooling rates between the surface and the

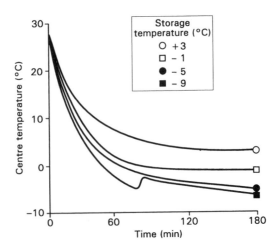

FIGURE 6.3. Effect of retarding temperature on changes in 60 g dough piece centre temperature.

centre are related in part to the retarding temperature but are mostly influenced by product size, or more specifically by product radius; the larger the dough piece radius the greater the temperature differential for a given retarding temperature over a given period of time.

Yeast activity in the dough continues during the initial cooling and subsequent storage periods as long as the temperature is not so low as to cause the dough to freeze. The activity is more vigorous in the warmer parts of the dough, such as the centre, because of the poor conductivity of the dough. In some circumstances considerable gas production and expansion of the dough can occur. The actual quantity of gas produced depends on the yeast level, initial dough temperature, retarding temperature, the ability of the dough to retain the gas being produced and several other factors. The influence of retarder storage temperature on the expansion of 60 g roll dough pieces with the same yeast level is illustrated in Figure 6.4. For dough pieces placed for storage in a unit set at −9 °C (16 °F) there is a small degree of expansion which stops once the dough is frozen. At higher storage temperatures the dough can continue to expand so that at 3 °C (38 °F) the dough piece can double in size during retarded storage.

Weight losses from the dough pieces during storage are also significantly influenced by the retarding temperature, with greater weight losses being sustained at higher temperatures (Figure 6.5). As discussed above, weight losses are linked with skinning of the dough pieces and subsequent quality losses in the final product. When stored at temperatures between 0 and −5 °C (32–23 °F) the surfaces of the dough pieces may be dry to the touch but are less likely to develop a thick skin, unless stored for more than 3 days.

Products made from dough pieces which have skinned frequently exhibit irregular shapes and always have a crust colour with a pale grey cast. In extreme cases, dough pieces may lean towards air inlets because the part of the dough facing the inlet air is subjected to greater moisture loss than the trailing edge of the piece, so that expansion during proof is uneven and is severely restricted by the skinning.

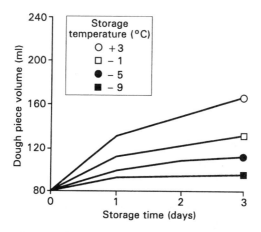

FIGURE 6.4. Effect of storage time and temperature on dough piece volume.

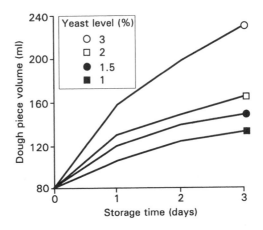

FIGURE 6.5. Effect of storage time and yeast level on dough piece volume.

Although many retarder–provers can operate at temperatures below −5 °C, they are not usually recommended for freezing dough because of their low freezing rates by comparison with blast freezers. While setting, retarder–provers at temperatures below −5 °C will increase the cooling rate because of the temperature differential with the dough, the poor conductivity of dough combined with the low freezing rate achieved in most units allows gas production to continue at the centre of the dough piece for some time after the surface has become frozen. Within the dough piece sufficient expansion can occur to cause the frozen surface to crack. These cracks in the surface do not close during the post-retarding phases and remain on the surface of the baked product. Dough pieces of large radius are more likely to display this problem than smaller ones.

## 6.2.4.  Storage Time

The longer dough pieces are stored under retarded conditions, the poorer the final product quality will be. The nature and magnitude of the decline in quality depends on factors such as dough piece size, yeast level and storage temperature. As a general rule the storage period for retarded doughs seldom exceeds 3 days mainly because of quality losses associated with dehydration of the dough pieces. The volume of dough pieces generally increases as storage time increases. The rate of increase of dough piece volume with time is greater for higher storage temperatures and yeast levels (Figures 6.4 and 6.5). Despite the increase in dough volume which occurs during storage, there is usually a progressive loss of volume in the baked product, the appearance of surface blemishes becomes more common and the cell structure becomes more open. These defects in product quality can be minimized, and in many cases prevented, by lowering the yeast level, storage temperature, or both, within the limits already discussed.

Weight losses from dough pieces increase with increasing storage time. This progressive weight loss from the dough pieces is mostly related to the loss of moisture. Although there is a contribution to weight loss from any escaping carbon dioxide gas resulting from the slow but continuing yeast fermentation of sugars, either natural or added in the dough, weight losses from this source are small despite the long fermentation times which are involved. Moisture lost from the dough pieces is held in the air in the retarder as the RH reaches equilibrium with the dough. After this point is reached weight losses should cease, but during the operation of the retarder some of the moisture in the air will form as ice on the coils, however efficient the design of the latter. The formation of ice removes moisture and lowers the RH of the air, which becomes lower than the equilibrium relative humidity (ERH) of the dough, and so further moisture losses occur as the system again tries to reach equilibrium. As well as condensation on the coils, some moisture will be lost from the retarder through the opening and closing of the doors and some smaller losses occur through door seals. Some of the moisture losses are offset in retarders which incorporate a defrosting cycle in their normal operation, since the water generated from the melting ice on the coils returns to the air within the retarder.

The rate of moisture loss with increasing storage time is greater at higher storage temperatures as shown in Figure 6.1, because the saturated vapour pressure (SVP) of the air is lower as its temperature falls. The lower SVP means that a smaller quantity of water is required to achieve saturation of the air. Weight losses with increasing storage time are also greater with higher yeast levels (Figure 6.2). The progressive loss of weight which occurs in all dough pieces must be taken into account when deciding on the initial scaling weight to be used, especially for products which are subject to weight control at the point of sale.

Enzymic activity in the dough also continues during storage, the rate of the action depending on the storage temperature used. One enzyme to be noted in this context is *alpha*-amylase which may occur naturally in the flour as cereal *alpha*-amylase or may be added by the miller or baker as malt flour or as fungal *alpha*-amylase preparations. *Alpha*-amylase breaks down the damaged starch to produce soluble carbohydrates and dextrins, some of which caramelize during baking to give the product a darker crust colour which may be undesirable. Other potential enzymic action in the dough during retarding can include the actions of proteinase and protease, the collective actions of which are to weaken the dough structure through interaction with the flour proteins. This may lead to a loss of gas retention and spreading or flow of the dough during retarding which will be accentuated during proof and the early stages of baking. Such enzymic actions will certainly contribute to the progressive quality losses observed with increasing storage times. In order to limit such effects of added enzymes, retarding temperatures should be kept as low as possible. Alternatively improver formulations or flour specifications may need to be modified to reduce enzymic activity.

## 6.2.5.  Proving and Baking

At the end of the retarding cycle the dough must be proved and expanded ready for transfer to the oven for baking. Even under scratch production the

FIGURE 6.6. Effect of proving temperature on bread quality, left, 40 °C and right, 21 °C (70 °F).

poor conductivity of dough results in a temperature differential between surface and centre by the end of proof. Since retarded doughs are much colder than scratch doughs when they enter the proof phase, it is important to raise the temperature of the dough piece gradually and uniformly during the warm phases of the proof cycle. There are two main ways in which this can be achieved; one is to use cool proving temperatures and the other is to modify yeast levels.

Proof to a constant dough piece volume is a time- and temperature-dependent operation so that any reduction in proof temperature requires a compensatory increase in proof time, and vice versa. When proving retarded doughs, uniform heating is probably more critical than overall proof time. This is especially true of retarder–provers, where the change from retarding to heating is automatically controlled and the absolute timing of the switch from one state to another is less critical. The effect of proving temperature on the quality of pan breads from retarded doughs is illustrated in Figure 6.6. Both dough pieces were proved to constant volume in the pan immediately before entering the oven, thus the pieces proved at 21 °C took longer to reach the required volume. The extra oven spring exhibited by the dough proved at 21 °C (70 °F) results from the more uniform temperature distribution within the piece. Although the dough piece proved at the higher temperature had reached the same volume, it had done so with a larger temperature differential in the dough. In effect, the outer layers of the piece were over-proved and the inner layers under-proved; the former would be over-expanded and exhibit reduced gas retention, and the latter would be too cold to achieve full gas production before the dough became set in the oven.

An alternative way in which to control the uniformity of proof is to prolong the proving time at any given temperature by manipulating the yeast level.

This approach has a dual benefit for the quality of the retarded dough in that it will also decrease the likelihood of quality defects such as white spot formation or skinning. An example of the benefits of lower yeast levels is illustrated with pan breads in Figure 6.7. Once again, all dough pieces were proved to constant volume before they were transferred for baking. Dough pieces with lower yeast level took longer to reach the required volume and the extra oven spring with the lowest yeast level arises directly from the more uniform temperature distribution within the dough achieved because of the longer residence time in the prover. Such results are largely the opposite with scratch doughs. A critical factor in determining the choice of proving conditions will be the initial size of the dough piece. For the same conditions of yeast level and proving temperature, differentials within the piece will be greater for dough pieces of large radius than those of small radius, just as they were in cooling.

In the case of processes where the retarder and the prover are separate pieces of equipment, it is helpful to introduce a 'recovery' stage between the two unit operations to help minimize dough piece temperature differentials. Traditionally recovery is carried out in the bakery to raise dough temperatures to around 16 °C (60 °F), depending on ambient temperature, the dough piece size and other factors. The doughs will need to be covered to prevent moisture losses and skinning.

Adjustment to the baking conditions when using retarded doughs is usually unnecessary, except with higher storage temperatures and longer storage times which may lead to increased enzymic activity and darker crust colours. Undesirably high crust colour can be lessened by lowering slightly the level of ingredients, such as skimmed milk powder, sugars and malt flour, which also contribute to crust colour.

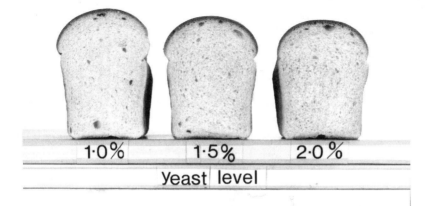

FIGURE 6.7. Effect of yeast level on bread quality.

## 6.2.6.   Guidelines for Retarded Dough Production

There are many potential combinations of yeast level, dough piece size, retarding temperature, storage time and proof time which will give acceptable retarded products, and so it is not possible to give precise 'rules' by which they should be produced. It is possible, however, to give some general guidelines on which to base choices. They include the following:

- reduce yeast levels as storage times increase;
- keep yeast levels constant when using separate retarders and provers;
- reduce yeast levels as the dough radius increases;
- reduce yeast levels with higher storage temperatures;
- the lower the yeast level used, the longer the proof time will be to a given dough piece volume;
- yeast levels should not normally be less than 50% of the level used in scratch production;
- for doughs stored below −5 °C, the yeast level may need to be increased;
- reduce the storage temperature to reduce expansion of and weight loss from dough pieces;
- lower the yeast levels to reduce expansion and weight losses at all storage temperatures;
- dough pieces of large radius are more susceptible to the effects of storage temperature;
- the low freezing rate achieved in most retarder–provers combined with the poor thermal conductivity of dough can cause quality losses;
- prove dough pieces of large radius at a lower temperature than those of small radius;
- lower the yeast level in the dough to lengthen the final proof time and to help minimize temperature differentials;
- maintain a high RH in proof to prevent skinning.

## 6.3.   Retarding Pizza Doughs

Some pizza restaurants store ready-made dough bases in a refrigerator or retarder, withdrawing them for topping and baking as required. Proof may be given to the bases either before entering the refrigerator or oven, or both. The composition of pizza bases is not too dissimilar to that of other fermented products and so the effects of yeast level and retarding temperature follow the same pattern. Since the base dough is usually thin, the cooling rate is very rapid, as is the warming rate but the large surface area of the base makes them susceptible to skinning unless precautions are taken to cover the surfaces, for example by stacking pans inside one another.

Novel methods for producing consistent pizza base quality using retarder–provers have been developed based on an automatically controlled cold–warm–cold cycling (Cauvain, 1986). The technique makes use of the narrow cross section of the bases, and after shaping and panning-up consists of three storage conditions:

1. an initial retarding period of up to 24 h at −4 °C (25 °F);
2. a proof period of 3–4 h at 21 °C (70 °F);
3. a return to retarding at −4 °C where the bases are held until required and then withdrawn for immediate topping and baking.

It is not usual to retard dough pieces which have undergone proof or significant fermentation before cooling because of potential problems with quality losses, especially the formation of white spots. However, in the case of the thin pizza base the cooling is so rapid that significant quality losses cannot occur. Once in the second retarded condition the bases remain relatively unchanged for up to 8 h.

## 6.4. Freezing Fermented Doughs

The problems of cooling dough associated with its poor conductivity have already been referred to above in the section on retarding. The low temperatures required to freeze fermented doughs and suspend yeast activity exacerbate those problems associated with poor conductivity, and restrict the choice of bread-making process which may be used for the manufacture of frozen dough even more than with retarded doughs. However, despite these drawbacks, the range of dough sizes which may be frozen and stored is somewhat greater than that usually encountered with retarded doughs.

### 6.4.1. Breadmaking Process, Recipe and Yeast

The instability of frozen doughs made with significant periods of fermentation before freezing has been shown by a number of workers (Godkin and Cathcart, 1949; Merritt, 1960; Kline and Sugihara, 1968). Straight or no-time doughmaking processes, such as the Chorleywood Bread Process (CBP), provide the most suitable processes for frozen dough production, as does sponge and dough, provided that the sponge is combined with a no-time doughmaking stage. The recipe for frozen doughs should be based on one which yields acceptable product quality in scratch production, and it is essential that the recipe formulation and the choice of fat and composite improver are compatible with the breadmaking process chosen.

The loss of gassing activity in frozen doughs is well documented (Cauvain, 1979; Dubois and Blockcolsky, 1986) and is most obviously manifested as an increase in proof time with increasing storage of the frozen dough (Figure 6.8). The illustrated results not only show the significant difference in proof time between the two yeast levels used but also suggest that the rate at which proof time increased with increasing storage time was lower with 2% than with 4%

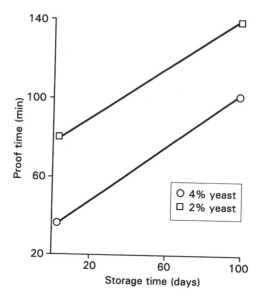

FIGURE 6.8. Effect of yeast level and storage time on proof time.

yeast. The loss in gassing activity is mainly due to the loss of viable yeast cells during the initial freezing and during subsequent storage and, as shown in Figure 6.8, can be compensated for by raising the yeast level in the recipe. In addition to the loss of gassing, there is a progressive loss of product volume with increasing storage as illustrated in Figure 6.9. Unlike the effect on proof times, the rates at which product volumes decreased with storage time were similar for both 2 and 4% yeast.

Volume losses with frozen dough typically occur even when dough pieces are proved to the same volume before entering the oven. Loss of oven spring

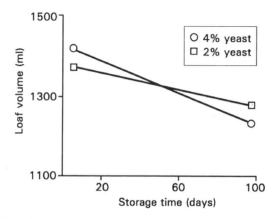

FIGURE 6.9. Effect of yeast level and storage time on loaf volume.

accounts for the volume difference and is probably linked with the release of enzymes and reducing agents from the contents of dead and disrupted cells (Hsu et al., 1979). Weakening of frozen dough can occur independently of any detectable loss of yeast activity (Bruinsma and Giesenschlag, 1984), although it is not clear just how significant this effect is in comparison with the effect of yeast cell contents. The same processes also most certainly account for the appearances of large blisters on some dough pieces during proof, a condition which is commonly associated with the use of weak flours and the lack of gas retention in scratch doughmaking.

Casey and Foy (1995) reviewed the effects of freezing and thawing on yeast cells and the mechanism by which cell damage occurred. In particular they considered the work of Mazur (1970), which showed that the yeast cell membranes lose their ability to block the passage of ice crystals below about $-10$ to $-15\,°C$ (14–5 °F). Usually frozen dough products reach temperatures around $-20\,°C$ ($-4\,°F$), if not during initial freezing then certainly during frozen storage. Ice crystals probably continue to grow with increasing storage time and under these conditions we can expect significant damage to cell membranes. The practical evidence reinforces the critical nature of the storage temperature on yeast activity with increases in proof time of thawed dough pieces being less if they have been stored at $-10$ than at $-20\,°C$ (Figure 6.10).

The compensatory increase in yeast level which is required depends in part on the length of the anticipated frozen storage time. For short periods of storage, say 3–4 days, an increase in yeast level may be unnecessary, while for periods of up to 30 days increases of up to 50% may be required. For storage periods in excess of 30 days, an increase in yeast level of 100% may be necessary.

The susceptibility of yeast cells to damage, both during the initial freezing and subsequent frozen storage, has led to the development of yeast strains with improved freeze–thaw resistance. Current developments and strategies for future developments have been thoroughly reviewed by Casey and Foy (1995). In some cases the new yeast strains have initially lower gassing rates compared

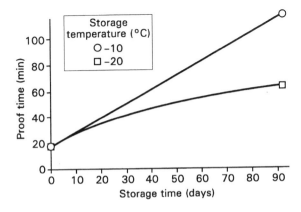

FIGURE 6.10. Effect of storage time and temperature on proof time.

to standard yeast which make them more suitable for the production of frozen dough (Hino et al., 1987; Baguena et al., 1991). Other approaches have included the use of trehalose-enriched yeasts, with the trehalose acting as a cryoprotective agent and the application of dried yeasts, although successes with the latter have been somewhat variable (Spooner, 1990; Neyreneuf and Van der Plaat, 1991).

## 6.4.2.  Processing and Freezing Doughs

As discussed above with retarded dough no changes to standard mixing techniques are necessary, and changes from the scratch dough norm for dough temperatures should be avoided wherever practicable. Once again, the temptation is to use doughs which are colder than normal in order to reduce gassing activity in the dough before it enters the freezer, but this will have an adverse effect on dough oxidation and enzymic action and the benefits of controlling gassing may well be lost in reduced bread volume and poorer cell structure.

Doughs should be processed as rapidly as possible in order to avoid excessive gas production before freezing, especially if yeast levels which are higher than normal have been used. It may be necessary to mix smaller quantities of dough than normal in order to achieve more rapid processing after mixing. The omission of intermediate proof between first and second moulding reduces gassing activity before freezing and slightly reduces the vulnerability of yeast cells. The resulting bread has a slightly finer cell structure and is less prone to the formation of small white spots on the surface of the product. These effects are related to improvements in gas retention in the dough, with shorter final proof times to constant height (Figure 6.11) and better loaf volume with increasing storage time (Figure 6.12).

A simple but often overlooked effect of the freezing process is that it fixes the dimensions of a dough piece, and it may be necessary for some adjustment

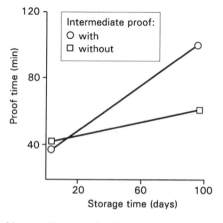

FIGURE 6.11. Effect of intermediate proof and storage time on proof.

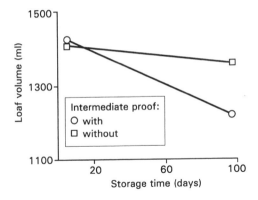

FIGURE 6.12. Effect of intermediate proof and storage time on loaf volume.

to be made to the moulder in order to ensure that the frozen pieces will fit in pans, or on trays, at the point of end use before defrosting. Dough pieces should be transferred to the freezing unit as quickly as possible after final moulding to reduce gassing activity and yeast vulnerability.

The freezing operation is best carried out as rapidly as possible to inhibit yeast activity. Of the types of freezer available, the blast freezer offers the best compromise between speed and operating temperatures. It is recommended that no part of the dough is subjected to temperatures lower than $-35\,°C$ ($-30\,°F$) to avoid excessive damage to the yeast cells present in the dough. Domestic-type freezers are generally not suitable for producing large quantities of frozen dough because the very low freezing rate (defined as the change of temperature within a dough piece in unit time, typically less than $-0.21\,°C/min$ in a domestic chest freezer) allows too much gas production to occur in the early stages of cooling with a subsequent loss of product quality. The low freezing rate that is achieved in most retarder–provers also largely makes them unsuitable for producing deep-frozen dough pieces, except for those products with a small radius, such as rolls. In a typical blast freezer, variations in air velocity have negligible effects on dough performance and bread quality (Cauvain, 1979).

The changes which occur in fermented doughs during freezing are mostly related to the size of the dough piece and the temperature at which freezing takes place. Dough pieces with large radii take longer to reach a given temperature at the centre than those with small radii. Figure 6.13 illustrates changes in the temperature at the centre of a dough piece for a typical range of bakery dough pieces during freezing. Plotting the freezing rates for these dough pieces against dough radius (Figure 6.14) confirms the close relationship between the two parameters. Further confirmation of the critical role of dough piece radius is given in Table 6.1, where the freezing rates of dough pieces each weighing 910 g (2 lb) but having different configurations are compared.

In addition to affecting freezing rate, the radius of the dough piece will also affect the temperature differential which will be set up between the surface and the centre. As discussed above with retarded doughs, pieces with large

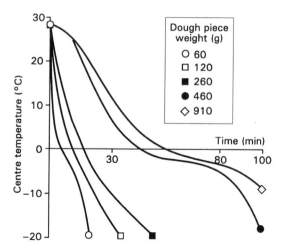

FIGURE 6.13. Changes of centre temperature with freezing time.

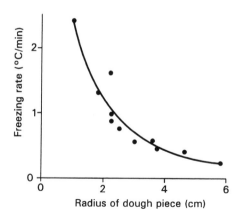

FIGURE 6.14. Effect of dough piece radius on freezing rate.

radii will cool with a larger temperature differential than those with small radii. An example of the changes which can occur during the freezing of a 460 g (1 lb) cylindrical dough piece is illustrated in Figure 6.15. In this example the centre temperature has taken nearly twice as long as the surface temperature to reach the frozen state, around −5 °C (23 °F), as shown by the inflexion on the centre temperature curve in Figure 6.3 after 80 min. In dough pieces with large radii the temperature at the centre remains relatively high for long periods of time, which allows considerable gas production to occur. The dough piece can, depending on yeast level, expand by up to 50% of its original volume. Typical times taken for a standard dough piece to reach three given centre temperatures in a blast freezer are given in Table 6.2 together with the calculated freezing rates.

TABLE 6.1. Effect of radius on dough freezing rate.

| Nominal shape | Length (cm) | Radius (cm) | Volume (cm³) | Freezing rate (°C/min) |
|---|---|---|---|---|
| Cylinder | 22.0 | 3.75 | 884 | 0.46 |
| Cylinder | 55.0 | 2.50 | 1080 | 0.75 |
| Sphere | – | 5.75 | 797 | 0.24 |

TABLE 6.2. Freezing times and rates.

| Centre temperature achieved (°C) | Time taken to cool from 30°C (min) | Average freezing rate (°C/min) |
|---|---|---|
| 0 | 47 | 0.62 |
| −10 | 86 | 0.44 |
| −20 | 110 | 0.44 |

Frozen products need to be held in storage below their glass transition temperatures in order to prevent change and deterioration during the storage period. Glass transition temperatures are formulation sensitive and so, in theory at least, every product will have its own unique minimum storage temperature. In commercial practice a storage temperature of about −20°C (−4°F) is usually applied, which is well below the glass transition temperature of fermented dough formulations. There is always a temptation for dough pieces to be transferred to frozen storage once the surface is frozen but, as the curves given in Figure 6.15 show, the central dough temperatures will be higher at any given time during the initial freezing.

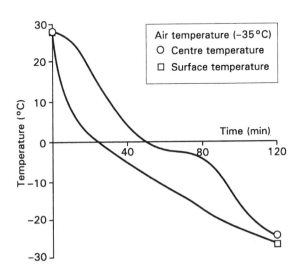

FIGURE 6.15. Effect of freezing time on dough temperature.

TABLE 6.3. Equalization times

| Centre temperature (°C) | Time taken (h) for centre to reach storage temperature | |
|---|---|---|
| | −10 °C | −20 °C |
| 0 | 5 | 11 |
| −10 | 0 | 1.5 |
| −20 | 1.75 | 0 |

In the storage period which immediately follows blast freezing, the dough pieces typically undergo a further cooling period. The length of that period depends in part on the particular storage temperature being used as shown from examples given in Table 6.3.

Changes in frozen dough performance occur with increasing storage time, the most obvious of which are lengthening of proof time (Figure 6.8) and a progressive loss of product volume (Figure 6.9). Even when stored at −20 °C (−4 °F), frozen dough pieces are able to lose water and thus for storage should be wrapped in polyethylene bags or boxes over-wrapped with polyethylene before transfer to the storage freezer.

## 6.4.3.  Defrosting and Proving

Because dough is a poor conductor of heat, the patterns of temperature change that occur in dough during defrosting are similar, in reverse, to those which occur during freezing, and a series of examples are compared in Figure 6.16. As with the freezing process, there will be a temperature differential between the surface and the centre of the dough piece during defrosting. In Figure 6.17 the temperature profile of a defrosted dough piece is compared with that typically observed in a scratch dough during proof. The curves show clearly the large temperature differential which exists between the surface and the centre of the defrosted dough. While the surface temperature after 50 min proof at 43 °C (110 °F) was only slightly lower than that of the scratch dough, the centre temperature was 25 °C (45 °F) lower. At the end of proof both dough pieces had the same volumes in the pan, but the bread baked from the frozen dough was the smaller of the two with a ragged crust break and a much denser crumb structure at the centre of the loaf.

Temperature differentials such as the one illustrated in Figure 6.17 can be minimized by gradual defrosting at more modest temperatures and under controlled conditions similar to those discussed in the section on retarding. Indeed retarder–provers offer a suitable means by which to raise gradually the temperature of the dough pieces and to minimize the temperature differential within them. However, standard retarders are designed only to cool dough and will be inadequate for defrosting when completely filled with frozen dough unless they incorporate provision for heating.

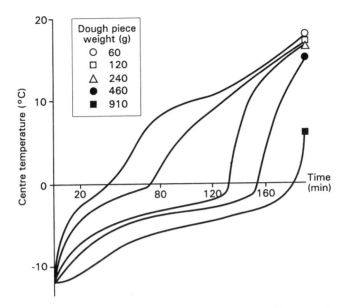

FIGURE 6.16. Changes of dough piece centre temperature with defrosting time at 21 °C (70 °F).

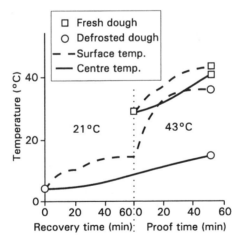

FIGURE 6.17. Changes in dough temperature during recovery and proof.

## 6.5.  Freezing Proved Doughs

It is possible to freeze doughs which have undergone a period of proof but these products will behave quite differently from unproved frozen dough during freezing, storage and baking. In proved dough pieces the yeast will have been fully activated and is more susceptible to damage from temperature shock. Large numbers of yeast cells will die during the initial freezing and more will die during the frozen storage period. It is likely that in a short time, probably only a few days, that there will be no viable yeast cells left in the dough. In such circumstances the dough will have a reduced ability to expand during the early stages of subsequent baking because no 'extra' carbon dioxide gas will be generated from yeast fermentation, such as would be seen in scratch or other frozen doughs. Any expansion of the dough during baking will come from the thermal expansion of gases trapped in the dough, supplemented by water vapour.

In all proved dough pieces the gas mixture comprises a mixture of nitrogen bubbles which provide the nuclei for carbon dioxide generated by the yeast. Any oxygen originally incorporated into the dough will have been rapidly used by the yeast during or immediately after mixing. The potential expansion of the dough during proof and the early stages of baking thus depends on these gases being retained within the dough matrix. At the warm temperatures experienced during proof, carbon dioxide is constantly being generated by yeast fermentation, going into solution and diffusing into the trapped nitrogen gas bubbles. When the dough is cooled, the diffusion process is reversed because of the high solubility of carbon dioxide at low temperatures compared with that of nitrogen (Sluimer, 1981). Without the inflating effect of the carbon dioxide gas the nitrogen bubbles shrink in size; some may become so small that they cease to exist and the nitrogen gas from them diffuses out of the dough system. In frozen proved doughs this diffusion process manifests itself as a collapse of the dough structure. During storage this collapse or shrinking continues, and with no yeast activity available at bake-off there is insufficient compensatory increase in gas expansion. In addition, the nitrogen bubbles which remain in the dough are those which had a larger bubble size, and the resulting bread is very open in cell structure.

Amongst the more successful of frozen proved products are those which contain yeast and which are laminated, for example croissant and Danish. In such products the final quality depends more on the ability of water vapour to force apart the dough layers during baking than on the generation of carbon dioxide from yeast. The vapour-resistant properties of the dough layers depend to a large extent on the integrity of the dough layers and the physical properties of the fat (Cauvain, 1995; Cauvain, 2001). When frozen proved laminated products are transferred to the oven, they still have the potential to expand and yield products of relatively 'normal' appearance.

## 6.6.  Factors Affecting the Formation of White Spots on Retarded and Frozen Doughs

Perhaps the most common of the quality losses associated with retarded and frozen doughs are the very small, translucent blisters or 'white spots' which sometimes occur on the top and bottom crusts of the baked product. They are not usually seen on the surface of the dough piece at any of the stages before baking, but very quickly become visible as the crust begins to colour. Observations of cross sections of suitable dough surfaces under a light microscope confirm their presence before baking.

The mechanism by which these white spots are formed still remains to be fully elucidated, but it is clear from research that any change which affects the production of carbon dioxide gas during the retarding or defrosting phases will influence their formation. The occurrence of white spots is associated with significant gas production before or during retarding. They are more likely to occur with high yeast levels, high storage temperatures, long storage times and even very slow defrosting in the case of frozen doughs. This evidence suggests that the significant factor in their formation is the high solubility of carbon dioxide gas at low temperatures as discussed above.

It is interesting to note that dough pieces which skin during the retarding phase do not normally exhibit white spots, which implies that high moisture content at the surface of the dough favours their formation. Some additional evidence for this view is provided when the bases of some retarded products, such as rolls, are examined. They show a similar formation of white spots, although their drier upper surface may exhibit skinning. The base surface of the dough is protected from the drying effects of the retarder atmosphere or air movement because of its intimate contact with the metal sheet or paper surface on which it stands. Pan bread can exhibit white spot formation on the upper crust and also a similar phenomenon on the side crusts. In the latter case the white spots appear to combine and form into patches having a 'waxy' appearance, which are sometimes referred to as 'condensation spotting' (Sluimer, 1978). Once again, the white spots seem to form most readily where the dough surface is protected from moisture loss, in this case by the sides of the pan.

This apparent relationship between white spot formation and higher moisture content may also help us to understand why the spots are translucent and contrast so readily with the darker crust areas. Under the microscope there is some evidence of water droplets suspended within the voids, which become white spots. This 'extra' water may act as a diluent for the colour-forming components within the dough, quite literally reducing the local concentration of sugars. The white cast that is observed when dough pieces skin occurs for quite different reasons. Here it is likely that the loss of moisture significantly reduces the Maillard browning reactions which normally occur.

It has been observed that the appearance of white spots is also related to the gas-retaining ability of the dough. White spots may occur when the dough has poor gas retention, and white spots are more likely to occur even when the retarding temperature and yeast level have been optimized. Because such a wide range of factors affect their formation it is difficult to predict their appearance with certainty.

Reductions in recipe yeast levels together with lower retarding temperatures are often the most effective means of eliminating or reducing the incidence of white spots. Other changes which may be beneficial include reducing the periods of bulk fermentation after mixing and minimizing delays after final moulding before transfer to the retarder or freezer. Traditional remedies for eliminating white spots have been to increase the degree of dough 'enrichment', usually sugar, fat or both. Sugar slows down (retards) yeast activity while increasing fat levels contribute to improved gas retention in the dough (see Chapter 3).

Not all process changes affecting the formation of white spots occur during the retarding phase. There is evidence that white spots are more likely to occur if the rate of warming of dough pieces after retarding is particularly low as a result of using a very cool proof temperature (e.g. 12 °C, 54 °F) or very low yeast levels (e.g. 1% of flour weight) or both. Supporting evidence was seen in work on frozen dough where a study of the defrosting technique needed for frozen Vienna bread doughs showed that relatively rapid defrosting at 21 °C (70 °F) did not give white spots while slow defrosting via a retarder at 3 °C (38 °F) did (Cauvain, 1983).

## 6.7.    Causes of Quality Losses with Retarded and Frozen Doughs

Variations in the quality of products made from retarded and frozen doughs may occur from time to time. There are many causes of such quality losses. Some arise from factors which have their origins in the dough formulation and processing conditions which are used before the doughs enter the retarder or deep freeze, while others arise directly from conditions within the refrigeration equipment or as the result of thawing and proving conditions. Identifying the causes of quality losses requires careful consideration of the whole process. A few possible causes of quality losses are given below, but the list is by no means exhaustive.

### 6.7.1.    Skinning

Low humidity or excessive air movement, even with sufficient humidity, may cause the surface of the dough piece to dry out or 'skin' during retarding or recovery (Cauvain, 1992). The upper crust will be hard and dry eating. If skinning occurs during retarding there will be a distinct tendency for the dough pieces to lean towards the air inlet. Skinning becomes more pronounced as the storage time increases. It is an irreversible phenomenon and will carry through to the baked products, which may be small in volume and have a 'pinched' appearance as the result of uneven expansion of the dough pieces during baking. Lowering the retarding temperature or the yeast level in the recipe, or both, reduce skinning.

### 6.7.2.    Crust Fissures

Products may sometimes exhibit small fissures or cracks, mainly on the upper crust. The depth of cracking may become significant on occasions. The cracks may be observed on the surface of the dough, and once formed will persist through to the baked product. Slow freezing, such as that encountered in retarder–provers, or excessively high yeast levels are the main causes. Such cracks are more prevalent with dough pieces of large radius. They can be minimized by lowering the yeast level used in the recipe.

## 6.7.3.  Ragged Crust Breaks

These occur most often when there is a large temperature differential between the centre of the dough piece and its surface. It can be a particular problem with frozen doughs of larger size where the poor conductivity of dough exaggerates the temperature differentials. When the dough piece enters the oven the full gassing potential of the dough centre is not realized until after the crust has set and the crust breaks at its weakest points. Generally this will be in the normal crust break area, although it may occur in other places. The only remedy is to give a longer proof using lower temperatures or lower yeast levels, or both.

## 6.7.4.  Small Volume

There is seldom a single cause of small-volume products and usually the problem is linked with other quality losses, such as skinning or ragged crust breaks. Improvements in volume may be achieved by reducing the length of time in storage (both retarded and frozen), optimizing yeast levels, and optimizing retarding and proof temperatures. In the case of frozen doughs, volume losses occur through the loss of activity and the release of proteolytic enzymes and glutathione from disrupted yeast cells.

## 6.7.5.  White Spots or Small Blisters

The origins of the white spots or small blisters on products baked from retarded or frozen doughs have been discussed above. In summary they have their origins in almost any change which affects gas production or retention. They can be eliminated or reduced by lowering yeast levels in the retarded dough recipe, adding fat where it will not adversely affect other product qualities, reducing bulk fermentation (floor-time) after mixing, and reducing delays before retarding and freezing. Lowering the retarding temperature or the yeast level in the recipe, or both, will eliminate white spots on the top crust and, to a lesser extent, the bottom. Solving a white spot problem in frozen doughs by adjustment of the yeast level is more difficult because of the need to maintain gassing activity in the dough after defrosting. Minimizing gassing of doughs before freezing is a better option, although using lower dough temperatures to achieve this may result in other problems.

## 6.7.6. Waxy Patches

Patches of uneven colour may occur on the side and bottom crusts of breads baked in pans. They have a shiny or 'waxy' appearance and are frequently observed on the lower corners of the loaves. This phenomenon is sometimes referred to as 'condensation spotting' and its incidence increases as storage time increases, even when the retarding temperature and yeast level have been optimized. The only remedy to this problem is to avoid storing dough pieces in pans at refrigerated temperatures for more than 36 h.

## 6.7.7. Black Spots

An unusual problem sometimes seen on the bottom of baked products is the occurrence of black spots, often mistaken for mould formation. This phenomenon arises from a reaction between the moist dough and the steel baking sheets and can even occur through silicone paper. It is easily overcome by switching to anodized aluminium trays or ones covered in non-stick material. The use of trays with damaged coatings should be avoided.

## 6.7.8. Large Blisters

Large blisters sometimes occur on products. They are of two sorts: either they protrude from the surface causing a distortion of shape or they appear as a large cavity under the top crust which only becomes apparent when the product is cut. The most common causes are poor moulding and damage to the dough piece during processing which may cause large gas bubbles trapped within the dough piece to become greatly expanded by the carbon dioxide gas released during the proof phase. Poor gas retention will also contribute to this problem, and it will be exacerbated by a short final proof, either as a consequence of using too high a yeast level or too high a proof temperature. Delays in transferring dough from the blast freezer to deep-freezer storage can lead to partial defrosting. Upon subsequent baking large blisters may be present under the top crust.

## 6.7.9. Dark Crust Colour

Enzymic action can continue in retarded doughs, and to a lesser extent with frozen doughs. *Alpha*-amylase activity is the main activity of concern under this heading as it can produce an increase in available maltose and dextrins, which can contribute to excessive darkening of product crust on baking. Excessively dark crust colours can be avoided by lowering the retarding temperature to reduce enzymic activity, or by reducing the level of added sugars, skimmed milk powder, malt flour or added fungal *alpha*-amylase.

## 6.7.10.    Uneven or Open Cell Structure

Sometimes doughs which normally produce a fine cell structure yield a more open one when retarded or frozen. More often than not, the origins lie in excessive gas production before the dough is chilled or frozen, or in the early stages of the retarding and freezing processes. The poor conductivity of dough, exacerbated with larger dough pieces, can make a significant contribution to the problem. There are a number of possible remedies which include lowering the retarding temperature and reducing delays before retarding or freezing. Controlling cell structure by manipulating the yeast level in retarded doughs presents some problems. To maintain a fine cell structure in retarded products it is necessary to lower the yeast level in order to minimize gas production during the retarding phase. However, lowering the yeast level requires a compensatory increase in proof time in order to maintain product volume, and excessively long proof times can also lead to more open cell structures. With frozen doughs the problem is potentially greater as yeast levels which are higher than normal may be used in order to avoid long proof times. In this case the best option is to keep the processing time between mixer and freezer as short as possible.

## 6.7.11.    Areas of Dense Crumb

This phenomenon may often be associated with ragged crust breaks. Uneven expansion of the dough piece during proof and the early stages of baking is the main cause, such as might be experienced by transferring cold doughs to a hot prover. There is a tendency for this phenomenon to manifest itself more readily as the storage time of frozen doughs increases.

## 6.8.    Principles of Refrigeration

In order to cool a product we must have some means of extracting heat from it. For a solid to become a liquid or for a liquid to change to a vapour, it must absorb a large quantity of heat. The heat involved in the transition processes is referred to as 'latent heat'. If we can find some means of using doughs to provide the necessary heat, we have the basis of a rudimentary refrigeration system. Large quantities of heat are also involved when a vapour changes back into a liquid and then into a solid, but in this case the heat is given out by the material instead of being absorbed.

The transition from liquid to vapour and vice versa forms an essential part of the refrigeration cycle which is used in retarding and deep freezing. In its simplest terms, a refrigeration system has an evaporator coil sited inside a cabinet and a compressor and condenser coil sited outside. All three components are connected and a pump circulates a liquid refrigerant between them. The refrigeration cycle is founded on four basic steps:

1. a reduction in atmospheric pressure on a liquid reduces its boiling point and increases its vaporization rate;
2. the change from liquid to vapour requires latent heat of vaporization;
3. increasing the pressure of a vapour increases its rate of condensation and the temperature at which it changes to a liquid;
4. latent heat of vaporization is given out in the change from vapour to liquid.

In our simple refrigeration system, liquid refrigerant is metered into an evaporator where it is vaporized under reduced pressure (step 1). The liquid cools as latent heat of vaporization is absorbed (step 2) and in turn cools the coils through which it is being pumped. The cooling coils have fins attached which considerably increase the potential cooling area of the coil, and since the coil is located within a chamber (e.g. retarder), the surrounding air is cooled and in turn any dough pieces contained within the cabinet. The efficiency of the air in cooling the product is improved by circulating the air by means of a fan. The vapour thus created in the evaporator coil passes to a compressor where the pressure is increased (step 3) and the latent heat is given up (step 4) to the atmosphere. Fins are also used in this part of the cycle to increase the efficiency of the condenser coil and rapidly pass the heat to the atmosphere.

Modern refrigeration equipment still uses the basic principles described above, but there have been many improvements in design and operating efficiency. There has been considerable change in the nature of the refrigerants which are used together with improvements in knowledge about their operating efficiency and impact on the environment. In designing a suitable retarder or deep freeze, the refrigeration engineer uses information on equipment loading, the specific heat capacity of the products and the materials used in the cabinet or room construction, and other less precise information on ambient operating conditions and the rate at which products will typically give up their heat. However, refrigeration is not a precise science, and while there have been many improvements in equipment design, the behaviour of fermented dough products during cooling still remains difficult to model accurately.

## 6.9.  Retarder–Provers and Retarders

The design of retarder–provers varies but most incorporate the basic elements illustrated in Figure 6.18. They can be programmed to follow a preset sequence of operations at given times. Initially the equipment is set in the retarding (cold) phase and remains so until, at a present time, it changes to the heating sequence. The unit continues in the proving (warm) phase until it is switched back automatically to retarding. This cycle of events can be programmed in advance for 7 days or longer.

In the retarding phase, warm air drawn from the cabinet passes over the refrigerating coils and is cooled. In smaller models a perforated metal screen in the cabinet ensures that the cool air is distributed evenly over the warm dough

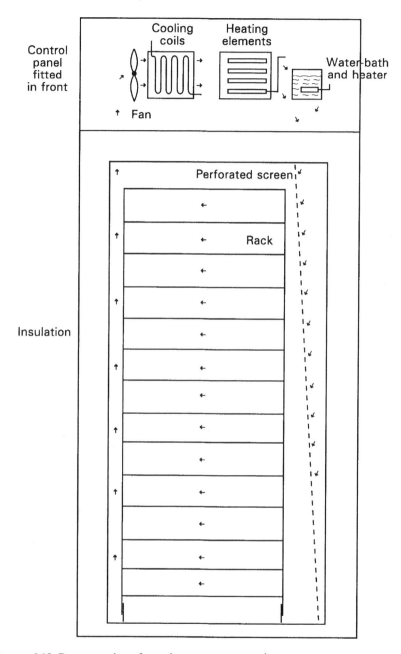

FIGURE 6.18. Representation of retarder–prover construction.

pieces. The air is re-circulated over the refrigerating coils and the cooling process continues until the temperature of the air and the dough pieces reaches that set on the thermostat.

The surface area of the refrigerating coils is made as large as possible to ensure rapid cooling and to help in maintaining a high RH in the cabinet during the retarding phase. Initially the air in the cabinet is relatively dry, but the dough pieces gradually lose moisture and the RH of the air increases. As the temperature of the surface of the refrigerating coils is close to, or below, 0 °C (32 °F), some of the water present in the air forms as ice on the coils and the RH of the air falls. In order to maintain RH at equilibrium, the products then lose more moisture.

As the operating temperature of a retarder–prover is reduced, the saturated vapour pressure of the air becomes lower and, therefore, a smaller mass of water is required to achieve a given RH. This lower moisture requirement for the air results in a smaller weight loss from the dough pieces. However, operating at temperatures below 0 °C increases ice formation on the coils and at other locations in the retarder–prover, with the result that the efficiency of the evaporating coils is impaired. Provided that the efficiency of the evaporator coil in keeping the cabinet at the desired temperature is not too badly affected, ice formation is of no great consequence. Indeed, when the retarder–prover changes to recovery and proof conditions, the melting ice can provide some of the water vapour necessary for proving conditions. In such cases the equipment must provide sufficient energy to overcome the latent heat of fusion of ice, as well as for warming the products.

Many retarder–provers include an independent defrosting cycle to keep the coils free from ice, and the water vapour produced helps to reduce the moisture lost from the dough pieces. However, it is important that the temperature rise within the cabinet during this defrosting period is limited; otherwise this will impair product quality.

At the end of the predetermined storage time the operation of the cooling coils ceases and the heating cycle commences. In general, retarder–provers are equipped with several different warming steps, the aim of which is to raise gradually the temperature of the unit and the dough pieces and to hold the latter until they are ready for baking. Humidity will be controlled during the proving phase to prevent skinning of the dough pieces.

In construction, retarders follow much the same lines as those described above with retarder–provers, except that they do not have proving facilities fitted. Retarders more commonly operate within the temperature range 2–4 °C (36–39 °F) and should be able to maintain a RH of about 85% but some are capable of operating at temperatures as low as −5 °C (23 °F).

## References

Baguena, R., Soriano, M.D., Martinez-Anaya, M.A. and Benedito de Barbre, C. (1991) Viability and performance of pure yeast strains in frozen wheat dough. *Journal of Food Science*, **56**(6), 1690–4, 1698.

Bruinsma, B.L. and Giesenschlag, J. (1984) Frozen dough performance. Compressed yeast – instant dry yeast. *Bakers' Digest*, **58**(6), 6–7, 11.

Casey, G.P. and Foy, J.J. (1995) Yeast performance in frozen doughs and strategies for improvement, in *Frozen Doughs and Batters* (eds K. Kulp, K. Lorenz and J. Brummer), AACC, St Paul, Minnesota, USA, 19–52.

Cauvain, S.P. (1979) Frozen bread dough. *FMBRA Report No. 84*, CCFRA, Chipping Campden, UK.

Cauvain, S.P. (1983) Making more from frozen dough, in *Proceedings of the 56th Conference of the British Society of Baking*, March, 5–10.

Cauvain, S.P. (1986) Producing fresh pizza bases. *FMBRA Bulletin* No. 3, June, CCFRA, Chipping Campden, UK, 126–32.

Cauvain, S.P. (1992) Retarding. *Bakers' Review*, March, 22–3.

Cauvain, S.P. (1995) Putting pastry under the microscope. *Baking Industry Europe*, 68–9.

Cauvain, S.P. (2001) Laminated products. *CCFRA Review No. 25*, CCFRA, Chipping Campden, UK.

Dubois, D.K. and Blockcolsky, D. (1986) Frozen bread dough. Effect of dough mixing thawing methods. *AIB Technical Bulletin Vol. VIII*, No. 6. American Institute of Baking, Manhattan, Kansas, USA.

Godkin, W.J. and Cathcart, W.H. (1949) Fermentation activity and survival of yeast in frozen fermented and unfermented doughs. *Food Technology*, **3**, April, 139.

Hino, A., Taken, H. and Tanaka, Y. (1987) New freeze-tolerant yeast for frozen dough preparations. *Cereal Chemistry*, **64**(4), 269–75.

Hsu, K., Hoseney, R.C. and Seib, P.A. (1979) Frozen dough II. Effects of freezing and storing conditions on the stability of frozen doughs. *Cereal Chemistry*, **68**, 423–8.

Kline, L. and Sugihara, T.F. (1968) Factors affecting the stability of frozen bread doughs. *Bakers' Digest*, **42**, October, 44–6, 48–50.

Mazur, P. (1970) Cryobiology: The freezing of biological systems. *Science*, **168**, 939–49.

Merritt, P.P. (1960) The effect of preparation on the stability and performance of frozen, unbaked, yeast-leavened doughs. *Bakers' Digest*, **34**, August, 57–8.

Neyreneuf, O. and Van der Plaat, J.B. (1991) Preparation of frozen French bread dough with improved stability. *Cereal Chemistry*, **68**, 60–6.

Sluimer, P. (1978) An explanation of some typical phenomena occurring during retarding. *Bakkerswereld*, **19**, January, 14–16.

Sluimer, P. (1981) Principles of dough retarding. *Bakers' Digest*, **55**, August, 6–8, 10.

Spooner, T.F. (1990) Hot prospects from frozen dough. *Baking Snack Systems*, **12**(11), 18–22.

# 7
# Application of Baking Knowledge in Software Systems

Linda S. Young

## 7.1. Introduction

In the baking industry the use of computer systems has become integral to the efficient operation of the plant and smooth running of the commercial aspects of the bakery. In the plant, computers have been used for the in-line control and reporting (for accountability) mechanisms built into equipment. The use of programmable logic cards (PLC) to ensure that the equipment performs as required is widespread. For example, PLCs are used in mixing machines to ensure that the ingredients are mixed at the correct speed for the requisite time period or, in the case of ovens, that each zone has the desired temperature and humidity profile. There is much software, for example spreadsheets, word processing and database packages, to assist the baker, large or small to perform the business function and to ensure the smooth running of materials requisition, stock control, accounting, production scheduling and communicating with the customer. An example of this type of software is SAP which integrates databases, applications, operating systems and hardware.

There is also another area where computer technology can be used by the baker. This is for assistance with the technology of the product.

Achieving quality products consistently, whether new or existing ones, must be the goal of any baker wanting to remain in or improve the business. Computer programs which advise on recipe formulation, on optimal equipment operation and assist in perfecting a faulty or substandard product can be useful tools to the bakery technologist in all aspects of product development, maintenance and production. They can help the baker to understand and apply the underlying technology of the products and their quality for existing or new products. They can assist bakers to meet their goals more effectively. However, such programs cannot be developed unless the knowledge exists about the technology of the product and is made available. Embedding knowledge into a computer program requires a different approach from that used for the development of spreadsheets and databases. The systems used are commonly called 'knowledge-based' or 'expert' systems.

It is useful to give some definitions:

- An **expert system** is a computer program that seeks to model the expertise of a human expert within a specific domain;
- A **knowledge-based system** embodies heuristic knowledge (rules-of-thumb, best guess, etc.) captured from intelligent sources.

For this chapter, I shall use the term 'knowledge-based systems' or 'KBS' to include both of these definitions as in reality the terms are used interchangeably.

It is easier to consider how KBSs differ from data and information processing, using an example as follows. If a baker has a recipe with ingredients and quantities, then using a data processing program the cost of the recipe can be determined simply by multiplying the unit quantity cost by the quantity of an ingredient and then summing all the costs of all the individual ingredients. Each time a new recipe is input the calculations can be done and the route or set of logic/numeric equations executed by the program is the same. If the baker then follows a recipe and processing method, say for pan bread, and finds that the product is faulty or below specification, then in order to obtain some advice for corrective actions the computer program that the baker uses must contain some rules and facts about the ingredients, the processing and the combination and interactions between them. When another recipe, perhaps for rolls instead of pan bread, is input to the program, the path and sets of rules are different. The route through such a program can take many paths. Each path is determined from information given and decisions made by the user and the knowledge contained and defined within the system. An explanation of the reasons behind the conclusions reached by the program and corrective action for rectifying the problem can be given to the baker.

In order to develop programs which can deal with the technology of the product, that knowledge must be available and be accessible. The integrity and validity of the knowledge must be sound. There are different sources from which the knowledge can be gathered. First, there is knowledge in the form of the written word together with the numerical data to support it. Second, there are the human sources — the experts. For example, the methods which are typically used to assess bread characters have been described in Chapter 1 but assessment of bread quality, whether subjective or objective, has no value unless the user has the ability to change bread character in a particular way. Traditionally this is a role for an 'expert'. Experts are hard to find and when they are found the very fact that they are recognized as experts means that they are busy people. Currently experts who thoroughly understand the technology of breadmaking and can apply their knowledge of breadmaking are in limited supply. Without them KBSs for the fermented products industry cannot be developed.

The knowledge, once captured and structured into a KBS program, can be used by technologists with different levels of skills. Whether used by an expert or novice, the KBS never forgets or becomes tired, and the advice or information returned to the user is consistent. From that information, users can learn and expand their own knowledge of the baking technology concerned.

The areas where KBSs have been most successful in the baking industry have been in advisory and fault diagnostic systems. The first important criterion is that the 'domains' or subject areas must be well defined before such systems are developed. Attempting to produce in one system a tool which covers the whole of a subject area is a mammoth task and will probably be doomed to failure. It is better to take a domain and split that domain, say bread technology, into manageable chunks. These chunks can be linked together to form the larger picture. The second important criterion when developing such systems is that the knowledge should exist and be sound. Where experts are used to provide the knowledge input and where there may be conflict between experts in the knowledge given, then care must be taken in sorting the 'fact' from the 'opinion'. In many instances there may be a case for performing further work to determine the facts to support the existing knowledge.

With advances in software programming languages and in the platforms in which they operate, encoding the knowledge has become much easier. However, the structuring of the knowledge for such purposes is still as important as ever. The systemization of knowledge has to be done before encoding can start and can be used independently of the software if the funds for programming are not available. The user interface or dialogue is designed to make the querying of that knowledge easy and meaningful to the user.

## 7.2.  Examples of Systems for Use in Bread Technology

The thrust in developing KBSs in the bread field has come from the Cereals and Cereal Processing Division of the Campden and Chorleywood Food Research Association (CCFRA). This division continued the cereals-based technological function of the Flour Milling and Baking Research Association (FMBRA) when it merged with Campden Food and Drink Research Association in 1995.

Aspects of the accumulated knowledge in bread technology which spanned 40 years were available for use in the development of the systems described in this chapter. The criteria required, that of sound knowledge and defined domain bounds, were met and all of the systems are in use in the industry helping bakery technologists provide quality products for their customers and so will serve as examples to help readers understand how knowledge can be used in these forms.

### 7.2.1.  Bread advisor

In 1990, a Bread Faults Expert System (Young, 1991) was released. Its domain was white pan breads made using the Chorleywood Bread Process (CBP) (Chapter 2). Its original research objective was to determine whether the knowledge-based systems area of computing science could be applied for the benefit of the baking industry. The potential for such systems was soon realized and the objective was extended to produce a commercially available computer program for the industry. The knowledge of faults was underpinned with baking trials and the manifestations of the faults were recorded photographically. These photographs were included in a manual which accompanied the system. When this system was developed in 1989, the expense of including these images in the computer-based system would have been prohibitive in cost and hardware memory.

Computer science has advanced tremendously since this first system and that along with the falling cost (in real terms) of computer hardware has seen development of such systems in other domains of baking technology, for example Cake Expert System – Fault DoC (Petryzsak et al., 1995), BALANCE (Young et al., 2001), ERH CALC™ (CCFRA, 2000). With these advances in computer science and the advent of the Internet, the potential for using knowledge systems was further realized by the development of a Bread Advisor, following the footsteps of the Bread Faults Expert System. However, the domain was widened to include knowledge about the technology of producing a variety of fermented bread types by the five main processing methods, making the software a tool for use in many countries around the world. The Bread Advisor was developed not only as a stand-alone software system to be used like any other computer program on a desktop or portable pc, but also as a 'back-end' program for use from a pc acting as a server with all that this enabled – multi-user access either from within a baking company or via the Internet. The latter, however, was only implemented as a demonstrator system. The former was made available commercially in 2000 (CCFRA) and is now in use in many companies worldwide.

## 7.2.2.    How the Bread Advisor is Used

First, the product type and processing method by which the product under examination was manufactured is chosen (Figure 7.1). This effectively defines the domain in which the Advisor will operate. During the consultation, these two choices define a product profile which is built and carried forward to other parts of the system. The product profile holds the information needed by each of the components of the system as they are reached. For example, if the problem in question was about un-lidded product produced in a pan (exhibiting collapse and lack of oven spring), mixed using a spiral mixer and a 'No-time dough' process, then these choices are made from the list of products and processes. The processes list includes 'Chorleywood Bread Process', 'No-Time Dough', 'Bulk Fermentation', 'Sponge & Dough' and 'Flour Brew'. Generic product types include pan breads, lidded or unlidded, freestanding (e.g. oven bottom/hearth) breads, sticks (e.g. baguettes), soft rolls, crusty rolls, twisted rolls or hamburger buns. The types of bread included are representative of the major product types produced throughout the world and represent products of different dimension types. Images can be viewed of the product type highlighted (Figure 7.1).

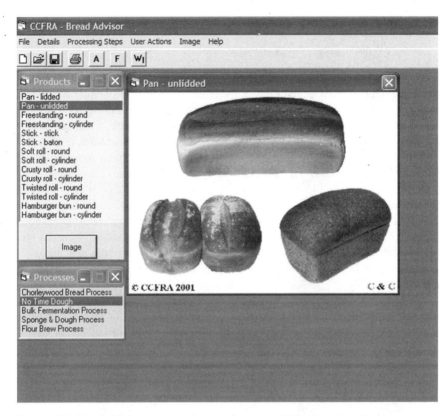

FIGURE 7.1. Bread Advisor screen – image of a pan, unlidded bread product.

## 7.2.3.  Fault Diagnosis or Quality Enhancement

If the purpose of the consultation is for fault diagnosis or for improving the product quality, then the more information about the product that can be given to the system, the more accurate the ultimate diagnosis can be. If conflicting pieces of information are given, the system will tell the user and ask for some information to be re-submitted.

The Bread Advisor 'mimics' the way a human might diagnose a fault or set out to improve the quality of the product. We identify the problem, try to ascertain the causes, gather information to eliminate them and then consider the options for corrective action or improvement for the most likely causes of the fault. On many occasions the manifestation of a particular fault or defect does not necessarily have a unique and identifiable cause and so there may be other intermediate steps to take into account in determining the real cause of the problem. This logic flow can be described schematically as follows:

problem → primary cause → contributing factors

→ corrective action

Or in more simple terms as:

What is seen → why → because of . . . → corrective action

Using this approach, the first step is to choose the fault or the quality attribute requiring improvement. In the Advisor the faults, in order to assist in selecting the fault quickly, are divided into categories. These are diverse and include aroma, crumb (faults which occur inside the product, e.g. holes, texture, structure, colour etc.), dough (faults which occur during the processing of the dough, e.g. sticky or soft dough etc.), eating qualities (for both crumb and crust), flavour, shape (e.g. concavity, low shoulders, lack of oven spring) and surface (e.g. crust colour, spots, blisters and wrinkles etc.). Faults such as low volume, collapsed product are included in a 'catch-all' category called 'General' (Figure 7.2). If the product is exhibiting several defects, then they can each be selected and viewed in a 'Selected' category. The software is intended for use internationally and as the naming or terminology of faults is often unique to the country and product, an image of a fault can be seen by clicking the image button for the fault in question.

Faults or product quality deficiencies rarely have a single cause. However, they can be split into those which are considered 'primary' or principle causes (the 'why') and those which have 'contributed' (the 'because of. . . .') to the faults in question. Consequently they can be obtained easily and quickly by clicking

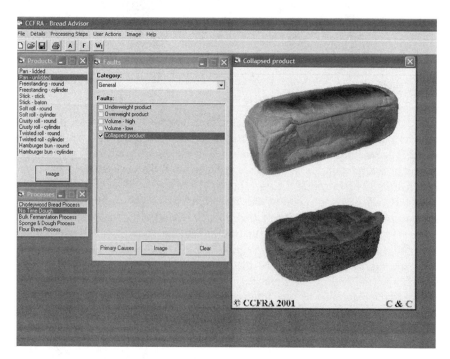

FIGURE 7.2. Bread Advisor screen – faulty product.

the 'Primary Causes' button. This action reveals a further pop-up window where such causes are again ranked in order of likelihood, the most likely being listed at the top (Figure 7.3). The list of primary causes can be considered and checked out with the processing conditions which occurred. Any factors which might have contributed to any of these primary causes can be displayed when the cause itself is checked and the 'Contributing factors' button selected. At the end of a consultation the baker has a 'suspects' list (in the case of the collapsed pan bread with lack of oven spring – lack of gas retention and three other factors that might have contributed to it) which can be considered along with the local conditions that the product underwent during processing in order to dismiss or confirm the cause or causes of the fault. Following this the necessary corrective action can be taken to improve the product quality.

For an experienced baker, the fault diagnostic aspect of the Bread Advisor considers all the necessary information and offers a quick and thorough investigation of the possible causes known to produce the faults in breadmaking. Unlike a human, the software never forgets or overlooks a possible cause. For novice bakers and technologists, the same aspect offers knowledge about the causes of faults from which they can build their own knowledge base about bread faults. Suspects can be eliminated quicker when processing conditions are checked or the 'once in a life time' fault is flagged for investigation.

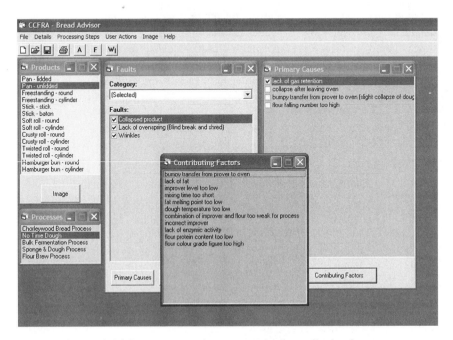

FIGURE 7.3. Bread Advisor screen – primary causes and contributing factors.

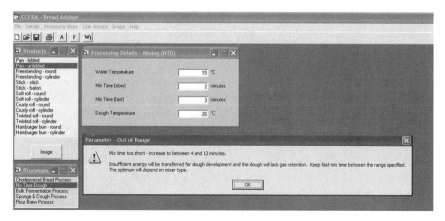

FIGURE 7.4. Bread Advisor screen – mixing check.

## 7.2.4. Processing Details

Investigating the 'suspects' can also be undertaken with the software. In the contributing factors list of the collapsed product example, 'mixing time too short' is flagged as a suspect. By inputting the known processing details about the mixing stage these can be checked for acceptability for the product in question. The generic settings given by the software are only those relevant to the process chosen. Where the input values for the mixing stage of the unlidded pan bread are at variance with the requirements for the process and product, information is displayed giving the range of values in which the parameter (in this case mixing time) should lie to achieve acceptable product quality. In addition a message is displayed giving details of consequences to the product quality (Figure 7.4). Such information is useful in isolating the cause of the fault and to the novice who may be unsure of the settings required for a successful product.

## 7.2.5. Questionioning Using 'What if?'

This KBS system provides an objective diagnosis of bread products, either faulty or below standard, made using any of the five common production methods. It can also be questioned using 'What if I...' experimentation in investigating changes in the process settings for the product, for example What if I increase the proof time?' This approach can be valuable when learning about the breadmaking process or attempting to change the quality of a product, perhaps to improve it or to explore a new product concept. Feedback for the questions posed can be considered. Questions can be posed about any of the processing steps and the parameters contained within the step. Comments about the effect on the product if the directional change is taken are displayed (Figure 7.5). This aspect of the software helps the novice user to learn quickly the consequences of any actions which might be taken on the plant.

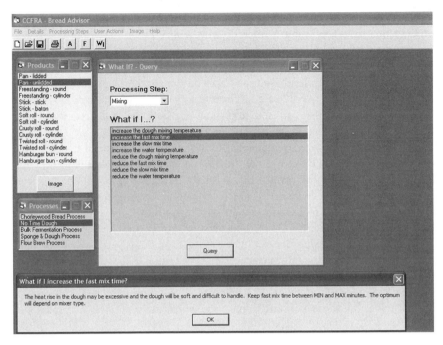

FIGURE 7.5. Bread Advisor screen – What if? query.

The ability to experiment about breadmaking at the PC provides a valuable training tool. The novice technologists/bakers can learn at their own pace, and can try out ideas before committing to full-scale development of new products.

## 7.2.6.  Retarding Advisor

Controlling production has been an area where knowledge-based systems has been less adopted in the food industry. Very few knowledge-based systems have been developed for online process control for baked products. At-line control has been attempted to a greater degree. The Retarding Advisor (Young and Cauvain, 1994) was the first in the baking industry to demonstrate the use of knowledge software to link to and to control the operation of bakery equipment– in this case a retarder-prover– for the production of optimal-quality products. The system was developed 4 years after the Bread Faults Expert System and in that time the leap forward in computing equipment was gaining momentum, in its availability (desktop personal computers), capability and speed, together with a realistic setting of price, making the technology accessible to many more companies and users. The technology required in order to link the PC to a particular piece of equipment was easier, and was tried and tested. The retarding process (see Chapters 6) for fermented goods allows bakers to 'time-shift' production to meet peak sales demands, eliminate night working and to give staff more sociable working patterns.

The Retarding Advisor enabled a baker to achieve acceptable and consistent product quality when using a retarder with a mix of product configurations, for example French sticks, rolls and pan breads. It encompassed the types of information on the interaction between product, recipe and process conditions as described in Chapter 6. It advised the baker on the appropriate settings required for the retarding equipment according to the mix of products to be made in the batch. It determined the start and stop times for both the cold (retarding) and the warm (proof) phases of the process in the specialized retarding cabinet. The complex relations between ingredients, formulations and between temperature, time, yeast level and bulk of dough in the cabinet are all taken into account to give the baker the optimum settings for the products to be retarded. These settings could then be downloaded to the cabinet. An intermediate step allowing a 'what-if' scenario of different equally viable settings to be examined by the baker before they were downloaded was also given (Figure 7.6). Display of the consequences of choosing one set of retarding conditions over another enables the baker to make a more informed decision. In addition to the at-line control, the Retarding Advisor could be used off-line to explore the reasons of poor quality by use of a detailed fault diagnosis with corrective actions indicated. The system has particular relevance not only to bakeries which use retarding equipment but

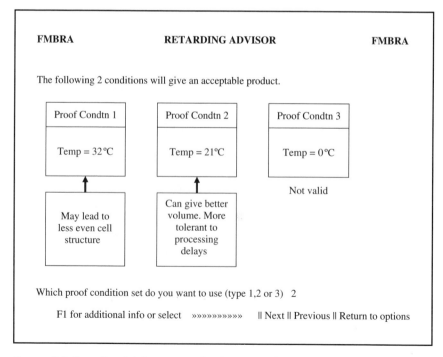

FIGURE 7.6. Retarding Advisor screen showing acceptable sets of conditions.

also to equipment manufacturers and ingredients suppliers, whose understanding of the principles of retarding might be less deep than that of bakers.

Information can be passed not only from the PC to retarding equipment, but also can be received by the PC from that equipment and further processed in the light of rules that apply to the equipment and the retarding technology in question. This processing information can provide indications of the performance of operation of the equipment (which has long been available in isolation), and as a monitor for preventative maintenance of the equipment. This can lead to less downtime for the equipment. Development of the system also led to improvements in the control systems of current retarding equipment. It can also be used for improving working practices within the bakery for the benefit of the product quality sought and the efficient operation of the bakery. Where throughputs for product are known, they can be linked to the scheduling requirements to supply quality products for a changing pattern of customer demands as is often experienced in supermarket in-store bakeries.

## 7.2.7. *Baking Technology Toolkit*

Making baking technology knowledge available in a software format does not have to be an onerous or extensive task. Sometimes smaller programs are equally useful and can help bakers to quickly reach their product goals. Knowledge encoded into a small program can assist in improving the accuracy of the baking process. Technologists and bakers have many small but vital calculations to make in order to ensure that the correct processing conditions apply to the product at its numerous production steps. For example, determining the temperature of the water to be added to the flour at mixing in order to get a dough at the correct temperature ex-mixer is important both for processing conditions and to the end quality of the product. Ensuring the appropriate water temperature, heat rise and energy input are essential to consistent product quality. With knowledge about the ingredients, calculations for these parameters have been included in a 'Baking Technology Toolkit' (CCFRA, 2004) software tool. For the water temperature calculator, the flour temperature, temperature rise during mixing and the desired dough temperature at the end of mixing are all input and, with the specific heats of the ingredients, are used to calculate the initial water temperature. Similarly in the energy calculator the quantity, temperature and specific heat of each of the ingredients is taken into account when the energy calculation is made (Figure 7.7). Often the raw data for such calculations are not held at the bakers' or technologists' finger tips and the calculations can be tedious and prone to error. Using software of this nature can help the technologist and baker to deliver more precise and controllable mixing to the benefit of consistent production. Also in the toolkit is a calculator to determine the modified atmosphere levels in packaging (MAP) invaluable when producers require an additional safety factor for the shelf-life of a baked product (Chapter 10). 'Toolkit' software of this nature can be made available to bakers, just as mechanics have a standard set of tools, for example screwdrivers, wrenches and so on, to help them in their work.

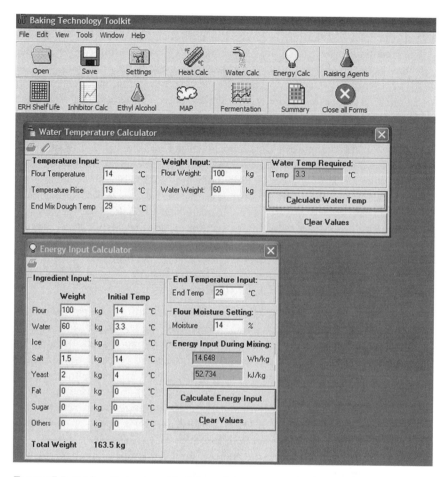

FIGURE 7.7. Baking Technology Toolkit – mixing energy calculator.

## 7.3.    Software for Complex Logistical Bakery Challenges

### 7.3.1.    Rollout

Every bakery, whatever its size, has to plan and schedule its production efficiently if it is to make the best use of its equipment, keep its costs under control and its customers happy by delivering products of the expected standard at the right time. Whether fermented products are being produced in a plant bakery or the more modest sized craft or in-store bakey, the timings and capacities of the equipment to be employed all have to be taken into consideration so that there is smooth transition, with few or no gaps, from one part of the process to the next. When the requirement for pans and trays, and for product to be ready at specific times of the day is considered the complexity of the task of scheduling is soon realized. Being able to visualise how the production flow dovetails together is no easy task and often conflicts arise during the processing sequence. For example, in an in-store bakery if scheduling is not accurate product quality can easily be compromised when oven capacity is not available and the product on exit from the prover is left in the bakery with only a rackcover to prevent it from spoiling. In plant bakeries, runs/batches might have to be re-scheduled at a moment's notice because of equipment breakdown, or a change in the priorities of the ordering sequence.

To address these problems a consortium of plant, craft and in-store bakeries worked together with cognitive scientists at Sussex University. A software system, known as ROLLOUT, was developed which 'visualised' the complex subject of bakery scheduling (Williams, 2006). The system comprised a 'planner' view and a 'scheduler' view. In the planner view the orders for customers or for a particular time window or deadline can be created. The orders specified the types and quantities of product required and are placed in the view at the approximate deadline time. The mixes/batches required to fulfil the orders can then be populated automatically into the scheduler view. Knowledge about the equipment capacities for each type and size of product along with process stages and their timings are brought into play in populating the scheduler view. The scheduler view comprises cascading horizontal bars (Figure 7.8) (the length depicts time, whilst height depicts capacity usage for the equipment at the process stage) imposed on a continuous timescale for the day's production. Conflicts which might occur, for example oven capacity insufficient, two products reaching the same process step at the same time, are flagged and highlighted in red. The scheduler/baker can re-arrange these 'cascading' production process runs so that the conflicts are eliminated. The powerful aspect of this software is the visualization on a pc screen of a complex problem and the opportunity for the

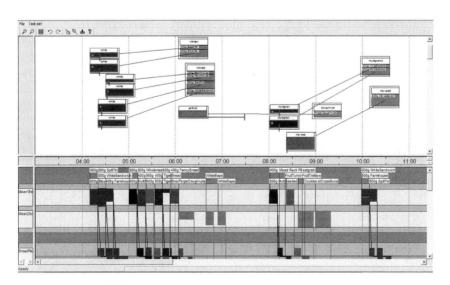

FIGURE 7.8. ROLLOUT example screen planner (top) and scheduler (bottom).

baker/planner to manipulate the schedule to meet the goals of the bakery business whether that be fresh products of the right quality available at specific times of the day, efficient use of valuable plant time or many other goals.

## 7.4.    Other Knowledge Sources

### 7.4.1.    Internet/Web Sources

The revolution in cheaper computing technology and access to global computing technology using the Internet and the World Wide Web has made the search for baking technology knowledge and information easier to achieve. However, the permanence, validity and accuracy of information need to be questioned before it is accepted and used. Often websites containing such information are transient and disappear. The information and knowledge they contain comes only in 'bite-sized' chunks with few threads of continuity linking one piece with another. Confidence in credible sources of information should be built up before the knowledge is applied in a commercial environment. Use of sites such as that provided by the professional bodies such as ICC (International Association of Cereal Science and Technology – www.icc.or.at), AACC (American Association of Cereal Chemists – www.aaccnet.org) and ASB (American Society of Baking – www.asbe.org) are good places to start as the information available on these sites has been 'peer-reviewed'.

## 7.5.    Conclusion

Knowledge-based systems are now an established computing technique which can be applied for the benefit of the baking industry. There are cases where they can be used 'solo' as diagnostic tools or as advisors (e.g. the Bread Advisor). They provide objective and supportive tools for use by an industry whose pool of expert resources has been much diminished since 1990s. They provide a lifeline to the hard-pressed product developers (Young, 2004) enabling them to innovate from a basis of sound knowledge and efficient software tools with which to achieve their goals. In a training environment they facilitate the uptake of and experimentation with knowledge about baking technology and processes. They provide easy access to focused and appropriate knowledge about fermented products to an emerging generation of bakery technologists who are computer literate and at ease when using computers as tools in the workplace.

Their greatest potential lies, however, with their integration with other computing environments, systems and techniques – for example with spreadsheets and databases, with online integration to equipment and with other advanced techniques, such as neural networks and fuzzy logic – to provide a limited element of artificial intelligence. In many computing scenarios in the baking industry the processing of 'data' in an ordered pattern is no longer enough to provide the solutions to the problems facing the industry. The data has to be linked to the technology of the product and to the rules, facts and techniques, that is the knowledge, which determine how that data can be best applied to provide the solutions for the production of baked products. For example, the patterns seen in the slice of bread captured by the C-Cell instrument (Chapter 1) and the data it determines might be linked via a knowledge system to detect trends in the product quality and to the correction of the settings on the plant to prevent faulty or sub-standard products reaching the customer. In some cases the knowledge is embodied within the system, such as when the physical link between the Retarding Advisor and the equipment was made to provide acceptable product quality in the batch operation. There are other cases where the knowledge can be called up and 'bolted in' when required. An example of this is when bakers might require advice in designing a new plant for a range of products. A search of the advertising literature, which can be found in bakery publications or in the newer electronic medium of the Internet, can yield many details about the engineering specifications of the equipment required. However, information concerning the technology of the product range and the important criteria and rules which apply to the fermented products to be made and their quality is seldom to be found. Where a user can interrogate a system in which the knowledge-based aspects applicable to the product are bolted to the data aspects applicable to the equipment, the true potential for the baking industry will be seen. The next few years should be an exciting time for the development of 'knowledge tools' for the baking industry. However, these tools can only be built from an understanding of the rules which govern the technology of the product and the quality required, and with close co-operation between bakery technologists (experts) and computer scientists.

## References

CCFRA (2004) Baking Technology Toolkit. A copyright product obtainable from CCFRA, Chipping Campden, UK.

ERH CALC™ (2000) A copyright product obtainable from CCFRA, Chipping Campden, UK.

FAULT DoC (1995) A copyright product obtainable from CCFRA, Chipping Campden, UK.

Williams, A. (2006) Production schedules undergo IT revolution, in British Baker, 5 May 2006, William Reed Publishing, UK, pp. 30–32.

Young, L. (1991) The FMBRA Bread Faults Expert System. *Manufacturing Intelligence No 6 Spring*, Department of Trade and Industry London, pp. 18–19.

Young, L. and Cauvain, S. (1994) Advising the baker, in *Proceedings of Expert Systems 94, Fourteenth Annual Conference of the British Computer Society Specialist Group on Expert Systems*, December, pp. 21–33.

Young, L.S., Davies, P.R. and Cauvain, S.P. (2001) Rise Again, Fair Knowledge, in *Applications and Innovations in Expert Systems IX, Proceedings of Expert Systems*

*2001, British Computer Society Specialist Group on Expert Systems*, Springer-Verlag, London, UK, pp. 89–99.

Young, L.S. (2004), We'll have it for T, in S.P. Cauvain, S.E. Salmon and L.S. Young (eds) *Using Cereal Science and Technology for the Benefit of Consumers*, Woodhead Publishing, Cambridge, UK, pp. 389–394.

# 8
# Baking Around the World

John T. Gould

## 8.1.   Introduction

Why is it that 'bread' means different things in different parts of the world? Place a loaf of white North American pan bread beside a loaf of Central European rye bread, Middle Eastern pita bread or Chinese steam bread and it could justifiably be said that they are all quite different products. Yet they are all bread. Why do bakers in different parts of the world use quite different equipment to make the same, or almost same, product?

To be able to answer these questions and hence understand baking in Europe, Africa, North America, Australasia and Asia, it is first necessary to appreciate

- the historical development of breadmaking;
- the factors which influence the process of breadmaking;
- customer expectations.

These factors are considered in the following sections.

## 8.2.   History

The breadmaking process is the interaction of raw materials, equipment, procedures and people in a particular environment. The end product which results from this interaction depends on the characteristics of each of the above process components. It will also depend on customer requirements and expectations.

Historically, the most dominant of the process components have been raw materials and environment. Over the centuries, and more particularly in Western countries, the nature of bread has changed over time with the evolution of modern wheats, the process by which they have been converted into flour, and technological advances in baking itself.

The evolutionary path of bread baking wheat from the primitive parents *Triticum boeuticum* and *Triticum dieoccoides* has been different in different parts of the world. These differences have resulted mainly from variations in climate, soil and culture, though in more recent years other environmental aspects, such as the economic, industrial and scientific factors, together with political development, have had a significant effect.

Because the evolution of wheat has occurred in different ways in different parts of the world, the nature of bread evolved differently in various countries and regions. Faridi and Faubion (1995) classified the different breads of the world in this way:

- those with high specific volume, such as the pan breads of the UK, North America and other Western countries;
- those with medium specific volume, such as French bread and many rye breads;
- those with low specific volume, for example flat breads of the Middle East and Eastern countries.

Each of these breads has its own particular characteristics which have resulted from the process of interaction referred to earlier.

A sample from a shipment of wheat from Danzig in the Baltic is attributed as the forerunner of the vast wheat industry of North America (Bailey, 1975). A wheat from an area not now known as being ideal for bread wheat found a true home in the North American countries and not only set the standards for future generations of wheat but also dictated the type of bread which was to become the standard for Canada and the USA, with of course some regional differences. Thus it is seen that an ordinary wheat from one part of the world became a much better wheat in the New World. This wheat was known as Red Fyfe, after the farmer who first grew it in Canada.

This transition is an example of the chance development of wheat which occurred from the earliest days until the latter part of the nineteenth century. Since then, scientists have used the principles of genetics and have developed variations of wheat to suit the needs of particular areas. More recently, particular types of bread and other flour-based foods have resulted from new wheat varieties.

Red Fyfe was to be one of the parents of what has been described by Professor Boyle of Connell University in the USA 'As the greatest single advance ever made in the United States.' Boyle was referring to the Maquis variety of wheat which had been bred in Canada by Sir Charles Saunders in 1903 (Bailey, 1975).

Special mention has been made of these developments in Canada because of the standing and influence of Canadian wheat on the world wheat market. For over half a century, Maquis became the standard by which bread wheats were measured, not only in North America, but also in most Western countries.

Cereal breeders in other parts of the world have not necessarily had the same ideal circumstances to work with. In particular, the climate and soil have imposed limitations on the extent to which hard-grained high-quality bread wheats can be consistently grown. The consequence of this has been a continuation of the production of bread-type wheats which originated in those areas. It should not be forgotten, however, that consumer preference is likely to have a significant

TABLE 8.1. Popular types of bread in Asian, Middle Eastern and North African countries (Qarooni, 1995)

| Country | Bread type |
| --- | --- |
| Afghanistan | Naan, Chapati and European bread |
| Algeria | Matlowa, French bread, Khobz El-daar, European bread |
| Bahrain | Tanoor, Arabic, Chapati, European bread, Samoon |
| Djibouti | Kisra |
| Egypt | Baladi, Shami, Samoon, French, Fatier, Shamsi, Bataw, European |
| Iran | Barbari, Tanoor (Taftoon), Lavash, Teeri, Suage, Sangak, European, French |
| Iraq | Tanoor (Khobz), Arabic, Samoon, European bread |
| Ethiopia | Injera |
| Israel | Sadj, Tarboon, Kimaj (Arabic), European bread |
| Jordan | Armani, Mafrood, Sauj, European bread |
| Kuwait | Tanoor, Arabic, European bread |
| Lebanon | Lebanese, Suaj, European bread |
| Libya | French bread, Arabic |
| Morocco | Moroccan Khobz El-daar, French, European |
| Oman | Arabic, Tanoor, Chapatti, European |
| Saudi Arabia | Samouli, Mafoud, Tannouri, Burr, Tamees, Korsan, European bread |
| Somalia | Injera |
| Sudan | Injera, Shamsi, Baladi |
| Syria | Mafrood (Arabic), Armani, Samoon, Suaj, European |
| Tunisia | Trabilsi, French bread |
| Turkey | Balzuma, Gomme, Yafka, French and other European bread |
| United Arab Emirates | Tanoor, Chapati, Arabic, Samoon, European |
| Yemen | Roti, Malouge, French, European bread |

influence on a wheat breeding programme. Sidhu (1995), when writing of wheat breeding in Pakistan, quotes 'Chapatis which have creamish white colour, soft silky texture and sweet wheatish aroma are preferred by the consumers.' The country's wheat breeding programmes are accordingly oriented to select wheat varieties conforming to these traits.

However, with the advent of modern communication, travel and trade, bread varieties originally confined to a particular area are now made in many countries of the world, and examples of this type of cross-culture in bread are provided by Qarooni (1995). This movement of bread types between countries and cultures can pose problems for bakers.

For example, the manufacture of French bread may be considered a challenge to bakers outside France. This is a challenge which can be overcome but does need an understanding of what makes French bread what it is. In Table 8.1 it is notable that those countries where French bread is popular are the countries with a French colonial presence in the late nineteenth and early twentieth century. Significantly, some of the best French bread in Australia is made by immigrant Vietnamese bakers who learned their craft from their former French colonial masters.

The special qualities of French bread are said to be due to the flour, the water, the yeast, the ovens, the atmosphere, the temperature, the Eiffel Tower, the aroma of coffee and Gitane cigarettes, the temper of the femme de boulanger and of current politics (Bailey, 1975). This statement will be examined later in the light of modern thinking about the baking process.

An historical aspect of bread baking which perhaps still has an impact on today's thinking about bread is that of adulteration. Particularly in Europe and the UK, the histories of bread make reference to various practices of economically hard-pressed millers and bakers adulterating flour and bread with chalk, rye and alum, bean meal, slaked lime and bone ash. In Roman times, bakers were accused of adding a whitening agent to their bread. It is thought that this may have been magnesium carbonate. The accuracy of these accusations is somewhat doubtful but then, as now, the public seemed to want to believe the worst of bakers. Adrian Bailey in *The Blessings of Bread* (1975) attributes most of the reasons for adulteration to the economic pressure placed on millers and bakers and the perception that white bread was seen as being preferable to the dark (by today's standards) bread produced from the so-called 'white flour' then available to them.

When attitudes towards bread are considered today, little has changed. White bread has to be white. Bakers still use additives to achieve this. It must be said, though, that soya flour is a much healthier ingredient than most of those listed above.

In most countries, regulatory standards, combined with today's sophisticated analytical techniques, ensure that the ingredients in bread are beneficial, not just to the baker, but also to consumers. Old attitudes die hard, though, and bakers and the baking industry still have an incomplete public relations task to convince consumers that today's bread is one of the safest food products on the shelves of the corner store or supermarket.

The claim by Bailey that adulteration in the early nineteenth century in the UK was the result of 'the desperate intense competition between bakers following the abolition of price fixing in 1815' also rings a bell when today's marketplace is considered.

To the environmental and cultural differences which have influenced the type of bread which is predominant in a particular country has been added the consequences of the various waves of refugees and immigrants which have moved between countries. Reference has already been made to the Vietnamese in Australia. Other examples are easy to find, not only in Australia, but in many other countries, of how people of different nationalities have influenced the bread types in their new homelands.

One of the most important examples of industrial progress which influenced the type of bread being produced was the introduction of roller mills to flour milling in the mid-nineteenth century. Only with this introduction could flour be produced which was almost completely free of the bran and germ which caused the darkness which was present even in the so-called 'white bread' previously produced.

Facilitating, as it did, the more efficient removal of bran and germ, the advent of this flour milling process was also the first significant step on the way to producing bread with the keeping qualities expected of today's packaged bread.

Other steps along this road have been the result of the application of science to breadmaking, which began in the late nineteenth century. These included better wheats following scientific breeding, oxidizing agents such as potassium bromate, refined fats, then emulsifiers, and followed later still by enzymes. All of these have only been made possible by improved scientific understanding of the breadmaking process. Folklore has gradually been replaced with scientific fact which, together with the ability to control the process, allows the production of the consistent product seen in most Western supermarkets.

Just as the Industrial Revolution provided the steel for the roller mills, so did it provide the steel for the first modern mixing machines and other bakery equipment. Of course, this equipment also needed the power to operate it – first this was steam and later electricity. To those people working in today's modern bakeries in Western countries, it is difficult to imagine that what is now seen as essential to make a product to satisfy present customers had not even been heard of 100 years ago.

Other environmental factors which have changed the nature of bread, particularly in Western countries, are methods of distribution and the advent of the supermarket and superstore. While the supply chain has lengthened and the time from oven to table has extended from minutes to hours or days, customers still expect their bread to be soft and palatable, that is 'fresh'. To achieve this the original flour, water, yeast and salt has, of necessity, been supplemented with a variety of 'improvers' and specialized packaging films.

With this background it is appropriate now to consider the breadmaking process in detail.

## 8.3.  The Breadmaking Process

Whatever type of bread is made around the world, to ensure that it is of acceptable quality to those who eat it, it is essential that the interactions between the various parts of the baking process are compatible and in balance with each other.

The extent of the balance established between the raw materials, procedures and equipment determines not only the quality of the end product but also the consistency of maintaining that quality. Craft bakers, whether they are in a peasant home in India or in the village high street in England, achieve this balance by subtle alterations to what they do as they proceed through the process. In contrast, the constraints inherent in automated mechanical baking dictate the need for sophisticated quality assurance systems to ensure consistent product quality.

Raw materials, particularly flour, which produce an acceptable or even superior product in one part of the world, may not do so when used under different circumstances in another country. Similarly, equipment developed for producing Western-type pan breads in one country may need considerable modification or the addition of supplementary equipment to produce the same type of bread using different flour, or indeed the same flour under altered environmental conditions.

There have been many instances where a failure to recognize these factors has resulted in unsatisfactory performance in bakeries and frustration for all involved. In order to understand why this is so, it is necessary to understand the basic processes of baking and how these apply in the situations which prevail in different countries, when different products are being made, or where different environmental conditions apply. When considering this approach it is necessary to start, not with the raw materials or the equipment, but with the question 'who are the customers and what do they want?' If the customer is a supermarket shopper in the UK looking for a loaf of bread with which to make sandwiches for lunch the following day, that person will be looking to buy a product which is already sliced, which will still be soft and palatable at lunchtime the following day, which can be spread easily with butter or margarine without coming apart (flexible crumb). Not only this, but any leftover bread should freeze and retain the above properties when thawed. In addition, if it is not used for sandwiches, it should brown evenly and quickly when toasted!

Compare this with the expectations of villagers in India and Pakistan when they make their own chapattis, or less commonly, buy them from a local baker. Sidhu (1995) states 'Probably the consumers of chapatti in Pakistan have not yet abandoned the age-old concept of using their product fresh from the hot plate (baking griddle)'. The same author also stated that 'the two most important quality parameters for chapatti (in India) are softness and flexibility'.

Thus two extremes of breadmaking sophistication have as their main quality characteristics softness and flexibility. At one end of the scale this is achieved by refining the raw material, processing to a high specific volume (giving a soft product to start with) and retarding the staling process by the addition of various anti-staling agents. At the other extreme these properties are achieved by eating the product immediately it is made. However, in both instances it is necessary to have the right type of wheat with which to start the process. Sidhu claims that more research is needed into the popularization of commercially produced chapatti and other baked products in Pakistan to produce a product with the same characteristics as that straight from the griddle.

In attempting to understand these extremes of breadmaking sophistication, it is also relevant to consider other factors which are basic to breadmaking, irrespective of the type of product. Sidhu gives a flow chart for chapatti preparation and after adding water to flour the next step is described as 'mix into smooth dough'. There is a clue here to one of the fundamentals of breadmaking – dough development.

## 8.4.  Flour and Dough Development

The extent to which dough should be mixed and developed depends on the type of flour available, the subsequent process steps, as well as the characteristics of the end product. Note there is no reference here to any testing method.

TABLE 8.2. Processing conditions for North American wheats (Faridi and Faubion, 1995)

| Process | Total mixing time (min) | Total fermentation time (h) | Cysteine added (ppm) |
|---|---|---|---|
| Sponge and dough | 15–17 | 3–5 | |
| Straight dough | 16–21 | 2 | |
| No-time straight dough | 16–21 | | 40 |

Rather it is end product quality which should determine the required amount of development.

North American breadmaking wheats require considerable mixing to achieve optimum development. This is true even if the mixing stage is followed by a period of bulk fermentation. If the fermentation stage is reduced, or even eliminated (no-time doughs), even more development is required at the mixing stage (Table 8.2).

It should be noted that, according to Faridi and Faubion (1995), the three different processes are used in different circumstances and the end products may be somewhat different. However, to achieve the same result in the same product, using the same raw materials, it can be assumed that the shorter the fermentation time, the longer the mixing time required. Bread baking flours in countries other than North America often require shorter mixing times than their North American counterparts.

Similarly, users of the Chorleywood Bread Process (CBP) find that the wheat varieties from different countries require different levels of work input to achieve optimum development. Difficulties can arise when this is not understood and major changes are made in a mill wheat grist. For example, if a mill changes from locally grown wheat to imported wheat without taking into account the requirements of the bakery, significant changes in the mixing requirement (work input) can result. Similarly, the writer has observed instances where the CBP has been installed in a bakery without an understanding of the available raw materials. The local flour was made from North American wheat with an optimum work input of 20 Wh/kg. As the temperature of the flour was usually about 30 °C (a tropical country) there was little chance that optimum development could be achieved while maintaining dough temperatures of approximately 32 °C. Significant flour cooling equipment needed to be installed to overcome the problem – the alternative was for underdeveloped doughs with a significant reduction in product quality. Collins (personal communication, 1993) has described such flour as 'awkward'. However, it is only awkward if it is being used in bakeries where the process has been designed without taking into account the nature of the available flour. A further example was observed by the writer, in Israel in the mid-1980s, where the importation of North American wheat to supplement local supplies caused real problems in a bakery using the CBP. In this instance the need to increase work input beyond 11 Wh/kg had not been recognized. A poor-quality product was the result.

The presence or absence of other ingredients in a dough can also affect the amount of mixing required to suit the process. This has been demonstrated where the use of potassium bromate has been discontinued in countries using the CBP. If ascorbic acid is used to replace bromate, a buckier, less developed dough results at the moulding stage. To overcome this, mixing, or development, need to be increased.

Related to the need to identify the correct mixing required for optimum development are the methods used for testing flour. Skilled craft bakers the world over may be able to adjust their processes to take into account variations in raw materials and the environment. However, in today's environment, where consumers expect consistency in product quality, such an approach is inappropriate for industrial plant bakeries. In these situations the ability to control the total process within clearly defined parameters is an essential prerequisite to producing bread of consistent quality. A logical consequence of this is that the performance characteristics of raw materials must be known before they enter the bakery. This being the case, attempts to establish international standards for flour testing, which do not take into account the process being used, have limited value as predictors of absolute quality or performance. Such international standards have their greatest value where wheat is being traded around the world since they allow both seller and buyer to make a reasonable assessment of the suitability of a given wheat variety for use on its own or as a component of flour millers grists.

While it is true that good bread baking flour can always produce good bread (Western style), it can only do this if it is treated in accordance with the functionality in the process being used. This has not always been recognized and the examples quoted above illustrate the difficulties which can be encountered when relying on the so-called 'traditional flour testing methods'.

New Zealand bakers and scientists recognized this soon after the introduction of the CBP into that country. Flours which were available there for making bread in the late 1960s and 1970s often only required half the work input that had been advocated by the originators of the process. The recognition of this led to a method of flour testing which closely approximated the CBP (Mitchell, 1982), a method which bakers in that country could confidently rely on to predict the performance of flour in their CBP-equipped bakeries.

The popularity of the CBP in New Zealand is, in no small part, due to this ability to predict flour performance accurately, but also to the fact that it was introduced to New Zealand for the same reasons underlying its adoption in the UK; namely, the ability to maintain existing bread quality while increasing the proportion of home-grown wheat in the flour miller's grist (Cauvain and Young, 2006). Today, in excess of 90% of packaged bread is produced by the CBP in New Zealand. The close relationship between the CBP and the home-grown wheat varieties has continued, and today such wheat varieties typically constitute 70% or more of milling grists.

Consistency of quality can also be obtained by introducing a very high degree of tolerance into the process. This, in turn, can be achieved in a number of ways.

One of these is to under-utilize the quality potential of the flour being used. This was dramatically demonstrated in South African bakeries observed by the writer in the mid-1980s. Potassium bromate was not a permitted additive, the baking industry was under strict government control and most of the bread observed appeared to have been produced from immature, or green, doughs. The result was loaf after loaf, looking and eating exactly the same. It was not bread of the highest quality, but it was absolutely consistent. In this case the tolerance of the process to variations in the raw materials and the environment was very wide. With deregulation of the industry and the availability of modern improvers, it has been observed that the general quality of South African white bread has improved dramatically, but it is no longer so consistent. The tolerance of the process has been reduced.

## 8.5.  Water

After flour, the next most important ingredient used in breadmaking is water. The impact of water addition on quality is often overlooked. Seibel (1994), in recognizing the importance of freshness in determining consumer acceptance, claims that 'higher water additions are being used world wide, resulting in more sensitive doughs needing special equipment'. This is probably only true for pan breads. In order to maintain their shape, varieties of bread baked on the sole of the oven or in shallow pans need to be made from firmer doughs than those used for pan bread. This will be a contributing factor to the way in which the keeping qualities of these breads diminish rapidly after the bread leaves the oven. A baker accustomed to making continental, or European, type hearth breads will have difficulty in adjusting to the need to increase the water to the extent necessary to produce longer-keeping pan bread.

However, English bakers are gaining a reputation in other parts of the world for overdoing the water addition. The sensitivity to dough damage during processing is very obvious in much of the pan bread produced in the UK. In the author's opinion, extra water addition appears to have been embraced for economic reasons rather than for improving quality. At the other end of the scale, in countries where bread weight has been determined by measuring the dry solids content, as, until relatively recently, in some Australian states, water addition has been limited. This had the opposite effect and bread keeping quality was adversely affected. A change in legislation in some Australian states in the early 1990s highlights the need to take into account the local legislative component of the environment when considering factors which influence the breadmaking process. In this case, bread weight legislation changed from a dry solids basis to net weight. This had a significant effect on bakers' expectations of flour water absorption, to which some mills had difficulty in adjusting. Also, water absorption became much more of a significant factor in determining the acceptability of new wheat varieties for the domestic market.

## 8.6.   Yeast

Modern bread yeasts are now available in most countries of the world. Compressed yeast and cream yeast are by far the most commonly used in plant bakeries. Active dry and vacuum-packed instant yeasts are now readily available and are often used in those areas where distribution is difficult and storage facilities are inadequate to maintain the viability of live yeast.

## 8.7.   Salt

The use of salt in bread is mainly influenced by consumer preference and varies from the extremes of zero for unleavened chapatti to a reported 3% for a popular Hungarian product (Lasztity, 1995). Most Western-type pan bread is usually made from doughs containing around 2% salt by weight of flour. In the UK, government pressure on the baking industry from the Foods Standard Agency and the Department of Health has increased. This is because of the potential links between salt levels in the diet and high blood pressure and other possible medical conditions. This pressure has resulted in a gradual reduction in salt levels in bread with the promise of more to come. It remains to be seen whether similar pressures for salt reduction will develop elsewhere in the baking world.

## 8.8.   Other Improvers

The use of dried vital wheat gluten, fats, soya flours, emulsifiers, malt flours and enzyme-active preparations varies greatly from country to country and from one type of bread to another. Maintaining product quality determines the type of improver required and the extent to which it is used. Also, variations in flour quality, environmental conditions and equipment idiosyncrasies all have an influence. For instance, one would not expect to find a widespread use of malt flour in northern Europe where wheats with low Falling Numbers are common (Salvaara and Fjell, 1995). This is in contrast to the wheat in many Australian states which commonly have Falling Numbers in excess of 400 and malt flour is often used as a base in improver mixtures.

In some instances legislative control determines the level to which various additives can be used in bread; for example, in Australia for many years there were legislative limits to many commonly used emulsifiers. In New Zealand there were no constraints for emulsifiers, and bakers there have tended to use higher levels than their Australian counterparts. The result is that the crumb characteristics of New Zealand white pan bread tended to be finer, but less resilient, than that found in most Australian bread. It will be interesting to observe whether or not the current review of the Australian Food Standards Code will allow the greater use of emulsifiers in bread. Of even greater interest will be whether Australian bakers would make greater use of emulsifiers to alter subtly the characteristics of the bread produced or if present consumer expectations will prevail.

## 8.9.   Dividing

Following mixing and development of the dough, either entirely in the mixer, for no-time doughs, CBP and continuous mixers, or by fermentation, the next stage in the breadmaking process is usually dividing the original bulk dough into the sized pieces needed for further processing and baking. This is just as true for the unsophisticated domestic production of chapatti as it is for the mechanical production of Western-style packaged breads (Sidhu, 1995). The bulk dough must be divided into pieces of specified weight to produce a loaf of the nominated net weight without damaging the bubble structure of the dough.

This double objective often requires a compromise between achieving accuracy of weight and minimal damage to the dough. This compromise is not often fully appreciated by some machinery manufacturers whose experience is frequently limited to one type of process. Australian and New Zealand bakers have had difficulty in convincing manufacturers of extrusion-type dividers that while that type of divider admirably meets the first part of the objective very well, it can cause severe damage to the delicate structure of CBP doughs. This damage can, to some extent, be reduced by the use of stronger flours or by the addition of dried gluten. However, this negates any cost advantage of improved accuracy. For this reason most bakeries in these countries still prefer to use suction dividers. As mentioned earlier, this sensitivity to dough damage during processing can be apparent in plant bread produced in the UK, though recent developments in divider technology have reduced its likelihood.

## 8.10.   Resting

This is not the term commonly used to describe the stage in bread production between moulding and dividing. Sidhu (1995) used it in his flow charts describing the production of unleavened bread. This is in contrast with the Western term, intermediate proof, which describes the stage of breadmaking where the dough is allowed to relax between moulding stages. Of course, intermediate proof, implying as it does yeast activity, would be quite inappropriate when referring to unleavened bread. The term 'intermediate' refers to the proofing stage following bulk fermentation (first proof) but before final proof prior to baking. Where no bulk fermentation takes place, for example in no-time doughs, first proof would be a more accurate term. However, the term 'intermediate proof' is firmly fixed in bakery terminology around the world.

The purpose of intermediate proof is to produce a piece of dough which is sufficiently soft, extensible and relaxed to allow optimum performance in the moulding stage. It also provides time for dough pieces to become properly mature.

Apart from equipment variations, the main difference found in this stage of breadmaking is that of time. This can vary from zero in the case of some continuous bread production methods to as much as 2 h used by some methods of producing continental types of bread. In the latter case the intermediate proof partially replaces bulk fermentation, but it also serves to create the very open texture preferred by the consumers of this bread. In this case it is not just relaxation which takes place but it also allows the considerable yeast activity necessary to produce the carbon dioxide required to create the preferred texture characteristics. This dough structure needs to be maintained during the next stage.

## 8.11.  Moulding

The purpose of moulding is to create a dough piece of the right shape while at the same time producing a dough structure which will result in the best possible texture in both the internal crumb and the external crust of the final product. In the example of continental bread referred to in the previous section, this means very gentle moulding so as not to damage the texture formed during intermediate proof. This is also true for the correct moulding of CBP doughs. In this process the required dough structure is formed at the mixing stage and should be protected right through to the actual baking.

Doerry (1995) claims that the purpose of the first stage of moulding, sheeting, is to expel the excess gas from the dough pieces, thus 'degassing' it. While this may be an appropriate description for what happens in North American bakeries, there is no explanation as to how this can be achieved without damaging the structure of the dough. It certainly does not apply to CBP doughs where, throughout the process, it is essential to maintain the structure of the dough created during mixing.

## 8.12.  Panning and Pans

It is a matter for discussion as to whether the pans used by bakers in different parts of the world determine or reflect many of the characteristics of the final product. In the mid-1980s, bakers in New Zealand introduced a new baking pan which significantly changed the size and shape of what had been the traditional loaf shape in that country. From being a relatively long narrow shape the loaf became rather shorter, broader and higher. The specific volume was increased by about 25% from approximately 4.0–5 ml/g or somewhat higher. This type of product, which was much softer, met with immediate consumer acceptance and is now the standard shape for packaged bread. In response to consumer demands for even softer bread, the specific volume is now approximately 6.0 ml/g.

It is interesting to note that in doing this, New Zealand bakers were able to increase the oven throughput and thus obtain an economic benefit. In contrast, as a result of a different oven tray and pan configuration, Australian bakers, in doing the same, were faced with a reduced throughput but, because of the consumer acceptance experienced in New Zealand, made the change regardless. The consumer acceptance in Australia has been similar to that in New Zealand. It is also relevant to note that Nagao (1995) reports that in Japan 'the introduction of a new white pan bread with a softer crumb than normal was welcomed by consumers and resulted in the recovery of total white bread consumption'. The writer has also observed that so-called 'premium packaged breads' in the UK have a higher specific volume, that is they are less dense than the 'standard' product. The premium product attracts a higher price. Such moves demonstrate the advisability, indeed necessity, of producing bread which meets consumer requirements.

The use of pans allows bread doughs to be softer and to flow more than when the dough piece is required to be self-supporting, and hence is a prerequisite for the production of 'soft' bread.

## 8.13.   Final Proof

With the possible exception of single-layer flat bread, the doughs for leavened bread are subjected to a period of raising before being baked. The purpose of this is to allow the moulded dough piece to relax and expand to produce an aerated piece of dough which, when baked, will be of the required shape and volume.

The warm, humid conditions which are necessary for this process to occur are well understood and practised by bakers everywhere. There is perhaps a lesser understanding of the time needed for this to produce the required results. Failure to allow enough time can result in the dough piece being unable to relax sufficiently even though it has risen to the required amount. This in turn can result in misshapen loaves and has been evidenced in both Australia and New Zealand when process and raw material changes have occurred. This is another example of the challenges which are faced when transferring technology from one environment to another. An exception to the humid proof conditions is that of double-layered flat breads. During the final proof of these products the surface dries out, resulting in the formation of a skin which assists in the formation of the double layer during baking (Qarooni, 1995).

## 8.14.   Baking

Baking is the final step in the making of a loaf of bread, during which heat is applied causing the rapid expansion of gas in the dough, water to be driven off, the gelatinization of starch and the coagulation of proteins. These reactions, together with the formation of a crust, transform an unpalatable dough piece into a (light) porous palatable loaf of bread which is nice to eat.

It is the last few words of the above definition which distinguish the different types of bread. Not only do the types of oven differ significantly, but baking times can vary from as low as 18–20 s for Arabic bread (Qarooni, 1995) to over 1 h for heavy, dense rye breads of Eastern Europe. The range of temperature is equally wide, between 150 and 650 °C. These extremes may seem way out to bakers used to baking Western-type breads at around 200 °C. Each situation can be explained by the fundamental principles of heat transfer and a knowledge of the gelatinization properties of moist starch. With the very thin Arabic breads the intense heat from the clay oven very quickly gelatinizes the starch and coagulates the proteins in the centre with minimal crust formation. The thicker the bread, the longer this process takes. In double-layered Arabic bread the rapid generation of steam and expansion of carbon dioxide during the first stage of baking forces the separation of the top and the bottom crusts and creates the characteristic pocket.

This is the same process which takes place in Western-style pan breads, but the lower temperature, slower heat transfer, and different volume and shape of dough produce a different result. Attempts to vary the process for making Arabic bread by retarding doughs result in a product which could be acceptable to Western consumers but may not find favour with those used to the traditional product (Inoue, 1995).

Bakers with an understanding of heat transfer principles will also recognize that differing oven conditions found in bakeries in Western countries are necessary because of different characteristics of the end products. In New Zealand and Australia the move in recent years to bread of higher specific volume has also resulted in reduced baking times. The heat transfer in the less dense dough piece means that the internal temperature can reach the required level, 96 °C, sooner than was previously the case.

## 8.15.  Cooling

In the markets where sliced and packaged bread predominate, bakers are required to give close attention to the conditions under which bread is cooled before it is sliced and packed. During this stage of bread production it is necessary to reduce the temperature of the bread to a level at which it can be sliced and packed without damage to the loaf either during slicing and packing or later due to excessive mould growth.

In markets where there are no restrictions on the moisture content of bread, the extent of moisture loss and hence weight loss during cooling needs to be taken into account. This moisture loss can be significant and should be avoided. Not only will excessive moisture loss adversely affect the cost of manufacture but it will also accelerate the staling process.

The design of a bread cooler which may be suitable for the climatic conditions experienced in the UK would be quite inadequate to deal with the humid conditions in a country like Singapore or the extremely dry conditions of, say, Perth in Australia. Each of these situations requires its own particular design criteria to enable bread to be cooled under conditions which will produce optimum results. Like every other stage of the breadmaking process, the correct balance of all the factors involved needs to be achieved. Machinery manufacturers and bakers alike need to adopt a much more theoretical approach to the specification of equipment used in different situations. Using first principles, process requirements need to be matched with equipment design to produce the required results under the conditions where the operation will take place. With regard to coolers the Australian Bread Research Institute has identified most of the problems and developed appropriate solutions (Adamczak and Kalitsis, 1992).

## 8.16.    Slicing and Packing

The final stage in the production of modern Western bread is also subject to variations from country to country. Depending on the nature of the product and consumer preference, the thickness of the slice can vary, as can the package. For instance, the thickness of sandwich bread in New Zealand is approximately 10 mm while in Australia for a similar type of loaf it is 13 mm.

Not only this, but the type of equipment will also differ considerably. The different type of bread found in different Western countries also dictates the type of slicing machine used. This difference has been demonstrated recently in New Zealand. As the specific volume of the bread has increased, bakers have changed from using the reciprocating slicers common in the UK to the band slicers usually found in North American bakeries. The band slicers handle the softer, higher volume bread much better than the reciprocating slicers. At the other end of the scale it is necessary to employ special spinning disc slicers to cut very heavy, dense rye bread.

## 8.17.    Packaging

The function of bread packaging for Western-style breads can be described as:

• to contain the slices and crusts together in a single unit for handling purposes;
• to maintain product quality by reducing crumb drying, minimizing the risk of contamination;
• to present an appealing and informative package to the consumer.

The most common material now used in Western countries is a low-density polyethylene, usually in the form of bags. This product provides a good barrier to water vapour, thus inhibiting drying, is easy to handle, and provides a clear background through which the consumer can see the product being purchased.

Many attempts have been made to package French-style country breads so that the crustiness can be maintained – these have not met with any great success. Similarly, attempts to maintain crustiness while microwaving have not been very successful. The migration of water vapour from the internal crumb to the crust has foiled all trials to date.

With the increasing emphasis on consumer information in many countries, the bread package, like other food packaging, is becoming more and more informative. The responsibility of bakers to ensure that the information they provide is accurate has also increased. Consumer laws regarding misrepresentation and false information mean that not only must the original source of the information be correct, but that products within the package must be consistently made to standard recipes. This all requires an extremely high level of quality assurance and traceability.

## 8.18.    Bread and Sandwich Making

Bread and other fermented products have always played a significant role in the development of the convenience food and snack market. In the past the use of sliced bread to prepare sandwiches was essentially confined to the home. Gradually as the advantages of having an edible 'package' were realised, bread-based convenience products proliferated. Today bread finds use in the preparation of hamburgers, sub-rolls, filled baguette, filled croissant and flour tortilla wraps. However, it is perhaps the re-discovery of the sandwich based on two slices of bread from a lidded loaf that best show both the versatility and the utility of bread.

In the UK it has been re-born as the classic 'meal on the move' and the sandwich making industry has grown to be worth an estimated GBP 5 billion. It is based on large-scale production of both bread and the assembly of the sandwich components in specialist units and sells in large retailers (Figure 8.1), gas stations, food halls and coffee shops. A wide variety of bread types are used in the manufacture of the sandwich with the standard white loaf being relegated to a supporting role in sales. In part, this arises because the sandwiches must be held at refrigerated temperatures (5 °C) at the point of sale for food safety reasons. This does not present the bread at its best since bread stales fastest at 5 °C (see Chapter 10) and if (as is often the case) the sandwich is consumed cold the bread 'lacks flavour'. Additions of fibres, cracked grains, malted products and even oatmeal all restore some of the flavour of the bread product.

The development of the UK sandwich market has significantly impacted on the bread baking industry. The sandwich maker sees the properties of every slice of bread during the sandwich assembly process and effectively quality controls every slice. The triangular pack has become the industry standard and this in turn has placed exacting standards on the bread baker. Tolerances to variations to lidded loaf height are small; too large a slice area and the bread triangle will not fit the pack, too small and the sandwich moves around in the pack and is likely to lose its filling. Even the internal features are carefully scruitnised. Holes are unacceptable since the filling falls out and makes a mess in the pack and a coarse cell structure can lead to the sandwich assembler using too much butter or margarine spread.

The exacting standards of the sandwich maker are passed back to the bakers who must now ensure better control of raw materials, recipe and process in order to meet the requirements of their customers. It is no longer a case of worrying about whole loaves meeting specification but of each slice having to be 'right

FIGURE 8.1. Sandwich display.

first time'. Such demands have led to the emergence of bakeries dedicated to making little else other than bread for sandwich making, not just a few loaves but often 12,000 units an hour, 24 h a day for 6–7 days a week. To meet the exacting standards, true 'partnerships' have been formed with bakers, ingredient suppliers and sandwich assemblers all working to a common goal – to make the humble sandwiches, but lots of them.

The sliced bread sandwich has not been confined to the UK with British bread for sandwich making being exported far and wide, including daily deliveries to the south of France and weekly frozen bread supplies to Japan. Production of the bread and assembly of the sandwich is also spreading with plants operating in France, Spain, Greece and Romania – further examples of the truly universal nature of bread and the consumer-led desire for new products.

## 8.19.   Crustless Breads

The contribution of the crust on bread and fermented products to flavour has been discussed in earlier chapters (see Chapters 1 and 5) but a significant part of the consumer base – children – apparently do not like the texture or flavour that comes with the crust. This seems to be especially true for the preparation of sandwiches. It is somewhat ironic to watch some consumers removing the crust

FIGURE 8.2. Invisible crust loaf made from a mixture of white and wholemeal flours.

and throwing it away (or feeding it to the birds) while others will tear out the crumb of baguette and simply consume the crust.

The marketing opportunity for crustless breads have been met in the USA, Mexico, Spain and elsewhere by removing the crusts from standard loaves baked in pans. This may seem a simple enough product to make but it is not without its technical challenges. In particular the dimension of the slices and their weights must meet the appropriate quality and legislative criteria.

A recent development has produced a genuine crustless or 'invisible' crust loaf (Figure 8.2). The product has thin but colourless crust. The outside of the product is soft to the touch and the crumb cell structure resembles 'conventional' bread. The technology used to make the product remains undisclosed and once again shows the ingenuity of bakers and the diversity of products that are classified as bread.

## 8.20.  The International Market

Western style bread is now made in many Asian, Middle Eastern and African countries. Arab-style pocket bread is made in many Western countries. French-style crusty bread is made throughout the world. Canadian, American and Australian wheats are exported all around the world.

The result of all this is that many traditional methods have been modified and traditional products have had their characteristics altered as they have been made with non-traditional flours and equipment. Tradespeople in some countries have had to be extremely versatile in adapting to an inconsistent supply of flour made from wheat which has been imported, stored and distributed with little or no regard for its breadmaking properties (Sidhu, 1995). Consumers who have not been exposed to an authentic loaf, made within its homeland, may become accustomed to a somewhat different product.

To overcome some of these problems, much work has been done in some laboratories in an effort to characterize the flour, and hence wheat, requirements for products made from wheat grown in one country and exported to another. The Australian Bread Research Institute has defined appropriate flour requirements for products such as Arab bread (Quail, 1996) and Chinese steam bread to be made from Australian wheat. Similarly the American Institute of Baking as well as the Canadian Grain Commission (Kilborn and Tipples, 1981) have studied the breadmaking processes used in countries importing wheat from their countries.

In many cases the challenge still remaining is for that knowledge to be transferred from the original publications to practical use. This challenge applies not only from country to country but within individual countries. The results of what is probably the most successful transfer of baking knowledge on an international scale can be seen in the ubiquitous Big Mac. The McDonalds hamburger chain has been successful in facilitating the worldwide adaptation of the baking process and raw materials to produce a product which is indistinguishable from country to country. This is an excellent example of the requirements of the end product dictating not only the control of the manufacturing process but also the specifications of all the raw materials. This has resulted in, at least in New Zealand, a special wheat being grown. This has only been achieved by a thorough and complete understanding of the principles involved.

One aspect of this, which applies to at least some Western countries, is the need to distinguish between craft skills and those required by operators of large plant bakeries. Most practical courses for bakers place a heavy emphasis on the former to the detriment of those whose skills should be in operating what should be the rigidly controlled process of automated bread manufacture. A step towards achieving this is, at the time of writing, being developed in Australia. In that country the Plant Baking Industry, in conjunction with the Australian National Food Industry Training Council developed a 'National Certificate in Food Processing – Plant Baking'. It was anticipated that this certificate would eventually replace traditional apprenticeships as far as most plant bakery operations are concerned. It is also interesting to note the fact that this project was being conducted under the auspices of the Cooperative Research Centre for Quality Wheat Products, a joint industry and government research initiative. Such an integration of research and training follows that adopted by organizations such as the American Institute of Baking, the Australian Bread Research Institute and the New Zealand Crop and Food Research Institute Ltd. This approach facilitates the uptake of research findings by industry and ensures that the most up-to-date knowledge is included in training programmes.

As a result of the research being carried out in cereal food laboratories around the world, the molecular basis of the baking process is gradually becoming clearer (Cauvain, 2003). With this level of understanding it is only a matter of time before a technological breakthrough occurs. As the basic building block in the process is dough formation, this breakthrough is likely to occur in that area of cereal science. Already the possibility of lower energy requirements for dough development is being mentioned (Larsen, 1996). Such an eventuality is likely to impact on all breadmaking methods with benefits accruing not only to Western countries but to the whole world of baking. We can but wait and see.

## References

Adamczak, T.P. and Kalitsis, J. (1992) Cooling control in the bread industry. *Paper presented to 42nd RACI Cereal Chemistry Conference*, Christchurch, New Zealand.

Bailey, A. (1975) *Blessings of Bread*, Paddington Press Ltd., London.

Cauvain, S.P. (2003) *Bread Making: Improving Quality*, Woodhead Publishing Ltd., Cambridge, UK.

Cauvain, S.P. and Young, L.S. (2006) *The Chorleywood Bread Process*, Woodhead Publishing Ltd., Cambridge, UK.

Doerry, W.P. (1995) *Breadmaking Technology*, American Institute of Baking, Manhattan, Kansas, USA, p. 175.

Faridi, H. and Faubion, J.M. (1995) Wheat usage in North America, in *Wheat End Uses Around the World* (eds H. Faridi and J.M. Faubion), American Association of Cereal Chemists, St Paul, Minnesota, USA, Chapter 1, Plate 1.

Inoue, Y. (1995) Bake pita bread from a standard MDD dough and at normal temperature. *Bakers' Bulletin*, New Zealand Institute for Crop & Food Research, Christchurch, New Zealand.

Kilborn, R.H. and Tipples, K.H. (1981) Canadian test baking procedures II. GRL–Chorleywood method. *Cereal Foods World*, **26**(II), 628–30.

Larsen, N.J. (1996) Dough development beyond 2000. *Bakers' Bulletin Issue No. 27*, New Zealand Institute for Crop & Food Research, Christchurch, New Zealand.

Lasztity, R. (1995) Wheat usage in Eastern Europe, in *Wheat End Uses Around the World* (eds H. Faridi and J.M. Faubion), American Association of Cereal Chemists, St Paul, Minnesota, USA, pp. 127–47.

Mitchell, T.A. (1982) The new MDD baking test. *NZ Wheat Review*, 15, 1980–82, 73–4.

Nagao, S. (1995) Wheat usage in East Asia, in *Wheat End Uses Around the World* (eds H. Faridi and J.M. Faubion), American Association of Cereal Chemists, St Paul, Minnesota, USA, p. 170.

Qarooni, J. (1995) Wheat usage in the Middle East and North Africa, in *Wheat End Uses Around the World* (eds H. Faridi and J.M. Faubion), American Association of Cereal Chemists, St Paul, Minnesota, USA, p. 226.

Quail, K.J. (1966) *Arabic Bread Production*, American Association of Cereal Chemists, St Paul, Minnesota, USA.

Salvaara, H.O.M. and Fjell, K.M. (1995) Wheat usage in northern Europe, in *Wheat End Uses Around the World* (eds H. Faridi and J.M. Faubion), American Association of Cereal Chemists, St Paul, Minnesota, USA, p. 158.

Seibel, W. (1994) Recent research progress in breadmaking technology, in *Wheat Production Properties and Quality* (eds. W. Busheck and V.F. Rasper), Blackie Academic and Professional, Glasgow, p. 67.

Sidhu, J.S. (1995) Wheat usage in the Indian sub-continent, in *Wheat End Uses Around the World* (eds H. Faridi and J.M. Faubion), American Association of Cereal Chemists, St Paul, Minnesota, USA, p. 210.

# 9
# Speciality Fermented Goods

Alan J. Bent

## 9.1. Introduction

'Foreign breads slice white sales' appeared as a headline of an article by Joanna Bale, in *The Times* on 13 August 1996. The article stated that 'While croissants and others increase in popularity the humble loaf of bread continues to fall out of favour. Consumers are becoming more adventurous and acquiring a taste for the new products available.'

The term 'speciality fermented goods' is rather difficult to define and falls into another nebulous category – 'morning goods'. The definition of morning goods can vary from 'anything that is not bread or cake – small bakery products exclusive of savoury products' to 'There is no legal definition covering this term. Morning goods range from very lean, crusty products such as Vienna breads to goods containing high levels of fat and sugar, and possibly fruit and spice, e.g. hot cross buns. They can range from laminated croissants and Danish to lean rolls.'

Taking the 'anything that is not bread or cake' category leaves thousands of baked products with infinite permutations and methods of finishing. Any attempt to cover the possibilities in one chapter would result in just a mention, if at all, of any one topic. In this chapter concentration has been placed on four major areas of speciality fermented goods, namely hamburger buns (Figure 9.1), part-baked bread, croissants and Danish pastry production.

## 9.2. Hamburger Bun

### 9.2.1. Production Rates

Hamburger bun plant come in a range of sizes with examples having outputs of up to 66, 240 buns per line per hour. This example comes from an eight-lane AMF Extrusion Bun Divider operating at 138 cuts/min (AMF, 1995, states production speeds of up to 138 cuts/min), that is $8 \times 138 \times 60 = 66, 240$. For a week's production of say 160 h (plus 8 h cleaning and maintenance) the output

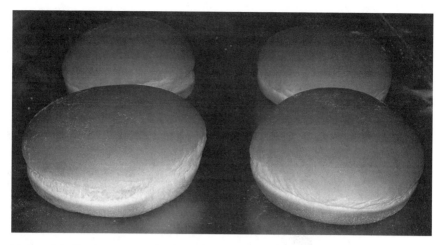

FIGURE 9.1. Hamburger buns (courtesy *BakeTran*).

of each line would be $66,240 \times 160 = 10,590,400$ buns. European lines of this type are said to be operating at a maximum of 125 cuts/min. Work in the USA reported by Steward (1990), Stevens (1991) and LeCrone (1994) also indicates commercial speeds as 125 cuts/min, which gives an output of 1000 buns/min on an eight-lane plant, that is $8 \times 125$. Even at this slightly reduced rate which equates to a weekly output of $(8 \times 125 \times 60) \times 160 = 9,600,000$ buns per line, the output is substantial. The following is a brief outline of the details covering large-scale hamburger bun production.

## 9.2.2. Formulation

Table 9.1 compares the formula for UK standard white pan bread with that of a hamburger bun. Two major differences can be identified. Higher levels of fat/oil are used in buns compared to bread, and sugar is added to hamburger buns. The reduced water in the hamburger bun formulation compensates for softening of the dough caused by the addition of the higher levels of fat and sugar. Without this reduction, doughs would not process through a plant. The flour for high-volume production of hamburger buns is required to be strong and relatively high in functional protein, to carry the high levels of fat and sugar. Readers interested in microwaveable, hamburger bun formulations based on high fat and high fibre/water ratios are referred to Neufeld (1991).

## 9.2.3. Liquid Brews and Fermentation

Before doughmaking it is common to include some prior fermentation. Water brews were at one time popular, in which most of the yeast, some of the salt and sugar are mixed and fermented with a brew buffer for 60–90 min. Water brews

TABLE 9.1. Comparison of a white sliced loaf recipe and one for hamburger buns.

| Ingredient | White pan bread | White hamburger buns |
|---|---|---|
| Flour | 100.0[a] | 100.0[b] |
| Salt | 1.8 | 1.8 |
| Water (approximate) | 60.0 | 55.0 |
| Yeast (compressed) | 2.5 | 3.0 |
| Fat | 1.0 | 3–5 (fat/oil) |
| Sugar | Zero (in the UK) | 5–10 or more |
| Improver | 1.0 | 1.0 |

It is difficult to give exact protein levels for a flour and whilst they are important, low ash figures and/or low Grade Colour Figures should also be taken into account.

[a] Flour with a protein level (as-is) of 10.5–11.5% suitable for pan bread made on the Chorleywood Bread Process.

[b] A higher functional protein level is required, < 13%. This higher level for hamburger buns is needed to maintain gas retention while supporting the additional fat/oil and the sugar, both of which have to be carried in the final product and do not contribute towards product volume. Without the higher functional protein acceptable product, volume and stability would be difficult to obtain.

based on 3.5–4.5% sugar solids and 3.0% yeast (based on flour weight) were used to develop the necessary amounts of acid to produce good buns (Sutherland, 1989). Other methods involve fermenting part of the flour, in the form of a sponge. A traditional sponge takes a proportion of the flour, yeast and salt, and a tight sponge is made with water which is allowed to ferment at specific temperatures. After fermentation of the sponge the balance of the ingredients in the recipe are added and mixing follows to form the dough.

A flour brew (sometimes referred to as a liquid sponge) is a variation of the traditional sponge in which more water is added, usually just a little more water than flour. A much slacker, pumpable, liquid sponge is produced. Traditional monitoring of brew development during fermentation is by measurement of pH and determination of total titratable acidity or TTA. [TTA, which is the term used to describe the amount of acid in a substance, is the number of millilitres of 0.1 N NaOH needed to raise the pH of 20 g of brew to a value of 6.6 (Sutherland, 1989).] In the UK these measurements are difficult because of the addition to all flour (for use in baked products sold in the UK) of not less than 235 and no more than 390 mg (per 100 g of flour) of calcium carbonate (Bread and Flour Regulations, 1996) which buffers the brew acidity. Sutherland (1989) states that 'Salt can also be used as a buffer, but it acts differently from calcium carbonate. Salt basically retards yeast action.' Presumably this describes the effect of salt in slowing down fermentation which affects the subsequent formation of acid. Dubois (1984) discusses further the effect of salt in the brew of a liquid ferment. After a 1–3 h fermentation the flour brew is degassed, filtered and cooled down, by passing it through a heat exchanger, to between 6 and 9°C(41–48.2°F) to arrest any further fermentation. The cooled brew can then be stored in bulk, in a water-jacketed reservoir or holding tanks, ready for addition to each mix. Correct sanitation procedures should be followed between brews, that is water rinsing between brews, and complete system CIP (clean-in-place) must occur at regular periodic intervals, that is at least once a week. Prior to the start of a mix, brew is pumped into a liquid batching tank, weighed via load cells and dropped into the mixer. The normal flour level in liquid brews was usually between 40 and 50% of the total. Watkins (1991) describes the advantages of liquid ferment systems which have the capacity to handle in excess of 70% of the flour in the brew, 'the continuous liquid ferment mixer was developed to improve the mixing action, eliminate mixing damage and gluten separation, and to facilitate the incorporation of higher flour levels'. The advantages to the product are described as a softer loaf, cleaner cut, better keeping quality, improved flavour and aroma, a more resilient crumb and an improved sheen. Charbonneau et al. (1978) describe bun manufacture with a 65% liquid ferment sponge system.

Many advantages are claimed for a prior fermentation of part of the flour. One of the practical advantages of using liquid flour brew is found to be improved pan flow (i.e. the ability of dough pieces to flow and fill the base of the pan indent), but bulk flour brew held in the water-jacketed reservoir or holding vessel becomes a mixture of batches of brew, which is excellent for long, continuous production runs of similar products and also gives consistency. However, major trials should only be undertaken after the holding vessel has been emptied and cleaned of old flour brew. After cleaning, the holding vessel will be filled with new brew containing the result of a major trial. Minor trials on a day-to-day basis are restricted to those which can be carried out on the ingredients that are added at the mixer. When traditional sponges are used for prior fermentation, an individual sponge is made for each mix. The traditional sponge system allows for both major and minor trials with each batch.

Sutherland (1989) gives the pH and TTA values shown in Table 9.2. In an earlier article Sutherland (1976) gave further values for ferments and doughs from flour brews, given in Table 9.3.

Further information on the changes taking place in brew systems are given by Sutherland (1974), Kulp (1983) and Doerry (1985). An outline of brew equipment is given by Matz (1992).

TABLE 9.2. Values of pH and TTA for various brews (Sutherland, 1989).

|  | pH | TTA (ml) |
| --- | --- | --- |
| 90 min water brew | 4.2 | 10–12 |
| 2 h, 50% flour brew | 4.95 | 6.2 |
| 3.5 h, 66% flour sponge | 5.05 | 4.7 |

TABLE 9.3. pH and TTA values for ferments and doughs from flour brews (Sutherland, 1976).

| Flour (%) | pH | TTA (ml) |
|---|---|---|
| 0 | 4.2–4.3 | 12–13 |
| 10 | 4.3–4.4 | 11.75–12.75 |
| 20 | 4.4–4.5 | 11–12 |
| 30 | 4.5–4.6 | 10.5–11.25 |
| 40 | 4.6–4.7 | 9.5–10.5 |
| 50 | 4.7–4.75 | 8.5–10 |
| Dough–developer | 5.1–5.15 | 4–4.5 |
| Dough–oven | 4.9–4.95 | 5–5.5 |
| Baked bread | 5.1–5.2 | 3.8–4.2 |

## 9.2.4. Mixing

At a scaling weight of, say, a modest 50 g, the dough requirements of some plants may be up to 3,310 kg (66, 240 × 0.05 kg) per hour; at 85 g the hourly dough requirement will be 5,630 kg. Horizontal bar mixers with mixing times of the order of 7–10 min are traditionally used. The machine mixing capacity is considerable and can be up to 1,300 kg per batch. Substantial refrigeration on the sides of horizontal bar mixers help in producing the required ex-mixer dough temperature of approximately 26–27 °C(78.8–80.6 °F). The exact ex-mixer temperature will be plant specific and will be dependent on the pipe length and the heat generated during dough transfer and handling on a plant. High-speed mixers suitable for producing dough using the Chorleywood Bread Process (CBP), with energy meters to control the mixing energy and time, are used in some hamburger plants. During a standard CBP pan bread mix, to 11 Wh/kg of dough, a temperature increase of the order of 15 °C is generated in the dough. The control of dough temperatures during warmer months of the year becomes very important. Thompson (1983) states that doughs mixed on high-speed mixers using normal-strength American wheat flours require somewhat longer mixing times (brought about by a higher energy requirement per kilogram of dough) than those common to the softer wheat flours of the UK. This will lead to a greater increase in dough temperature during mixing. The addition of refrigerated liquid flour brew to a high-speed mixer will contribute towards controlling dough temperatures (further discussed by Thompson, 1983). When a 40% flour brew is used nearly half the weight of each mix is refrigerated flour brew.

The use of the pressure–vacuum, CBP, high-speed mixer (described in Chapter 4) may have benefits in the future of hamburger bun manufacture, especially when producing them in the absence of potassium bromate.

## 9.2.5. Dough Transfer

After mixing, the dough is emptied into a dough pump. A pair of augers gradually move the dough through an outlet at the base of the dough pump or hopper.

The dough can then be moved onto transfer conveyors or through a pipe for delivery to the divider. At maximum output of 50 g rolls, an eight-lane extrusion divider will use $0.05 \times 138 \times 8 = 55.2$ kg dough/min. If a large mixer is used to produce, say, 1,300 kg of dough, it will then take $1,300/55.2 = 23.56$ min before all the dough is divided. During this time some fermentation will occur and some chemical reactions will take place; the action of the augers in the dough pump is said to homogenize and degas the dough during transport. This supplies a dough of a more uniform density to the divider, from the beginning to the end of each dough (AMF, 1995).

## 9.2.6.  Do-Flow Unit

The Do-flow unit is sited above the divider and is a kneader that degasses dough before it enters the divider. The Do-flow unit is said to allow precise control of product grain structure (AMF, 1995). Dough from the dough pump/hopper is delivered into the Do-flow unit onto a helical feed screw which controls the rate of dough delivery onto the developer paddle. The developer paddle further develops the dough, which is extruded as an even curtain of dough into the divider hopper. The Do-flow unit supplies the divider with a dough of uniform density from the beginning to the end of each batch, which enhances scaling delivery. The speeds of the feed screw and the developer paddle are adjustable.

## 9.2.7.  Dividing

The bun and roll make-up systems of AMF offer two types of dividers, the Model K divider head and the Extrusion Bun divider (similar types of Do-flow units, dividers and first provers are also made by other manufacturers, e.g. Dawson and Cummins Eagle, Inc.).

### 9.2.7.1.  Model K divider

The Model K consists of a stainless steel rotating drum which has two rows (of four or six across) of cut-out cylinders. The rows are on opposite sides of the drum. Dough enters the divider hopper and is drawn by vacuum into one of the rows of cylinders. As the drum rotates, a knife cuts off the dough. In the down position, a piston is activated and forces the dough to the edge of the cylinder. A wire then releases the dough from the edge of the piston bore and the dough pieces drop onto the bed of the moulder. The Model K divider can operate at up to 100 cuts or strokes per minute. The Model K uses divider oil to prevent dough from sticking.

### 9.2.7.2.  Extrusion bun divider

The method of operation of this divider is given in the name; dough is drawn in a vacuum chamber containing rotary screw augers. It is fed into a metering pump which then goes to manifolds containing adjustable valves on each outlet.

Altering the valves changes the weight of dough from each outlet port. The dough is extruded from four, six or eight extrusion ports where a knife or blade cuts off or guillotines the extruded dough into equal pieces onto the bed of the moulder. Steward (1990) states that 'the device used to cut off the dough piece must operate at extremely accurate, highly repeatable frequencies in order to divide the extruding product in equal portions'. The speed range of this type of divider is from 40 to 138 cuts/min. An accuracy of ±1% is guaranteed by AMF. The operation is such that no divider oil is required.

Stevens (1991) states that 'dough goes into the degasser and is automatically pumped through a metal detector feeding the eight-across guillotine bun divider'.

### 9.2.8.  AMF Pan-O-Mat

The Pan-O-Mat rounds, first proves, forms and pans dough pieces. From the divider the dough pieces drop onto the moulder which consists of an endless belt. Across the moulder belt four, six or eight rounder bars are accurately placed. The dough pieces travel down specially concave rounder bars, which are designed to round each dough piece into a smooth, surface-sealed ball.

Dough pieces, which originally left the mixer at 26–27 °C (78.8–80.6 °F) and have passed through a dough pump, Do-flow unit, and a Model K dividing head leave the rounder bars at a temperature of approximately 30 °C (86 °F). Steward (1990) states that with the extrusion type of divider, because no oil is used on the dough or the rounder belt, the dough balls begin to round immediately upon contact with the rounder, which improves indexing gate transfer efficiencies and reduces the number of doubles in the intermediate prover.

From the moulder bars a driven kicker bar transfers the dough pieces onto the zig-zag board. On the zig-zag the dough pieces pass down an incline from the moulder. A fine dusting of flour falls onto the zig-zag and each dough piece is evenly coated in a thin film of flour. The dough pieces are assembled again in a row and transferred into a tray of prover pockets by a synchronized rotary gate system.

The first proof allows the dough pieces to relax sufficiently for sheeting or forming. At the end of proving a row of balls of dough are inverted and are then discharged into what are known as closed clam shell gates. The latter ensure that the dough balls enter (when the clam shells open) the adjustable sheeting rollers at the same time, which allows an even row of sheeted or slightly flattened dough pieces to be transferred into the indented pans (after the sheeting rollers, additional equipment can be fitted, e.g. a finger pressure board, which moulds the dough pieces into a hot dog or finger shape). The pans are fed by conveyor under the sheeting unit and a pan stop and indexer system synchronizes each row of indented pans just under the discharge conveyor.

### 9.2.9.  Flour Recovery and Pan Shaker Units

The tray of buns then passes through a unit which employs targeted jets of compressed air to just lift each row of buns, whilst an extraction unit removes unwanted excess flour from the surface and importantly from the base or heel of each dough piece. An adjustable pan shaker is then employed to ensure that the dough piece is in the centre of each indent. At the start of a run the pans are likely to be cold and will require a higher setting on the pan shaker to centralize each dough piece. Once the pans have been round a plant and through the oven and become warm, the dough piece tends to move more easily in silicone-treated pans, and consequently lower pan shaker settings are required.

## 9.2.10.  Proving

Large-scale high-volume hamburger bun manufacture usually employ a proof and bake system. Baking pans are moved by a track into a conveyorized prover. With this system, pans are supported either by a device in the centre of each tray or by a magnet. A track then transfers the pans into the oven. Stevens (1991) describes this type of system as an endless conveyor system. Prior to entry to the prover the space between each baking pan is monitored. Within the prover the track bends in an oval and gradually travels to the top, crosses over and then down again. Pan spacing is essential if the edges of the pans are not to collide as the track bends. The prover temperature and relative humidity are adjustable. Typical settings for hamburger buns are of the order of 38 °C (100.4 °F) and 78% relative humidity, for a proof time of between 50 and 55 min depending on the weight of the bun. Proving time is altered by changing the speed of the track.

## 9.2.11.  Seed Application

After an appropriate proof time, the track with indented pans full of proved dough pieces leaves the prover. If the buns are to be covered with sesame (Figure 9.2), other seeds or grits, the pans of proved dough pieces are passed through a water mist to give a fine and even covering of water. Immediately after this, seeds are evenly distributed onto the top surface of the proved buns. Two systems are available for adding seeds. In the first, after water misting the pan activates the vibration of a tray of sesame seeds held at an adjustable height above the pans. The base of the flat hopper of sesame seeds is perforated and the vibration releases seeds onto the top of the tray of buns. The time during which the tray vibrates and the degree of vibration are adjustable. These adjustments will alter the weight of seed applied. The second system uses the pan indents to activate a mandrel or a grooved spindle, housed in the base of a hopper of sesame seeds. As the spindle revolves, seeds are released onto the top surface of each row of dough pieces. Whichever type of seed applicator is used, it is important that they are positioned correctly and level and that the correct settings are made appropriate to the seed and product to ensure an even coating on the top surface of the buns.

Finger rolls can be partially slit down the centre by a water splitter. The pressure of the water in the jets is adjusted to alter the depth of each cut or split.

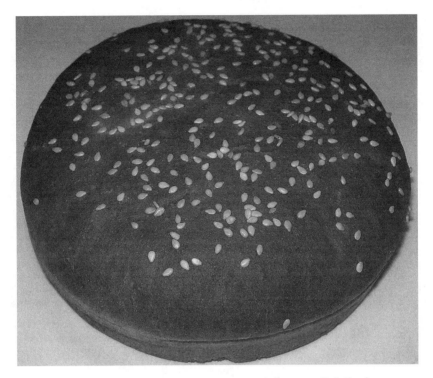

FIGURE 9.2. Hamburger bun coated with seasame seeds (courtesy *BakeTran*).

## 9.2.12. Baking

The pans of fully proved, seeded buns are then transferred onto an oven track, and the spacing or distance between each pan is again metered to ensure the correct spacing for the bends in the oven. The metered pans then enter into a conveyorized oven. Within a gas oven the pans pass over adjustable ribbon burners. The various layers of track are divided into six controlled temperature zones. In each zone the temperature is adjustable. In addition to the ribbon burners, fans are used to increase hot air circulation to ensure an even heat distribution. Typical baking temperatures for hamburger buns are of the order of 220 °C (428 °F) for 7.5–9.5 min depending on the weight of the product on the line (Figure 9.3). Further details on conveyorized proofing and baking systems are given by Wells (1983) and Grogan (1980).

## 9.2.13. Depanning and Cooling

After baking, the track passes under a depanner unit. Compressed air is used to loosen the buns from the tray, and a vacuum unit removes some of the surplus seeds. A belt which is as wide as the tray of buns, made up of special soft

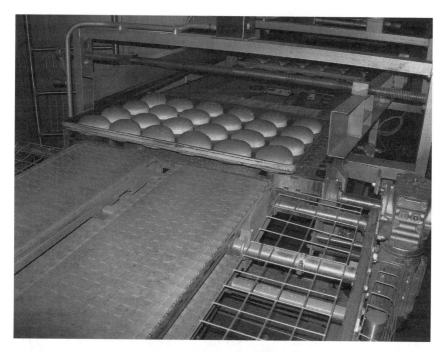

FIGURE 9.3. Hamburger buns leaving the oven (courtesy *BakeTran*).

depanning cups, is employed to remove the buns from the tray. Through each cup an adjustable vacuum is employed. The cups gently suck the buns off the tray and place them onto a cooler. Typically, a stainless steel spiral cooling conveyor is used. The cooling of buns takes in the order of 20 min, during which they are cooled to 35–38 °C (95–100 °F) before slicing.

From the cooler the buns are aligned into six lanes, and bunches of five rows are then sliced and packed into a pillow pack possibly with a central seal to give $2 \times 15$ buns in each pillow pack. The date-coded pillow packs are then passed through a metal detector and placed on cleaned trays which may be automatically stacked, ready for dispatch from the production area. For most products, a double layer of buns in pillow packs are added to each tray. The bottom layer should have the same characteristics as the top layer. The bottom layer should not show any signs of compression (typically known as 'crow's feet'). Further details of slicing and packaging are given by LeCrone (1994).

The baking pans are then conveyed back to the Pan-O-Mat via an in-line tray cleaning system. As an aid to keeping these high-volume plants clean, the baking pans are not greased at all but are treated with a non-stick material. Various non-stick pan coatings are available. Typical materials are silicone glazes, rubberized silicone and fluoropolymers. Schneeman (1995) discusses the advantages and disadvantages of each material, while Moss (1987) reported on a survey of pan coatings and baking surfaces.

## 9.3.   Part-Baked Breads

### 9.3.1.   Introduction

'Part-baking' is a term used to describe a method of bread manufacture which involves two stages of baking. Other terms include partially baked, par-bake, brown-and-serve and two-stage baking. During the first stage, dough pieces are baked just to the point when they are 'set', traditionally without colour formation of the crust (Figure 9.4). After cooling, packaging and distribution, the uncoloured partially baked dough pieces are given a second bake (Figure 9.5). During the second bake, a crust is formed and coloured. Additionally, the characteristic aroma of baked bread is produced. The second bake can be carried out close to the point of sale or consumption. Products which are usually part-baked are yeasted buns and rolls and laminated Danish pastries and croissants. Kamel and Stauffer (1993) state that 'part-baking is most successful with products which have a large surface area in relation to crumb area, such as baguettes, where the heat can penetrate quickly to the crumb centre to set it without setting the crust'. Valentino (1994) reports the suitability of part-baked pizza crust. Pyler (1988) describes how General Mills, Inc., introduced a new baking process called 'Brown-N'-Serve' in 1949. Turner (1970) describes the high-volume production of partially baked rolls in the late 1940s.

FIGURE 9.4. Part-baked rolls after the first bake (courtesy *BakeTran*).

FIGURE 9.5. Part-baked rolls after the second bake (courtesy *BakeTran*).

## 9.3.2.  Use of Part-Baked Bread

The manufacture of freshly baked crusty bread from a traditional single baking process produces a very desirable product. The facilities required to carry this out are, in outline:

- delivery of raw material – access and storage;
- manufacturing floor space;
- breadmaking equipment – weighing, mixing, scaling, moulding, pans, proving, baking and cooling;
- a trained workforce.

Establishing all these facilities near to the point of sale or consumption is not always feasible. Part-baked bread, manufactured in a bakery sited near an industrial area and distributed close to the point of sale, can be an attainable way of producing a regular supply of hot crusty bread. The minimum equipment requirements to produce coloured crusty loaves and the aroma of fresh bread are a small bake-off oven, often sited behind the counter and in view of the consumer. With minimum effort, new batches of hot bread can be produced every 20 min throughout most of the day.

The convenience of part-baked breads allows small retail shops to offer hot bread for sale, for example butchers and even mini shopping stores attached to garages. Brittain (1996) states that 'Different outlets use bake-off in different ways. Its rise in non-bakery related outlets such as garage forecourts and butchers continues unabated.' Baking off part-baked bread can be a solution for catering establishments wanting to offer customers crusty hot bread and rolls, whatever the demand, throughout their hours of opening. Some hot bread is also enjoyed in the home. Supermarkets and food shops offer for sale part-baked breads for the consumer to purchase and finish in domestic ovens.

## 9.3.3.   Manufacture of Part-Baked Breads

Similar plant operations are required to make both 'traditional' bread and part-baked bread. Some modifications to the formula to increase the stability or rigidness of the part-baked dough pieces are recommended. For part-baked French bread Bonnardel et al. (1990) suggested a slight reduction in the amount of water added at the dough stage. A slightly tighter dough is recommended to encourage the product to maintain a circular cross section as opposed to one which is oval in appearance. Stear (1990, p. 707) states that dough water absorption must be reduced to give a stiff dough which will remain rigid on removal from the oven. Additionally, Stear suggests a moderate reduction in yeast and yeast food to avoid excessive oven spring. These recommendations are also given by Pyler (1988). Both Pyler and Stear state that straight doughs require higher mixing temperatures, in the range 32.2–35 °C (90–95 °F). Presumably, in order to facilitate plant handling the reduced dough water absorption suggested must offset the dough softening which is associated with such high ex-mixer temperatures. Stoecklein (1995) gives an up-to-date definition of part-baked as 'baking beyond the point of starch gelatinization to some 90 to 95% completion in terms of desired crust colour and crispness'. Stoecklein states that absorption, as a general rule, will be slightly higher in part-baked products compared to frozen doughs (except in traditional hearth-baked items). The definition given by Stoecklein would appear to be similar to three-quarters-bake, part-baked bread which is described later. The bread improver employed could be one with a higher level of ingredients which will offer greater dough piece stability, for example DATA ester.

## 9.3.4.   Baking Nets

Bonnardel et al. (1990) state that the use of baking nets with deep, flexible, rounded corrugations, in which the dough pieces can sit, are found to contribute towards maintaining the shape of a dough piece, as opposed to inflexible baking pans with shallow ridges. Baking nets can be used to support part-baked French breads during cooling and freezing.

## 9.3.5.   Proving and Baking

Kamel and Stauffer (1993) describe two baking methods to achieve minimum crust formation and colour. First, a low temperature for the normal baking time may be used. This requires a reduction in proof time (or a reduction of yeast level) to take account of the resultant increase in loaf volume caused by the delay in the dough piece 'setting'. Second, a high temperature bake may be used to set the dough structure quickly, followed by removal from the oven before the crust has set or coloured.

For French breads a shorter final proof time is suggested (Bonnardel et al., 1990). A less aerated, lower volume dough piece is found to favour the maintenance of the part-baked bread's structure. Longer final proof times produce part-baked breads with higher volumes and lower stability. It is recommended that the top surface cutting of dough pieces made from tighter doughs, with less final proof time, should be a little deeper. The objective of the first bake is to obtain a sufficiently rigid semi-finished product which has an uncoloured surface. Two types of baking are given: (1) high temperature and short time and (2) low temperature and long time. The ideal first bake for French breads is a two-stage process. For example, in a rack oven, during the first stage, dough pieces are placed into a steam-filled oven at high temperature for a short time, typically 3 min at 270 °C (518 °F) to open up the cuts. The second phase is a lower temperature bake for a longer time, typically 7 min at 250 °C (482 °F) to ensure that the centre of the dough pieces have been transformed from dough into 'set' bread. The temperature profile of 170 g half-baguettes was followed during part-baking or just to the beginning of crust formation. They found that just after 8 min of baking at 235 °C (455 °F) the centre temperature had reached approximately 87 °C (188.6 °F), whilst the temperature on the edge of the dough piece had nearly reached 100 °C (212 °F). Pyler (1988) states that in order to avoid shrinkage and collapse on cooling the internal temperature ex-oven must exceed 76.7 °C (170 °F) and should preferably reach 82.2 °C (180 °F). To achieve this, products must be baked at between 121.1 and 148.9 °C (250–300 °F) for as long as possible without the formation of crust colour. Turner (1970) stated that 90.6 °C (195 °F) was the proper internal temperature of part-baked rolls. For maximum rigidity and volume, without crust formation, a 10–15 min bake at a steady 140 °C (284 °F) to a core temperature of 82 °C (179.6 °F) is given by Stear (1990). Fetty (1966) describes the use of microwave energy in the production of part-baked products. A reduction of final proof time to 30 min is required. The optimum internal temperature of microwaved part-baked breads required to set up the cellular structure was 88 °C (190 °F).

The average of the core temperatures above stated is 84 °C (183.3 °F). The exact temperature of an oven for baking off part-baked bread will vary with each type of oven and the size of each batch. Oven conditions during the first bake will vary additionally with the size and shape of the dough pieces. A pragmatic approach to determining the optimum conditions during the first baking period is to adjust the baking time and temperatures until the disappearance of an unbaked ring in the centre of the dough piece, without the formation of crust and colour.

## 9.3.6.  Depanning and Cooling

Part-baked bread contains a much higher moisture level than fully baked bread. Bonnardel et al. (1990) report weight losses for part-baked bread after the first baking and cooling of only 13.5% compared to up to 22.5% for fully baked

French bread. The higher moisture content of part-baked products makes them susceptible to mould growth (Kamel and Stauffer, 1993). Strict plant hygiene procedures should be implemented to reduce mould contamination. Turner (1970) states that it is possible to minimize mould problems by using 0.25% (based on flour weight) mould inhibitor in the dough. Additionally, the use of an oil spray containing 1.3% calcium propionate is described. Vacuum cups on depanning units are recommended to be cleaned several times per day with a good germicide solution. Hickey (1980) suggests spraying a 1–1.5% potassium sorbate solution on hot part-baked products as a method of doubling or tripling mould-free shelf-life.

## 9.3.7.   Storage of Part-Baked Breads

The shelf-life of part-baked bread without any further processing is limited. With storage at room temperature, surface drying, rapid staling and an increased risk of surface mould development limit the shelf-life to 24–48 h (Bonnardel et al., 1990), depending on the storage temperature.

Modified atmospheric packaging (MAP) or gas flushing considerably extend the mould-free shelf-life of part-baked bread stored at room temperature (Figure 9.6). This method is commonly employed in the retail sector. MAP involves sealing the part-baked product in a moisture-proof, airtight packaging material. Before the package is sealed, the air (and thus oxygen) is replaced or flushed out with an atmosphere of carbon dioxide and nitrogen. The resultant anaerobic sealed environment cannot sustain aerobic metabolism of microorganisms. During storage at ambient temperature an increase in crumb firmness or staling occurs. The second bake is used to reverse the staling process and the crumb regains the softness associated with fresh bread. However, it should be noted that after re-heating the rate of staling increases (Schoch and French, 1947) so that crumb firming takes place in a matter of a few hours after the second bake. Pool (1991) gives a shelf-life expectation for brown-and-serve rolls packed in MAP of up to 3 months and for croissants of 15–25 days.

FIGURE 9.6. Part-baked products in gas flushed-pack (courtesy *BakeTran*).

## 9.3.8.  Freezing of Part-Baked Bread

Freezing of part-baked bread stops the process of staling and the development of mould. Additionally, freezing hardens the product, giving stability to an otherwise fragile product. Freezing can be carried out in a blast freezer operating at $-40 \pm 5\,°C$ ($-40 \pm 9\,°F$). 'Shock freezing' can be used in which a liquid carbon dioxide or nitrogen spray is used to freeze rapidly a thin layer of bread. This forms a barrier against moisture migration and increases rigidity before slower freezing down to $-20\,°C$ ($-4\,°F$) (Bonnardel et al., 1990). After the freezing process the part-baked products are packaged in cardboard boxes with an inner sealable polythene liner. Frozen storage follows at $-18\,°C$ ($-0.4\,°F$). No moisture loss was reported in part-baked bread stored under these conditions for 7 weeks.

## 9.3.9.  Second and Final Baking

Final baking brings about the following four changes:

- a dry crisp crust is formed and coloured;
- the aroma of baked bread is produced;
- the crumb of the product is refreshed reversing the process of staling (Kamel and Stauffer, 1993);
- the volume of French bread is reduced by 13–15%.

Bonnardel et al. (1990) first reported loaf shrinkage during final baking after their investigation into part-baking of French bread. A final bake-off temperature of $210\,°C$ ($410\,°F$) for 10 min is given for a frozen part-baked half-baguette. The recommended baking temperature for part-baked bread in the UK is almost universally given as $220\,°C$ ($428\,°F$), and variations are given in time depending on size and whether the product is fresh or frozen.

## 9.3.10.  Quality of the Final Product

When correctly baked, quite acceptable bread and rolls can be produced after the second bake, and products are enjoyable when eaten before they firm up after a few hours. With appropriate training of the person carrying out the second bake, it is possible to maintain an evenly coloured crusty product. Not following closely the recommended baking procedure can lead to pale-looking unappetizing bread and rolls.

In addition to the reduction in volume of French bread reported above, other quality factors are altered after the second bake. Kamel and Stauffer (1993) state that it is widely recognized that the crumb of re-baked part-baked bread stales quickly. The refreshed crumb stales at a higher rate than the crumb of a freshly baked product. They recommend that part-baked bread should be consumed fairly quickly after re-baking.

Stoecklein (1995) states that 'the overall quality of par-baked products is not necessarily superior to that of scratch products, or of products made from frozen dough. Because a reheating stage is required for par-baked products, the most significant quality problem facing par-baked goods is their moisture loss and its impact on the appearance and eating quality of the finished product. Its shelf-life is very limited, and its consumption should take place shortly after reheating.'

The final bake of part-baked bread is out of the control of the bakery that produced it. Concern over wide variations in the interpretation of when the second bake is complete has led to the introduction of what is known in the UK as three-quarters-baked products. The baking of three-quarters-baked products is greater than conventional part-baking and is sufficient to give some crust colour. Three-quarters-baking reduces the possibility of pale-looking under-baked bread being produced after the second bake. The final bake of three-quarters-baked items is really a reheating process.

Stoecklein (1995) states that the final preparatory stage prior to product use is more in the nature of reheating (of products part-baked to 90% completion) at temperatures of 204 °C (400 °F) for 10 min, for example, depending on the type of product and the consumer's preference.

## 9.3.11.  The Milton Keynes Process

A variation on the theme of part-baking was the development of a process which evolved after a request from a UK supermarket to see 'if it was possible to provide ambient preformed bread with qualities that would be enjoyed by the customer'. A partnership of four companies evolved: the supermarket, a plant baker, a machinery manufacturer and a bread improver and yeast manufacturing company. The process was described as being composed of 'the uniting of several elements' (Kear, 1995). The main difference between this process and traditional part-baking is the claim that the ambient shelf-life extension was, before the final baking, from 5 to 12 days (Anon., 1995; Kear, 1995; Grindley, 1996; Macdonald, 1996; Milton Keynes Process).

Prior to dough mixing, part of the flour could be fermented in either a liquid brew or as a plastic sponge. Kear (1995) stated that 'Tweedy high speed pressure/vacuum mixers are used to mix pan breads while French sticks and morning goods are produced on spiral mixers.'

The process involved subjecting dough pieces to baking conditions of low temperature and a long time duration, so arranged that the crumb of each piece was fully baked, but crust formation was incomplete. Steam was being applied to maintain humidity and prevent dehydration of the crust (Anon., 1995). Kear (1995) described this as 'a crumb conditioning stage'. During this stage, crust caramelization was minimized as the dough piece was fully formed.

Immediately after the first baking or period of heat transfer, the pale/white loaves, whilst still in the pans (in the case of pan bread), were cooled under vacuum. Kear (1995) described this process thus, 'the bread then goes to a stabiliser where the negative pressure stabilizes the product against collapse when cooled'. The principle of vacuum cooling of baked products involves the reduction of pressure or the evacuation of the atmosphere which contains the hot baked product, down to a relatively high vacuum. As a result of this reduced or sub-atmospheric pressure, the temperature at which water is converted into steam is greatly reduced. The heat of the product provides energy to convert water into steam; as this happens the temperature of the baked product is rapidly reduced. Cooling rates are rapid and an 800-g loaf can reach a core temperature of approximately 30 °C (84 °F) in a time of 4–5 min.

After vacuum cooling and depanning, the surface of the part-finished product could be sprayed with a preservative to suppress mould growth. The patent which covered the 'Manufacture of baked farinaceous food stuffs' describes how this preservative could be removed by volatilization in the final baking step in the process (Anon., 1995).

It is possible that the rapid cooling associated with vacuum cooling has a beneficial effect on starch crystallization, in terms of altering the rate or progress of starch retrogradation (APV, personal communication, 1996). A combination of the use of enzymes that reduce the rate of firmness of the crumb of bread and vacuum cooling could possibly contribute to the extension of shelf-life at ambient temperature, from the staling point of view, and the preservative spray would increase the mould-free shelf-life.

Kear (1995) stated that the preformed product could be packed and sent out, under ambient conditions, to stores where baking takes place in an oven and the crust was caramelized. The finished loaf, according to Kear, was of superior quality to part-baked breads, and was slow to stale.

Newbery (1996) described, in an article on 'vacuum cooling', an in-store baking breakthrough which fulfils the demand for bakery foods that can be baked within minutes at the point of sale with simple equipment. A revolutionary new technology allowed a wide variety of high-quality breads and rolls to be produced without in-store batching, mixing or make-up. No freezer storage, thawing or proving was required. After 10 min in the in-store oven, the foods, it was claimed, retained their freshness for a normal 3–4 days.

It has been reported (Anon, 1996) that 'the equipment manufacturer and the supermarket have developed a small forced convection oven specially for baking Milton Keynes products in-store. The system provided consistent final bake-off conditions. Simple colour-coding allowed the in-store oven to be operated at the predetermined temperatures and time settings.'

Following its initial introduction the Milton Keynes Process created considerable interest in the UK and many other parts of the world. However, its life as a 'new' process was short-lived and as originally conceived it is not longer in use. There were a number of reasons for its demise but the main reason was that the final product quality did not match consumer expectations. In particular the moisture content of the products after the second bake was lower than a scratch-baked product which gave a drier mouth feel. The second bake also accelerated staling and this meant that any product not consumed within a couple of hours after second bake became inedible. Such quality changes were in stark contrast to scratch-baked products normally enjoyed by the customers of retail in-store bakeries.

## 9.4.    Yeasted Laminated Products

### 9.4.1.    Introduction

The main yeasted laminated products are croissants and Danish pastries (Figure 9.7). Both would appear to have their origin in the seventeenth century. Rijkaart (1984) describes how in 1615 the law in Vienna only allowed bakers of bread to make products containing yeast. Bread bakers made Danish pastry, which is described as puff pastry with added yeast. In Denmark, Danish pastries are called 'Wienerbrot' or the 'pastry from Vienna'. Asklund (1965) gives a further account of the history and evolution of Danish pastries.

Montagné (1961) states the origin of the croissant was in Budapest and dates from 1686. Bakers working during the night gave alarm when they heard the noise of what were to be Turkish invaders who had dug underground passages to reach the centre of the town. The bakers who had saved the city were awarded or granted the privilege of making a special pastry. The pastry had to take the form of a crescent, in the shape of the emblem on the Ottoman flag.

### 9.4.2.    Formulations

Published formulations of croissants and Danish pastries indicate that both usually contain nine major ingredients: flour, salt, water, yeast, shortening, sugar,

FIGURE 9.7. Pan-au-chocolat, mini croissant and pecan Danish (courtesy *BakeTran*).

egg, milk solids and laminating margarine or butter (note that egg may be one exception, which may not be added to some croissant formulae). Cauvain and Telloke (1993) surveyed 36 Danish pastry and 14 croissant formulations and concluded that the two products have similar 'average' essentials, although, they say, the proportion of individual ingredients can vary substantially.

Table 9.4 shows a comparison of the mean or average proportion of ingredients from the survey carried out by Cauvain and Telloke (1993).

Examination of Table 9.4 indicates some general trends. Danish pastries contain lower proportions of salt and higher proportions of yeast, sugar, egg and laminating margarine/butter. The same general trend is found in Table 9.5 which compares croissant and Danish pastry formulations in standard American and UK textbooks.

## 9.4.3. Ingredients

Fully automatic make-up plants are available that can produce 12,000 curved croissants and 14,000 straight croissants/h (Sasib, 1996). Rijkaart (1984) describes croissant machines producing 800–24,000 croissants/h (followed by manual bending and placing the croissant on trays with the slot down). For such plants it is very important to have a knowledge of and control over the raw materials.

## 9.4.4. Flour Protein

Rowe (1985) states that the best results are consistently obtained from croissants manufactured with a flour with good protein at a level of about 11.5–12.5%. He concludes that most bread flours can be used successfully by making proper adjustments. Rowe found that substituting 30% of the bread flour with lower protein cake flour produced croissants with lower volume. The interior was finer and had a more bread-like grain, and the layering was not as well defined.

TABLE 9.4. The mean or average proportions of ingredients from a survey of 14 croissant and 36 Danish formulations (Cauvain and Telloke, 1993).

| Flour (%) | Croissant | Danish pastry |
|---|---|---|
| Flour | 100 | 100 |
| Salt | 1.8 | 1.3 |
| Water | 52.2 | 43.6 |
| Yeast (compressed) | 5.5 | 7.6 |
| Shortening | 9.7 | 9.6 |
| Sugar | 6.1 | 9.2 |
| Egg | 2.6 | 12.4 |
| Skimmed milk powder | 6.5 | 5.4 |
| Laminating margarine/butter | 50–57 | 62–64 |

TABLE 9.5. Comparison of an English and an American formulation for Croissants and Danish pastries.

| % flour | Croissant | | Danish pastry | |
| --- | --- | --- | --- | --- |
| | Sultan[a] | Brown[b] | Sultan[a] | Brown[b] |
| Flour | 100 | 100 | 100 | 100 |
| Salt | 2 | 1.8 | 1.56 | 1.1 |
| Water | 52 | 55.4 | 50 | 52 |
| Yeast (compressed) | 5 | 4 | 6.25 | 6 |
| Shortening | 8 | 2 | 12.5 | 6.3 |
| Sugar | 10 | 2 | 25 | 9.4 |
| Egg | 24 | nil | 25 | 5 |
| Skimmed milk powder | 5 | 3 | 6.25 | 4 |
| Margarine/butter for lamination | 32 | 45 | 50 | 50 |

[a] Sultan (1989), USA.
[b] Brown (1985), UK.

(Many references discuss the addition of lower protein flour for Danish pastry; this may be due to a requirement to dilute or adjust the protein of very strong North American flours). Rijkaart (1984) states that the optimum type of flour (containing potassium bromate) has approximately 12–13.5% protein. Cauvain and Telloke (1993) examined the effect of gluten fortification of flour and found that weak low-protein flours are unsuitable for Danish pastry and croissant production. Weak flour doughs were rather extensible and proved to a lower height and spread more during proof, thereby producing flatter pastries. Stronger flours had a greater ability to withstand the effect of paste lamination and the adverse effects of yeast. They concluded that medium to strong flour, or weaker flours with added dry vital wheat gluten, can be used to make Danish pastry and croissant with acceptable product qualities. Telloke (1991a) concluded that puff pastry doughs made with weaker flours pass through the optimum elasticity ranges faster and require shorter resting periods than those made with stronger flour. Doughs made from stronger flours need longer resting periods to avoid excessive pastry shrinkage and distortion, unless their dough has received intensive mixing.

It would appear that flour protein levels for laminated products will be plant specific. Higher protein flours might be more suited to make-up plants with appropriate methods of relaxation employed.

## 9.4.5.  Fat Addition to the Base Dough

Cauvain and Telloke (1993) reported the effects of dough fat addition. Based on flour weight, up to 15% of butterfat or plain shortening or 16.6% laminating margarine were mixed into the base dough. Mixing 3–6% fat into the dough was found to improve the paste machinability by increasing the elastic properties of dough slightly, which in turn increased the paste recovery after sheeting. Higher fat levels caused the croissants to be flatter and the crumb structure to become more closer grained and 'bun-like'. The only difference found between the fats was that the butterfat imparted a yellow colour to the crumb and a pleasant butter flavour to the pastry, especially at higher levels. Rowe (1985) states that increasing the amount of margarine or butter will result in a softer croissant. Smith (1990) states that if it is desirable to add quantities of fat higher than 25% to a sweet dough, then it must be rolled and folded into the finished dough rather than adding it with other ingredients at the mixing stage. This they state is the main difference between regular sweet dough and Danish pastry.

Telloke (1991a) found that certain emulsifiers could be used to reduce the fat level in puff pastry. He states that the addition of diacetyl tartaric acid esters (DATA esters), sodium stearoyl lactylate (SSL) or glycerol monostearate (GMS) to the base dough, or a blend of GMS, lecithin and sorbitan monostearate to the laminating fat, increased the tender eating quality of puff pastry.

### 9.4.6.  Sugar Levels

Cauvain and Telloke (1993) found that the addition of sugar at levels between 5 and 12% (flour weight) gave optimum pastry quality. Higher levels inhibited yeast activity and reduced pastry lift. Danish pastries with 18% sugar tended to distort in shape, and croissants tended to uncurl and show signs of internal collapse.

### 9.4.7.  Yeast Levels

Optimum pastry quality was produced with a yeast level of 7.5% (flour weight) with a dough temperature of 19–20 °C (66–68 °F) and a proof time of 60 min at 31 °C (86 °F). Increasing yeast levels (and proof time) caused croissants to assume a flatter external appearance with a more open crumb structure. Yeast contributes to a pleasant flavour and gives a tender eating character to the pastry (Cauvain and Telloke, 1993). Rowe (1985) gives three typical croissant formulae with 7% yeast addition in each.

Poehlman (1979) states that much Danish pastry is made using active dried yeast. He states that reconstituted active dried yeast seems to have a beneficial reaction within the Danish dough system. In the discussion of the paper he stated 'you have a larger percentage of inactive dry yeast in it (dried yeast) due to the drying process which seems to have a beneficial effect on machinability'. This was assumed to be due to the presence of glutathione (a reducing agent) in yeast which gives more dough extensibility.

### 9.4.8.  Laminating or Roll-In Fat

Vey (1986) suggests a classification of 'enrichment' of Danish pastry based on the level of roll-in or laminating fat. To every 100 parts by weight of base dough the following parts would be added for:

| Lean Danish | 12.5–17 | (2–2.75 oz/lb of base dough) |
| Medium Danish | 18–25 | (3–4.5 oz/lb of base dough) |
| Rich Danish | 26–35 | (4.75–5.5 oz/lb of base dough) |

Smith (1990) gives the following guide based on flour as 100%:

| Lean Danish | 30 |
| Medium-rich Danish | 40 |
| Rich Danish | 50 |
| Very rich Danish | 70 |

Rowe (1985) states that the selection of proper roll-in fat is very important, taste and processing being the two important aspects in croissant production. The storage or conditioning of the fat at the correct temperature is important. If the roll-in fat is too warm and soft, it will not spread evenly and will tend to be absorbed in the dough. The resultant croissants have reduced volume and a poorly defined interior crumb and layering. Roll-in fat which is too hard will not spread evenly and results in variable shape, size and layering. The roll-in fat should be stored in a cool place. Consistent results are obtained when the temperature of the roll-in fat before lamination is about 15.6 °C (60 °F) with a dough temperature of 12.6–15.6 °C (55–60 °F). Rijkaart (1984) states that the butter temperature should be no more than 14 °C (57.2 °F) and shortening up to 18 °C (64.4 °F) with a dough temperature from a high-speed mixer of 24–25 °C (75.2–77 °F). Vey (1986) states that regular bakers' or pastry margarine containing between 16 and 19% water has become the most popular roll-in shortening for Danish pastry production, produced by hand or semi-automatic systems. Problems have developed with these shortenings containing moderate amounts of water, with newer extruding and laminating systems. Anhydrous margarine containing minimal water and the same plastic range as regular bakers' or pastry margarine are favoured. The improved functionality of the anhydrous margarine allows usage levels of up to 25% less than regular bakers' margarine. Vey (1986) reported that roll-in shortening for Danish pastry must have a putty-like consistency.

Telloke (1991b) states that in order to avoid irregular layer formation in puff pastry and variation in pastry lift and shrinkage, the roll-in or laminating fat or margarine should be plasticized prior to lamination of the paste. The best paste and pastry qualities were obtained with uniform fat crystals no greater than about 5 $\mu$m in diameter within an optimum solid fat index (SFI) of between 38 and 45%. Cauvain and Telloke (1993) state that for croissant and Danish pastries, the firmer the laminating fat the more resistant the paste to deformation after sheeting. If the solids content and the firmness of the laminating fat are too high, pastry expansion is restricted during proof and the final specific volume is low. They found that a fractionated butter was soft, greasy and difficult to machine at 20 °C (68 °F). Fat layers in the paste were partially destroyed during lamination and croissant specific volume was low. When used

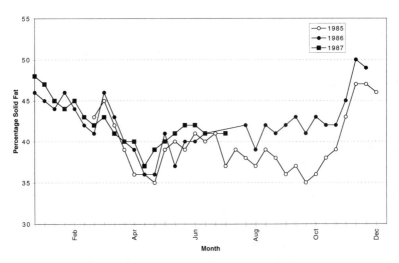

FIGURE 9.8. Seasonal variation in percentage of solid fat at 10 °C (50 °F) for UK-produced butteroil (courtesy D. Lawrence, Dairy Ingredients (UK) Ltd, Slough, UK).

at 10 °C (50 °F) the butterfat gave a firmer paste, layering remained intact and pastry-specific volume remained high. Their investigation suggested that for fermented laminated products it is more important that the roll-in or pastry margarine is reasonably firm and plastic during lamination but relatively soft during proof to assist pastry expansion.

Smith (1990) states that butter is considered to be the best of all shortenings from the flavour standpoint but, due to its cost, low melting point and short plastic range, it finds limited use as the total roll-in fat. Blends of butter with high melting point fats make very good Danish pastries. Asklund (1965) states that for Danish pastry 'summer butter, also called grass butter, is not suitable for rolling'.

The amount of solid fat in butter varies throughout the year and even during some months, as feeding regimes change. Figure 9.8 is a plot of the percentage solid fat at 10 °C (50 °F) of butter throughout the year, for the period 1985–87. It shows the considerable variation in the percentage solid fat between summer and winter months. Typically during summer months, butter has a lower content of solid fat at 10 °C, that is between 35 and 42%, whereas in the winter months the percentage of solid fat is higher at 42–50%. Figure 9.9 shows the percentage of solid fat determined by NMR on samples of winter and summer butteroil at temperatures between 0 and 35 °C. A similar trend of higher content of solid fat is found in winter butter throughout the temperature range. The important practical interpretation of these graphs is that winter and summer butter should be stored, conditioned and used at different temperatures.

Kazier and Dyer (1995) describe an investigation into 'reduced-fat pastry margarine for laminated dough in puff, Danish and croissant application'. Attempts at fat reduction by lowering the amount of laminate were not feasible.

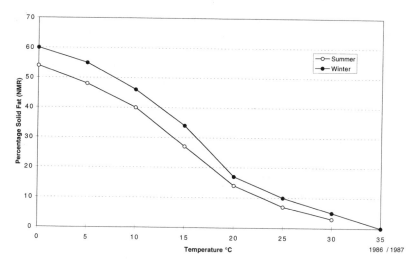

FIGURE 9.9. Variation in percentage of solid fat with temperature for UK-produced butteroil (courtesy D. Lawrence, Dairy Ingredients (UK) Ltd, Slough, UK).

However, by reducing the fat content of the laminate roll-in margarine, the system developed allowed a 1:1 replacement of the control roll-in shortening. Reduced-fat versions of croissants and Danish had slightly higher post-bake moisture and slightly larger volumes. The finished products using the reduced-fat system appeared to be softer and more tender compared to their control.

## 9.4.9. Dough Mixing

The basic methods available for mixing doughs for fermented laminated products are (1) low-speed straight dough, (2) high-speed straight dough and (3) the blitz system or Scotch or Dutch method. In any of these three systems, proportions of the flour can be made into a sponge and fermented prior to final dough mixing. The degree of dough mixing and development will be plant- and process-specific and will depend on the characteristics required in the final product.

Rijkaart (1984) found from experience that vertical high-speed mixers with mobile bowls were ideal for croissant dough. A 1-min mix without too much gluten development to the dough is described, further development being carried out during the lamination process. A mixer with a capacity of 136 kg (300 lb) of dough gives 2 t of finished product per hour. The high-speed mixer is described with a 30-kW (40 hp) motor which drives the mixing arms at speeds up to 380 rpm. High-speed doughs after lamination are stated as being less prone to shrinkage.

Vey (1986) described a conventional, retail Danish pastry dough after mixing as being very slack and underdeveloped. In a second method, the 'blitz system' (Scotch method in the UK, Dutch in some parts of Europe) roll-in shortening is combined in the mixer with semi-mixed dough. The roll-in shortening is mixed with the dough to give a random dispersion.

Rowe (1985) states that the low-speed mixing time for croissant doughs is generally 5–15 min and depends on the characteristics desired in the croissant.

Pyler (1988) describes how, when doughs are to be extruded, some adjustments to doughs are required to counteract the adverse effects of the greater abuse that is inherent in the process. These include making a sponge of up to 75% of the flour, increasing absorption by 1–3%, extending dough fermentation time by 30 min, maintaining a yeast level of 5% in the sponge and an additional 3% at the dough stage.

Poehlman (1979) states that there is controversy about the amount of mixing and development and the desirable dough temperature of Danish pastry. Some advocate long mixing and warm doughs, while others advocate short mixing and cold doughs. He says both are right, depending on the product line. For a mixed production (in terms of size, 56.5 g (2 oz) and large 340–453 g (12–16 oz) Danish coffee cake), he believes that a short, cool mixing provides the best overall performance. For best performance at make-up, the proper temperature for Danish dough is 12.6–15.6 °C (55–60 °F).

Asklund (1965) emphasizes that, in the production of traditional Danish, the dough should literally be mixed and not kneaded.

## 9.4.10. Methods of Adding the Roll-In or Laminating Fat to the Dough

Five methods of adding the roll-in or laminating fat to the dough are described as:

- French or English;
- Blitz, Scotch or Dutch method;
- Co-extrusion;
- Sandwich; and
- Curling.

With the French or English method, roll-in fat is placed on part of the sheeted-out dough. The remaining dough is then folded over the roll-in fat to give, in the French method, one layer of fat surrounded by dough. The English method, after the first fold, gives two layers of fat and three layers of paste.

In the blitz or Scotch method, previously described, optimum pastry quality is influenced by the amount of mixing of the complete paste. Pastries produced by this method after lamination have lower lift than those made by the English or French method (Telloke, 1991a).

Co-extrusion described by Rowe (1985) and Vey (1986) involves the extrusion of a cylinder of paste. A layer of roll-in fat is co-extruded on the inside of the cylinder of paste. The co-extruded cylinder is then passed through sheeting rollers to flatten the cylinder into a sheet of one layer of roll-in fat surrounded by dough.

With the sandwich system a continuous sheet of dough is extruded onto a moving conveyor. Roll-in fat is then extruded onto the base layer. A top layer of dough is then extruded onto the roll-in fat.

The curling process as described by Rijkaart (1984) employs just one layer of dough onto which a layer of roll-in fat is extruded. Diagonally across the belt on which this is travelling, a curling roller causes the paste to form into a cylinder The cylinder takes the form of a giant Swiss roll with six layers of roll-in fat and seven layers of dough. The number of layers can be altered by decreasing or increasing the thickness of the extruded base dough and roll-in fat. Increasing the depth of paste and roll-in fat gives fewer layers in the Swiss roll.

## 9.4.11.  Lamination

Cauvain and Telloke (1993) examined photomicrographs of pastes with and without yeast. They found that, after final sheeting, small bubbles were retained in the layers of the yeasted paste. During proof these bubbles expanded and the layered structure was largely destroyed. The bubbles reduced pastry lift by creating holes in the gluten sheets in the dough layers. Bakers try to avoid this destruction of dough layers by giving less lamination to Danish and croissant pastes, compared with puff pastry. Danish and croissant pastes are usually laminated to give between 20 and about 50 fat layers, while in puff pastry there are between 130 and 250. Cauvain and Telloke recommend, for Danish and croissant doughs, lamination to give between 18 and 32 fat layers.

Rijkaart (1984) states that 36 layers of butter is more than enough for croissants and that in Europe the number of layers varies from 16 to 48. Rowe (1985) showed that croissants with 16 layers have a larger volume, and are flakier with an open grain with a decided lacey effect and pronounced layers. Increasing the number of layers to 48 reduces flakiness and volume and the grain becomes more bread-like.

After lamination has been carried out, individual pastries are produced. Useful references on the numerous varieties of Danish pastries are given by Poehlman (1979) and Asklund (1965) and major books covering practical baking.

## 9.5.    Frozen and Fully Proved Frozen Laminated Products

As discussed in Chapter 6, laminated doughs may be readily frozen. They can be either in the unproved or proved state. If frozen unproved, then thawing and proof are required before bake-off. More convenient is the production of frozen fully proved laminated products since they can be taken from deep freeze storage and placed immediately in the oven for baking. The loss of potential gas production by the yeast which is present in the dough is not a major problem since the steam pressure which is generated during baking is sufficient to force the dough layers apart. Good quality croissant and Danish pastries can be produced this way provided that the product recipe and manufacturing processes have been carefully controlled in the stages before the products pass into the freezer.

## References

AMF (1995) *Bun and Roll Make-Up Systems* (leaflet).

Anon. (1995) *Manufacture of Baked Farinaceous Food Stuffs*. Patent WO 95/30333.

Anon. (1996) The Milton Keynes Process. *Food Trade Review*, **66**(8), 546.

APV (1996) Personal Communication.

Asklund, D.O.A. (1965) Danish pastry production. *Proceedings of the 21st Conference of The British Chapter*, 22–32.

Bonnardel, P., Maitre, H. and Poitrenaud, B. (1990) Part-baked technology of French bread. Notes from FMBRA conference, *Using Refrigeration in the Bakery*, December 6.

Bread and Flour Regulations (1996) HMSO, London, UK.

Brittain, J. (1996) Bake-off boom. *Bakers' Review*, September, 24.

Brown, J. (1985) Recipe and processing. *The Master Bakers Book of Breadmaking*, Turret-Wheatland Ltd, Herts, UK, Chapter 8 (2nd edition 1995).

Cauvain, S.P. and Telloke, G.W. (1993) Danish pastries and croissants. *FMBRA Report No. 153*, August.

Charbonneau, T., Gonzales, R. and Gorton, L.A. (1978) Liquid ferment system feeds double bun line at central bakery. Produces 6000 lbs/hr of 65% sponge. *Baking Industry*, April, 24–6.

Doerry, W.T. (1985) Liquid preferments: A study of factors affecting fermentation parameters and bread quality. *American Institute of Baking, Research Department Technical Bulletin*, VII, June, 1–9.

Dubois, D. (1984) Salt in the brew of a liquid ferment, effect on processing and quality of white pan bread. *American Institute of Baking, Research Department Technical Bulletin*, VI, March, 6.

Fetty, H. (1966) Microwave baking of partially baked products. *Proceedings of The American Society of Bakery Engineers*, 145–52.

Grindley, E. (1996) Made in Milton Keynes. *Bakers' Review*, January, 16–17.

Grogan, P.E. (1980) Conveyorized proofing and baking systems. *Proceedings of The American Society of Bakery Engineers*, March 4, 113–20.

Hickey, C.S. (1980) Sorbate spray application for protecting yeast-raised bakery products. *Bakers' Digest*, **54**(4), 20–2, 36.

Kamel, B.S. and Stauffer, C.E. (1993) *Advances in Baking Technology*, Blackie Academic & Professional, Glasgow.

Kazier, H. and Dyer, B. (1995) Reduced-fat pastry margarine for laminated dough in puff, Danish and croissant applications. *Cereal Foods World*, **40**(5), 363–5.

Kear, H. (1995) The Milton Keynes Process. *Proceedings of the 79th Conference of the British Society of Baking*, November.

Kulp, K. (1983) Technology of brew systems in bread manufacture. *Bakers' Digest*, November 8, 20–3.

LeCrone, D.J. (1994) Slicing and packaging 1,000 buns per minute. *Proceedings of the American Society of Bakery Engineers*, 145–51.

Macdonald, S. (1996) Milton Keynes patent in the public domain. *British Baker*, January 26.

Matz, S.A. (1992) *Baking Technology and Engineering*, 2nd edn, Van Nostrand Reinhold/AVI, New York, 603–7.

The Milton Keynes Process (Undated Leaflet) The Milton Keynes Process, Mandeville Drive, Kingston, Milton Keynes, MK10 0AP, UK.

Moss, T.M. (1987) A survey of pan coatings and baking surfaces. *Proceedings of the 64th Conference of the British Society of Baking*, March, 5–9.

Montagné, P. (1961) *Larousse Gastronomique, The Encyclopedia of Food, Wine and Cooking*, 1st edn, Hamlyn Publishing Group, London, UK.

Neufeld, K. (1991) Microwave baking. *Proceedings of the American Society of Bakery Engineers*, 77–84.

Newbery, D. (1996) Vacuum cooling. *Proceedings of the American Society of Bakery Engineers*, 81–6.

Poehlman, R.W. (1979) Premium Danish pastry. *Proceedings of the American Society of Bakery Engineers*, 91–105.

Pool, H. (1991) Modified atmosphere packaging. *Proceedings of the American Society of Bakery Engineers*, 159–65.

Pyler, E.J. (1988) *Baking Science and Technology*, Sosland Publishing, Kansas City, KS, USA.

Rijkaart, C. (1984) Croissant production. *Proceedings of the American Society of Bakery Engineers*, 137–44.

Rowe, C.S. (1985) Croissants. *Proceedings of the American Society of Bakery Engineers*, 154–64.

Sasib Bakery (undated, received 1996). Rijkaart croissant curling and bending machine (leaflet).

Schneeman, H.L. (1995) *Proceedings of the American Society of Bakery Engineers*, 65–7.

Schoch, T.J. and French, D. (1947) Studies on bread staling. 1.Role of starch. *Cereal Chemistry*, **24**, 231–49.

Smith, R.Z. (1990) Sweet dough and Danish. Baking Production and Technology, *Seminar, American Institute of Baking*, Honolulu, January 29–31.

Stear, C.A. (1990) *Handbook of Breadmaking Technology*, Elsevier Science Publishers, Amsterdam.

Stevens, E. (1991) 1,000-per-minute bun line. *Proceedings of the American Society of Bakery Engineers*, 125–37.

Steward, C.D. (1990) The latest in bun dividers. *Proceedings of the American Society of Bakery Engineers*, 100–10.

Stoecklein, R.C. (1995) Frozen par-baked products. *Proceedings of the American Society of Bakery Engineers*, 49–54.

Sultan, W.J. (1989) *Practical Baking*, 5th edn, Van Nostrand Reinhold, New York.

Sutherland, W.R. (1974) What is pH and TTA? *Proceedings of Southern Bakers' Conference*, Atlanta, Georgia, September 8.

Sutherland, W.R. (1976) pH and TTA in baking. *Proceedings of the American Society of Bakery Engineers*, 85–98.

Sutherland, R. (1989) Hydrogen ion concentration (pH) and total titratable acidity tests. *American Institute of Baking, Research Department Bulletin No. XI*, May, 5.

Thompson, D.R. (1983) Liquid sponge technology applied to high speed dough mixing. *Bakers' Digest*, November 8, 11–17.

Telloke, G.W. (1991a) Puff pastry I. Process and dough ingredient variables. *FMBRA Report No. 144*, January.

Telloke, G.W. (1991b) Puff pastry II. Fats, margarines and emulsifiers. *FMBRA Report No. 146*, August.

Turner, J.E. (1970) Partially baked foods. *Proceedings of the American Society of Bakery Engineers*, 74–83.

Valentino, F. (1994) Par-baked pizza. *Proceedings of the American Society of Bakery Engineers*, 153–8.

Vey, J.E. (1986) Danish. *Proceedings of the American Society of Bakery Engineers*, 111–20.

Watkins, F.H. (1991) High-flour liquid ferments. *Proceedings of the American Society of Bakery Engineers*, 168–78.

Wells, B. (1983) Proofing and baking systems. *Proceedings of the American Society of Bakery Engineers*, 119–25.

# 10
# Bread Spoilage and Staling

Irene M.C. Pateras

## 10.1. Introduction

Bread is the most important staple food in the Western world and it is recognized as a perishable commodity, which is at its best when consumed 'fresh'. Unfortunately, bread remains truly 'fresh' for only a few hours after it leaves the oven. During storage it is subjected to a number of changes which lead to the loss of its organoleptic freshness. The factors that govern the rate of freshness loss in bread during storage are mainly divided into two groups: those attributed to microbial attack and those that are result of a series of slow chemical or physical changes which lead to the progressive firming up of the crumb, commonly referred to as 'staling'.

## 10.2. Microbiological Spoilage of Bread

The most common source of microbial spoilage of bread is mould growth. Less common, but still causing problems in warm weather, is the bacterial spoilage condition known as 'rope' caused by growth of *Bacillus* species. Least common of all types of microbial spoilage in bread is that caused by certain types of yeast.

### 10.2.1. Mould Spoilage

Mould spoilage of bread is due to post-processing contamination. Bread loaves fresh out of the oven are free of moulds or mould spores due to their thermal inactivation during the baking process (Ponte and Tsen, 1978). Bread becomes contaminated after baking from the mould spores present in the atmosphere surrounding loaves during cooling, slicing, packaging and storage.

The environment inside a bakery is not sterile because dry ingredients, especially flour, contain mould spores, and flour dust spreads easily through the air. It has been estimated that 1 g of flour contains as many as 8000 mould spores. In some bakeries a similar number of spores settle on $1 \, m^2$ of surface every hour (Doerry, 1990). Production operations such as weighing and mixing of ingredients increase the mould count in the air. In larger bakeries where segregation is possible, the flour handling areas are separated from the cooling and packaging area of the finished bread.

Bread crust is rather dry (Cauvain and Young, 2000) and if the relative humidity of the atmosphere is below 90%, moulds will not grow on it. Also, moulds are relatively slow to develop, so that in dry climates the surface of a slice of bread may dry before mould growth is sufficient to be visible. In a humid atmosphere, however, and especially on a loaf inside a wrapper, moulds will grow rapidly. This is true especially if the bread is wrapped hot from the oven so that droplets of water condense on the inside surface of the wrapper. When bread is cut, the inner, more susceptible surfaces are exposed to mould infection. Sliced, wrapped bread is more at risk, because the moist, cut surfaces are an ideal substrate for moulds to grow on and the packaging prevents the moisture loss.

The rate of mould growth in various breads depends on the recipe and the processing method (Seiler, 1992). Brown and wholemeal (wholewheat) breads appear to become mouldy rather earlier than white breads because mould growth is often more clearly visible on the darker surfaces. Cultured breads, such as rye bread (Chapter 13), tend to have a slightly longer shelf-life because of their increased acidity and lower pH. The processing method has also been shown to have an effect on the rate of mould growth. For example, bread made from no-time dough, that is Chorleywood Breadmaking Process (CBP) and Activated dough development (ADD), has a slightly shorter shelf-life than bulk-fermented bread. This difference is considered to be largely due to the higher alcohol content in fermented breads.

The most common bread spoilage moulds are *Penicillium* spp., although *Aspergillus* spp. may be of greater significance in tropical countries (Legan, 1993). In wheat-breads a wide range of spoilage moulds including *Penicillium, Aspergillus, Cladosporium, Mucorales* and *Neurospora* have been observed (Table 10.1).

*Rhizopus (nigricans) stolonifer* is the common black bread mould. It has very fluffy appearance of white cottony mycelium and black sporangia. *Neurospora sitophila* is another type of mould which is reddish in colour, and is found in bread stored at a high humidity or wrapped while still warm. Storage temperature has an effect on the type of moulds growing in bread. *Aspergillus* spp. was reported to be the dominant mould spoilage of bread in India, while the 90% of moulds isolated from a range of bread in Northern Ireland were *Penicillium* spp. (Legan, 1993).

In addition to spoilage, some moulds present a severe risk to public health because they can produce mycotoxins. Exposure to mycotoxins can occur either directly by eating bread spoiled by mycotoxigenic moulds or indirectly as a result of people consuming the products of animals fed contaminated bread. Mycotoxins are very resistant and can survive the heating process designed to kill moulds. It has been reported that 10% of *Aspergillus* spp. and *Penicillium* spp. are toxic to mice (Silliker, 1980). As a significant amount of mould growth is needed to form mycotoxins in bread, the risk to public health from mycotoxins

TABLE 10.1. Characteristics of bread moulds.

| Mould | Colony colour | Colony appearance | Comments |
|---|---|---|---|
| *Penicillium* spp. | Blue/green | Flat, spreads rather slowly | The most common type of bread mould |
| *Aspergillus niger* | Black | Fluffy, spreading with spore heads often clearly visible | Frequently present |
| *Aspergillus flavus* | Olive green | | |
| *Aspergillus candidus* | Cream | | |
| *Aspergillus glaucus* | Pale green | | |
| *Cladosporium* spp. | Dark olive green | Flat, spreads slowly | Often present on damp bakery walls, commonly encountered |
| *Neurospora sitophila* | Salmon pink | Very fluffy and fast spreading | Will grow very rapidly on moist bread |
| *Rhizopus nigricans* | Grey/black | Very fluffy and fast spreading | Will grow very rapidly on moist bread |
| *Mucor* spp. | Grey | | |

Source: Seiler, 1992.

in developed countries is minimal (Legan, 1993). The consumer today tends to reject the whole loaf rather than cut away the visible mouldy portion and eat the remainder, as was common in the past. The indirect risk via animals fed on mouldy bread is also very low (Osborne, 1980).

## 10.2.2. Bacterial Spoilage

Rope is a spoilage problem of bread and other bakery products that have high equilibrium relative humidity (ERH), that is greater than 90%. It is caused by a mucoid variant of *Bacillus subtilis* or *Bacillus licheniformis*. The causative organism, *B. subtilis*, appears naturally in the soil and thus rope bacteria may be present on the outer parts of grains and vegetables. It can also be present in the air and may be carried as an aerosol in dust in the bakery environment. The primary source of contamination is from raw ingredients that may be present in or on the equipment.

Almost any of the ingredients used in the production of bread may contribute the organisms, but flour and equipment that previously have been in contact with contaminated dough are the greatest offenders. According to Clark (1946), rope occurs during hot and humid weather. The spores easily survive baking, and germinate and grow within 36–48 h inside the loaf to form the characteristic soft, stringy, brown mass with an odour of ripe pineapple or melon. This is due to the release of volatile compounds including diacetil, acetoin, acetaldehyde and isovaler-aldehyde which contribute to the typical sweetish odour of rope (Legan, 1994). The bacteria are heavily encapsulated, which contributes the mucoid nature of the material. As spoilage progresses, bacterial amylase and proteases degrade the bread crumb, causing discoloration and development of stickiness. At this stage the crumb stretches into long silky threads which give rise to the name 'rope'. Conditions favouring the appearance of rope are (1) a slow cooling period or storage above 25 °C (77 °F), (2) pH above 5, (3) high spore level and (4) a moist loaf. The water activity inside a loaf is marginal for *B. subtilis* growth so that rope may appear in localized areas where the moisture content is high. The development of rope has become rarer in many countries because the addition of calcium propionate, good hygiene (sanitation) and good bakery practice keep it under control. However, in the parts of the world where salt levels are being reduced (e.g. the UK) there is an increased risk of rope growth, especially if inhibitors are not used in the dough recipe.

## 10.2.3.  Yeast Spoilage

Many of the complaints regarding off-odours in bread, when not due to rope, are associated with yeasts. Contamination with wild yeasts is rare in bread made using a short process, but can sometimes occur when long fermentation sponges or doughs are employed (Seiler, 1993). Yeasts, like moulds, do not survive the baking process, but bread can become contaminated with yeasts during the cooling and slicing operations. The main sources of contamination are through physical contact with dirty equipment or infected high-sugar foods, which are a perfect substrate for osmophilic yeasts.

There are mainly two types of yeasts involved in the spoilage of bread (Legan and Voysey, 1991).

1. Fermentative yeasts. Sugars present in bread are fermented by these yeasts. The spoilage is manifested by the development of an 'alcoholic' or 'estery' off-odour depending on the species of yeast present. Many different types of yeast are capable of growing on bread and causing such defects but *Saccharomyces cerevisae*, which is the bakers' yeast, tends to be encountered most often in bread.
2. Filamentous yeasts. These are commonly referred to as 'chalk moulds' because they form a white, spreading growth on the surfaces of bread which can readily be confused with mould growth. They are considered as yeasts rather than moulds since they produce single cells and reproduce by budding. There are a number of different chalk moulds, but the most common and troublesome is *Pichia burtonii* which has the ability to grow very fast on bread and has proved to be more resistant to preservatives and disinfectants than many other moulds.

## 10.2.4.  Control of Microbiological Spoilage

### 10.2.4.1.  Preservatives

The preservatives most commonly used in bread to prevent or minimize microbial growth are the propionates, that is propionic acid and its salts (Chapter 3).

TABLE 10.2. Anti-mould agents most commonly used in bread.

| Anti-mould agents | Recommended level of use (%)[a] |
|---|---|
| Propionic acid | 0.1 |
| Calcium propionate | 0.2 |
| Sodium propionate | 0.2 |
| Sodium dipropionate (70% solution) | 0.2 |

[a] % of flour weight
Source: Seiler, 1983.

Organic acids such as propionic acid act by distorting the pH equilibrium of microorganisms (Wagner and Moberg, 1989). The depression of the internal cellular pH by ionization of the acid molecules leads to the elimination of substrate supplies into the cell, causing inhibition of the microbial growth (Jay, 1992). The anti-microbial activity of propionates is mainly against moulds and the bacteria responsible for the development of rope in bread. Because their effectiveness on yeasts is minimal, propionates can be used in bread without disturbing the leavening activity in the dough (Beuchat and Golden, 1989). The types of propionates used in bread with their recommended levels of use are listed in Table 10.2.

Another substance with effective preservative action in bread is ethyl alcohol. The addition of ethanol at levels between 0.5 and 3.5% of loaf weight leads to a substantial extension of the shelf-life of bread (Legan, 1993). Seiler (1984) has demonstrated that the mould-free shelf-life of loaves treated with alcohol increases with ethanol concentration, since a 50% extension in life is achieved with a 0.5% addition of ethanol based on loaf weight (Figure 10.1). Similar increases in shelf-life were obtained when the same amount of ethanol was sprayed over all surfaces of the loaf prior to packaging and sealing as when it was added to the base of a bag of the same size before adding the product and sealing. This finding confirms that ethanol acts as a vapour pressure inhibitor. Sensory tests have indicated that a level of addition higher than 1% by product weight might be unacceptable by the consumer (Seiler, 1984). Despite the high cost of duty paid, ethanol has real potential as a bread preservative because of not only its anti-microbial activity but also its ability to delay staling.

## 10.2.4.2.   Modified Atmosphere Packaging

The storage of food in atmospheres of increased concentration of carbon dioxide is referred to as modified atmosphere. The application of modified atmosphere packaging (MAP) in bakery products became of interest in the late 1970s, mainly in Europe because new labelling regulations demanded that the presence of preservatives should be declared. In contrast, carbon dioxide and nitrogen do not need to be declared. In addition, there was a need for greater increases in shelf-life than were possible using propionates. Packaging in carbon dioxide has the advantage that it can increase the shelf-life of bread without affecting its flavour,

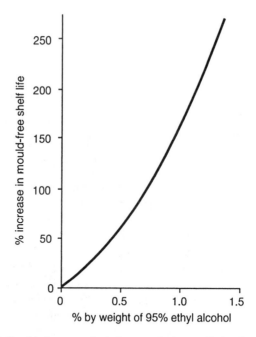

FIGURE 10.1. Relationship between alcohol concentration applied and percentage increase in mould-free shelf life (source: Seiler, 1984).

aroma or appearance. A carbon dioxide atmosphere is effective in retarding the development of moulds when present in concentrations greater than 20% (Seiler, 1984; Inns, 1987). At concentrations near to 100% the anti-microbial effect of carbon dioxide is maximized through anaerobiosis.

It has been shown that the higher the concentration of carbon dioxide in the atmosphere within the package, the greater the extension of the mould-free shelf-life (Seiler, 1984). Studies carried out on British bread have indicated a shelf-life extension of 100, 200 and 300% in packed atmospheres of 40, 60 or 90% carbon dioxide by volume (Seiler, 1984, 1989). The effect of carbon dioxide on the mould-free shelf-life of products is ERH dependent. Bakery products such as cakes, with ERH values of 85% or below, can obtain 400% extension in their shelf-lives in atmospheres of 75% or above of carbon dioxide. In bread-type products, which have an ERH higher than 90%, when a high concentration of carbon dioxide is present the achieved extension is about 250% (Figure 10.2).

A mixture of two gases, carbon dioxide and nitrogen, has also been used as a means of preventing package collapse, which can happen as carbon dioxide is absorbed into the product when it is used in its pure form. A gas mixture of 60% carbon dioxide and 40% nitrogen has been found to be the most suitable (Ooraikul, 1982).

The main methods of wrapping products under modified atmosphere conditions are as follows: (Seiler, 1984):

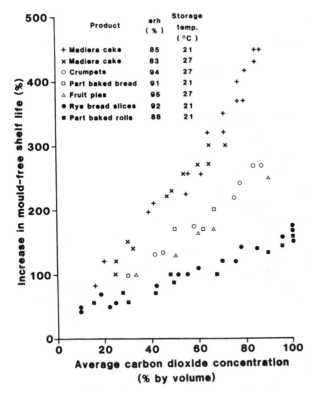

FIGURE 10.2. Increase in mould-free shelf life of various products with different carbon dioxide concentrations in the headspace of the pack (source: Seiler, 1984).

- Form-fill sealing, where the air is replaced by a continuous stream of gas and prior to the sealing of the package. The advantages of this system are the speed of the operation, versatility and the ability to adjust to different product sizes and wrapping materials. The disadvantage is a significant tendency towards gas leakage at the point of seal. The control of sealing pressure, time and temperature is absolutely necessary to avoid this gas leakage.
- Vacuum packaging, where the air is removed by creating a vacuum inside a partially sealed package, followed by the addition of the desired gas mixture. This is a less flexible system which requires a two-step approach to replace the air. However, this method can attain an equilibrium carbon dioxide concentration within the package approaching 100% and can even produce gas-tight seals between the lid and the base.

The application of gas packaging in bread-type products is mainly limited in high-value bread-type products intended to have long storage life because of the cost of gas, special wrapping equipment and the laminated films used in the process.

### 10.2.4.3.  Irradiation

Irradiation is used as a means of destroying mould spores which may be present on the surfaces of bread. The types of irradiation of primary interest in bread preservation are ultraviolet, microwave and infrared (Seiler, 1983).

### 10.2.4.4.  Ultraviolet Irradiation

Ultraviolet (UV) light is a powerful anti-bacterial agent with the most effective wavelength being about 260 nm. UV light is used to control the occurrence of mould spores on bread. The poor penetrative capacities of UV light limits its use to surface applications. Direct UV irradiation is used to treat the surfaces of wrapped bakery products, leading to useful increases of the mould-free shelf-life. The important advantage of UV light is that it does not generate heat to destroy wrapping films or promote condensation problems. The disadvantages are that it is difficult to treat a multi-surfaced product as mould spores in the air cell walls within the bread surface are protected from the irradiation. There is also the risk of exposure and it is necessary to shield the UV light from workers' eyes (Seiler, 1989).

### 10.2.4.5.  Microwave Radiation

Microwaves lie between the infrared and radio frequency portion of electromagnetic spectrum. The main advantage of microwave heating is that microwaves heat rapidly and evenly without major temperature gradients between the surface and the interior of homogeneous products (Potter, 1986). Seiler (1983) reported that wrapped bread can be rendered mould-free by heating in a microwave oven for 30–60 s, which is the time required for the surface to reach 75 °C (167 °F). The use of microwaves as a means of bread preservation is limited by the heating effect which results from their use. This can cause condensation problems which can adversely affect the appearance of the product.

### 10.2.4.6.  Infrared Radiation

Infrared treatment can be used to destroy mould spores on bread by heating surfaces to the desired temperature of 75 °C without adversely affecting the quality and appearance of the product or the integrity of the packaging material (Seiler, 1968). The time required to reach this temperature depends on the thickness of the packaging material, the nature of the product and the distance between the infrared projector and the surface of the product. The advantage of infrared radiation over microwaves and UV is that it is necessary to heat only the outside surfaces, which minimizes problems due to condensation or air expansion. One disadvantage is that it is quite costly for multi-sided products which are required either to rotate between heaters or to be treated in two separate ovens (Seiler, 1983).

## 10.3.    Bread Staling

The term 'staling' refers to the gradually decreasing consumer acceptance of bread due to all the chemical and physical changes that occur in the crust and crumb during storage excluding microbial spoilage (Bechtel et al., 1953). The result of these changes is a product which the consumer no longer considers 'fresh'. Staling is detected organoleptically by the changes in bread texture, as well as in taste and aroma.

The processes that cause staling actually begin during cooling, even before starch has solidified sufficiently for the loaf to be cut. Bread staling is mainly associated with the firming of the crumb. During storage, the crumb generally becomes harder, dry and crumbly, and the crust becomes soft and leathery. Quite often, these changes are solely attributed to the drying out of the crumb. However, the mechanism of crumb firming during storage is far more than a simple moisture redistribution from crumb to crust. The overall staling process is mainly part of two separate sub-processes (Guy et al., 1983): the firming effect caused by moisture transfer from crumb to crust, and the intrinsic firming of the cell wall material which is associated with starch re-crystallization during storage.

### *10.3.1.    Crust Staling*

During the storage of bread, the moisture content of the crust increases as a result of moisture migration from the crumb to the crust. With an initial moisture content of only 12%, the crust readily absorbs moisture from the interior crumb, which has a moisture content about 45%. It has been reported (Czuchajowska and Pomeranz, 1989) that during a storage period of 100 h, the crust moisture increased from 15 to 28%, while the crumb moisture loss was only from about 45 to 43.5%. In a zone near the crust the decrease was much more pronounced, from about 45 to 32%.

Wrapping of bread loaves in moisture-proof film accentuates crust staling by preventing evaporation of moisture that has migrated to it from the centre crumb. It must be said, however, since packaging slows down the rate of crumb staling by preventing excessive drying and minimizing the total moisture loss from the product into the atmosphere, that loaves are usually wrapped as soon as they cool. Generally, crust staling is less objectionable to consumers than the age-related firming of the crumb.

### *10.3.2.    Role of the Main Bread Components in Crumb Staling*

#### 10.3.2.1.    Starch

The changes that especially starch polymers undergo during the baking process and during storage determine the structure, textural properties and keeping quality of bread. A loaf of bread consists of about 50% starch, 40% water and 7%

protein. The main ingredients of bread are flour and water, with small quantities of fat, salt, various improvers and yeast. Starch is the major component of flour, being about 75% of the material, and plays an important role in the structure formation, physical properties and keeping quality of bread.

Bread will stale even when there is no net loss of moisture from the loaf. The mechanisms of bread staling have been studied for over 150 years. As early as 1852, the Frenchman Jean-Baptiste Boussingault, a pioneer in the study of nitrogen fixation, showed that bread could be hermetically sealed to prevent it from losing water, and yet still go stale. He further established that staling could be reversed by reheating the bread to 60 °C (140 °F), the temperature region at which, we now know, starch gelatinizes in bread.

Early researchers emphasized the role of starch in the staling process. The most comprehensive studies on changes in ageing bread were carried out by Katz (1928), in his effort to counteract staling to eliminate the need for night baking in Dutch bakeries. Katz utilized an X-ray diffraction technique to demonstrate that starch in bread retrograded with time in a manner similar to that of a starch gel. He concluded that water was involved in increased starch crystallinity, and thus starch was mostly responsible for crumb firming.

Since the work of Katz, much of the work on bread staling has concentrated on the gelatinization and retrogradation behaviour of the starch fraction. A brief description of these processes is in order at this point.

## 10.3.2.2.   Starch Gelatinization

Starch occurs as roughly spherical granules in wheat flour. The starch polymer consists of two structurally distinct polysaccharides: amylose and amylopectin. Amylose is an essentially linear polymer, apparently amorphous, containing approximately 4000 glucose units. Amylopectin – the partially crystalline component – is a multi-branched polysaccharide composed of approximately 100,000 glucose units. In the starch granules the polymer chains are held by crystalline junction points in a rigid network. Schoch (1945) separated and described the properties of the amylose and amylopectin fractions of starch.

Native starch granules are not soluble in cold water but, when heated in an aqueous medium, will absorb water and swell. Initially, swelling is reversible but it becomes irreversible as temperature is increased, and the granule structure is altered significantly. As temperature increases, the starch polymers vibrate vigorously, breaking intermolecular bonds and allowing their hydrogen bonding sites to engage more water molecules. The water penetration leads to an increased separation of starch chains, which has an effect in increasing the randomness and decreasing the number and size of crystalline regions. Continued heating results in complete loss of crystallinity. At this stage the viscosity of the system is very close to that of near-solid system, as the melting temperature ($T_m$) value is exceeded. This point is regarded as the gelatinization point or gelatinization temperature. In bread it is the setting temperature, when the viscous dough changes to a solid sponge. In most baked products the starch crystallites melt at between 60 and 90 °C (140–194 °F).

On cooling, all the polymers begin to lose mobility as the baked crumb cools and the viscosity increases. In bread, textural properties such as firmness will be determined by the concentration of polymers, particularly starch and the temperature.

The state of starch granules in the final loaf may range from being almost fully gelatinized to a complete loss of order and with variable degrees of exudation of granular polysaccharide (mainly amylose). The exuded polysaccharide would behave physically as a rubber while the amylopectin internal to the granule would likewise be in rubbery (flexible) state. At water contents greater than 20–30% w/w, such as that of bread, this would lead to retrogradation.

### 10.3.2.3.   Starch Retrogradation

The predominant mechanism of staling is the time-dependent re-crystallization of amylopectin from the completely amorphous state of a freshly heated product to the partially crystalline state of a stale product. Starch retrogradation is a process in which gelatinized starch molecules re-associate to form a double helix crystalline structure. An implicit requirement of starch re-crystallization is the availability of sufficient moisture, at least locally within the matrix, for mobilizing long polymer chain segments. The rate and extent of starch re-crystallization are determined primarily by the mobility of the crystallizable outer branches of amylopectin (Schoch and French, 1947; Ring et al., 1987; Russell, 1987). Water plays a very important role because it acts as a plasticizer. A plasticizer is a 'material incorporated in a polymer to increase polymer's workability, flexibility or extensibility' (Levine and Slade, 1990). Water is unique in that role because of its low molecular weight.

Schoch and French (1947) proposed a model that describes the heat-reversible aggregation of amylopectin as the principal cause of bread staling (Figure 10.3). They suggested that granule swelling is restricted by the limited water available in bread dough, so that swollen granules retain their identities as discrete particles. With increasing swelling, the linear amylose fraction becomes more soluble and diffuses into the aqueous phase, forming a concentrated solution. Immediately on cooling, this solution of amylose molecules associates by hydrogen bonding and rapidly retrogrades to set up an insoluble gel, contributing to the loaf structure. This gel is considered to remain stable during further storage and not to participate in the staling process. As illustrated in the Schoch and French model (see Figure 10.3), amylose quickly associates in bread soon after baking, affecting initial firmness, but plays no further role in crumb firming.

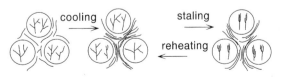

FIGURE 10.3. Mechanism of bread staling (source: Schoch and French, 1947).

This firming is attributed to changes in the physical orientation of the branched amylopectin molecules of starch within the swollen granule. In fresh bread, the branched chains of amylopectin are unfolded and spread out within the limits of available water. These chains of the amylopectin polymer gradually aggregate, aligning with one another by various types of intra-molecular bonding. This effect results in increasing rigidity of the internal structure of the swollen starch granules causing crumb hardening.

Ghiasi et al. (1984) investigated the effect of varying the ratio of amylose to amylopectin on the bread firming rate. They found that on day 1 of storage the high amylopectin breads (ratio of 83.4% amylopectin to 16.6% amylose) were less firm than the control (ratio of 75% amylopectin to 25% amylose), but by days 3 and 5 the firmness of high amylopectin breads was equal to that of the control. This demonstrates that amylose retrogradation occurs rapidly and that any subsequent firming does not involve the amylose fraction. When waxy barley starch (100% amylopectin) was used to replace normal maize starch in a reconstituted bread dough, bread loaves collapsed on cooling (Hoseney et al., 1978). This finding suggests that the amylose fraction is predominately responsible for the setting of crumb structure. The findings that the amylose in wheat starch is the component contributing most to bread staling has led to speculation that waxy wheats may be used to extend shelf-life (Graybosch, 1998) but bread staling is a series of complex changes to many components in the bread formulation.

Although a qualitative correlation exists between the extents of starch retrogradation and the increased firmness of the bread crumb (Eliasson, 1985), the correlation changes significantly when different bread recipes are considered. Factors that complicate the real situation include the degree of starch granule swelling, hydration of starch, the level and type of saccharide ingredients, moisture content, the presence of solutes (salt and sugar), the presence of flour hydrocolloids (pentosans) and the effects of various lipid materials. This confirms that coexistent phenomena contribute to the overall increase of firmness.

### 10.3.2.4.  Gluten

Although starch re-crystallization is the major component of bread crumb firming, there is reported evidence that some other mechanisms may also have a role in the staling phenomenon. Many investigators have supported the view that gluten proteins participate in the staling process. It has been suggested (Martin et al., 1991; Martin and Hoseney, 1991) that hydrogen bond interactions between the $-OH$ groups of protruding starch chains with the $-NH_2$ groups of protein fibrils increase in number and cumulative strength through storage time as the crumb loses kinetic energy, resulting in increasing crumb firmness.

According to Kim and D'Appolonia (1977a,b) there is, in general, an inverse relation between protein content and staling during storage of bread. Bread made from strong flour, forming gluten of adequate quantity and quality, has a higher volume and a slower rate of staling compared to bread made with weak flour. This finding was later confirmed by He and Hoseney (1991), who reported that gluten from poor-quality flours interacts more strongly with the starch granules than does the gluten from good-quality flours. This means that bread from poor-quality flours would be expected to firm at a more rapid rate. Willhoft (1973a) suggested that the anti-firming effect of increased protein is due to the dilution of the starch component and an increase in loaf volume affected by gluten enrichment.

### 10.3.2.5.   Water Redistribution

The redistribution of moisture during bread storage and its contribution to staling has been a controversial subject. Willhoft (1973b) suggested that crumb firming involved a loss of moisture from the gluten to the starch phase, whereas Cluskey et al. (1959) supported the view that moisture must migrate from starch to gluten in staling bread. Wynne-Jones and Blanshard (1986) demonstrated that 'bound' water increased slightly upon ageing of bread while 'free' water decreased. This finding was taken as an indication that moisture from gluten is moving to the crystalline starch structure. Leung et al. (1983) suggested that as starch changes from the amorphous state to the more stable crystalline state, water molecules become immobilized as they are incorporated into the crystalline structure, resulting in decreased water mobility during the ageing process. Slade and Levine (1987) proposed a possible explanation to the controversy of the direction of moisture migration within the bread crumb. He suggested that the re-crystallization of amylopectin within swollen starch granules leads to the development of a partially crystalline structure with B-type crystalline regions. Water must migrate within the crumb to the developing crystalline regions. The B-type starch crystal is a higher moisture crystalline hydrate than is A-type starch and requires the incorporation of more water molecules into the crystalline region while starch chain segments are realigning. Once incorporated within the crystal, water is no longer available as a plasticizer of the starch–gluten network, nor can it be perceived organoleptically. As a result, bread loses softness and develops a drier mouth feel.

Rogers et al. (1988) have shown that retrogradation (due to amylopectin) could be affected by the presence of water, which would result in a variable firming rate. They found that bread crumb firms more rapidly under conditions where starch does not retrograde and concluded that starch retrogradation and crumb firming do not correlate with each other across varying moisture contents.

The re-distribution of water at both the macroscopic and the microscopic level has a profound effect on the staling of bread. In general terms the higher the water level that remains in bread after baking, the slower the staling. The moisture content of the baked product is profoundly influenced by the level of water added during mixing and is profoundly influenced by the subsequent processing of dough to bread. A major impact at the macrolevel is the relationship between the product crust and crumb. Cauvain and Young (2000) showed how an increase in the thickness of the crust on a UK-style sandwich loaf from 1 to 2 mm would at equilibrium result in an extra 1.2% moisture being lost from the crumb. This increased loss of water from the crumb would yield not only a slightly firmer/drier eating product but one which would stale slightly faster.

Chinachoti (2003) discussed the mobility of water in bread and the conflicting theories that exist. The addition of glycerol (Baik and Chinachoti, 2001) caused more rapid firming of bread crumb despite potential plastization and reduced amylopectin re-crystallization. The addition of glycerol increased the diffusive mobility of water and the impact of the glycerol was to draw water from the starch and gluten macromolecules. Despite extensive studies on the role of water in bread staling, the underlying mechanisms involved remain relatively poorly understood. In part this is because of the complex nature of bread crumb and its associated characteristics. In this context the impact of the role of cell structure and moisture diffusion at macro-level and micro-level has not been addressed.

## 10.4.   Staling Inhibitors

Losses which result from bread staling are of great economic importance, especially under conditions of industrialized and centralized production. Thus considerable attention has been focused on this problem and much research effort has been expended on understanding the mechanisms by which certain groups of ingredients retard the staling process or minimize its effect.

### 10.4.1.   Enzymes

The addition of *alpha*-amylases to the dough has been shown to retard crumb firming (Miller et al., 1953). During baking, amylases partially hydrolyse starch to a mixture of smaller dextrins. The *alpha*-amylase enzymes cannot access intact starch granules, thus they essentially hydrolyse damaged starch by attacking *alpha*-(1,4) linkages along starch chains. This action is stopped at *alpha*-(1,6) branch points of amylopectin. Below 55 °C (131 °F) the activity of *alpha*-amylase is minimal and depends on the amount of damaged starch in the flour. Between 58 and 78 °C (136–172 °F) gelatinizing starch is rapidly attacked, with the rate of conversion slowing down above this temperature range because of enzyme denaturation.

Bacterial *alpha*-amylase affects bread crumb firmness more than fungal or cereal *alpha*-amylase because it is more heat stable than the other two types. The thermostability of added *alpha*-amylases decreases in the following order; bacterial > cereal (malt) > fungal (Herz, 1965; see also Chapter 3). As much as 20% of the initial bacterial enzyme activity can survive the baking process. The use of *alpha*-amlyase enzymes results in the formation of increased amounts of dextrins which can be extracted from the bread crumb. The amount of soluble dextrins present in the crumb is an indicator of the thermostability of the enzymes used.

The shorter-chain dextrins of a particular low degree of polymerization (DP 3–9) are presumably responsible for the anti-firming effect (Martin and

Hoseney, 1991). These mobile dextrins may interfere with and prevent the development of hydrogen bonds between starch granule remnants and the continuous gluten network. This action prevents the development of entanglements between starch and protein which increase the rigidity of the crumb.

Lineback (1984) observed that bread made with bacterial *alpha*-amylase remains soft for longer than 'normal' bread. He suggested that the lower molecular weight branched dextrins have a decreased ability to retrograde or interfere with retrogradation in some manner, thus reducing the extent of firming.

Comparison of the X-ray patterns of fresh and stored breads has indicated that the order of decreasing degree of starch crystallinity was bread with bacterial *alpha*-amylase, bread with fungal *alpha*-amylase, bread with cereal *alpha*-amylase, and un-supplemented bread (Dragsdorf and Varriano-Marston, 1980). Dragsdorf and Varriano-Marston (1980) observed that in bread supplemented with bacterial *alpha*-amylase, an A-type crystal pattern was obtained, whereas in the other supplemented breads the B-type crystal pattern was formed. They also noted that the crystallization of starch did not follow the change in firmness and reported that starch crystallinity and bread firming were not synonymous. The discrepancies between X-ray diffraction data and firmness results arise because the different types of crystals influence the distribution of water within the crumb differently. The A-type crystal contains eight water molecules, whereas the B-type crystal contains 36 water molecules. As a result, in breads amylopectin recrystallization develops B-type crystalline regions and the crumb is firmer because more water has migrated into the crystalline region. This water which participated in the formation of the crystal is no longer available as a plasticizer of the starch–gluten. Macroscopically, the lack of the plasticizing effect from water results in firmer bread which is perceived as being drier (Slade and Levine, 1987).

## 10.4.2.  Emulsifiers

Emulsifiers are used in bread as a means of maintaining crumb softness for longer by retarding the process of staling. The emulsifiers that are widely used for their ability to reduce the staling of the bread crumb are distilled, saturated monoglycerides (see Chapter 3). Other types of emulsifiers that are effective as crumb softeners are sodium stearoyl-2-lactylate (SSL) and diacetyl tartaric acid esters of monoglycerides (DATA esters), although generally not to the same degree as monoglycerides (Tamstorf et al., 1986).

The anti-staling ability of emulsifiers is mainly due to their interaction with starch but the exact mechanism is still unexplained. Emulsifiers complex with linear amylose, and they may do some complexing with the outer linear branches of amylopectin (Knightly, 1977). It appears that the formation of an emulsifier and amylose complex contributes to a decrease in the initial firmness of the crumb, while a complexing with amylopectin results in a distinct reduction in the rate of firming during storage (Knightly, 1988). Research studies have shown that when emulsifiers are added to a starch gel or to bread, the differential scanning calorimetry (DSC) endotherm due to re-crystallization of amylopectin is decreased (Russell 1983a,b; Eliasson, 1983). However, it has been demonstrated that the final extent of retrogradation is not changed by the addition of emulsifiers, but the staling process proceeds at a much slower rate (Krog et al., 1989).

According to Rao et al. (1992), although emulsifiers inhibit the process of amylopectin re-crystallization, they have no effect on the cellular mechanical properties of bread as measured by recoverable work. This means that emulsifiers have an effect on crumb softness but do not interfere with changes in the rigidity and elasticity of the cell walls in baked bread.

Distilled monoglycerides are used as anti-staling agents in the form of hydrates. These are suspensions of monoglyceride crystals (*beta*-form), usually 20–25% in water, with a creamy texture which make them easily dispersible in dough. Blends of saturated and unsaturated monoglycerides in the form of fine powder are also available and can be added directly to the dough.

## 10.4.3.  Pentosans

Pentosans, which are non-starchy polysaccharide materials, are a minor component of wheat flour present at the 2–3% level, roughly half water-soluble and half water-insoluble. The presence of pentosans reduces the tendency of the starch components towards re-crystallization because they are highly hydrophilic and absorb six or seven times their weight in water (see Chapter 11).

A number of researchers have investigated the effect of pentosans on staling rates and concluded that they retard starch retrogradation, especially the water-insoluble fraction (Kim and D'Appolonia, 1977c,d; Jankiewicz and Michniewicz, 1987). They suggested that pentosans affect the extent of retrogradation by reducing the amount of starch available for retrogradation in the system. The water-soluble fraction appeared to interact with the amylopectin, and the insoluble fraction with both the amylose and the amylopectin (Kulp and Ponte, 1981).

## 10.4.4.  Alcohol

It is generally accepted that breads produced from processes involving a bulk fermentation step have longer shelf-life than do products from 'no-time' accelerated processes (Chapter 2). This is mainly due to the presence of larger quantities of alcohol which is a product of the longer periods of yeast fermentation in bulk-fermented doughs though the moisture content of breads made by bulk fermentation is often slightly lower than that from no-time doughs.

Studies carried out on the effect of ethanol on the firming of bread have shown that the crumb modulus of bread which has been treated with ethanol increases during storage at a slower rate than that of untreated 'control' loaves (Figure 10.4). DSC has been used to follow changes in the rate of development

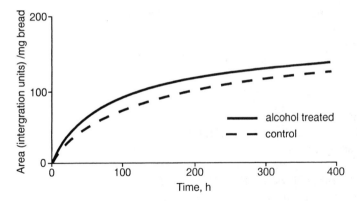

FIGURE 10.4. Effect of alcohol treatment on the extent of starch recrystallization during storage (as expressed by the endotherm area per mg of bread) (source: Russell, 1982).

of an endotherm associated with the starch fraction of bread during ageing, as a result of ethanol treatment (Russell, 1982). The loaves were treated by painting their surface with ethanol after they had been baked and allowed to cool. The rate of development of the staling endotherm caused by the re-crystallization of starch was lower for bread treated with alcohol than for the untreated control. For comparison, crumb compressibility measurements were carried out and the results confirmed that alcohol-treated bread exhibits a reduced rate of crumb firming paralleled by changes observed by DSC in the starch fraction of the bread crumb.

The manner in which the alcohol exerts its 'anti-staling' effect is uncertain, although it has been suggested that ethanol can complex with amylose and the linear exterior branches of amylopectin to prevent retrogradation (Erlander and Erlander, 1969).

## 10.4.5.  Sugars and Other Solutes

Sugars are also known to function as anti-staling ingredients in starch-based baked goods (Cairnes et al., 1991). I'Anson et al. (1990) investigated the effect of sucrose, glucose and ribose on the retrogradation of wheat starch gels at the nominal starch:sugar:water ratio of 1:1:1. Their findings suggest that the effectiveness of these sugars in reducing firmness and crystallinity follows the order ribose > sucrose > glucose. They also reported that for the level of added ribose no retrogradation or crystallinity was detected.

The exact mechanism by which sugars suppress the crystallization of starch is not yet fully understood. It has been suggested, however, that sugars locally increase the glass transition temperature ($T_g$) and thereby dramatically reduce diffusion of polymer molecules to a crystal nucleus (I'Anson et al., 1990). Levine and Slade (1990) have proposed that sugars exert an anti-staling influence by raising the glass transition temperature of the amorphous amylose matrix, thus suppressing the re-crystallization of amylopectin.

Increasing salt levels from 0 to about 2–3% has been reported to decrease the firming rate of bread crumb (Maga, 1975).

## 10.5.    Freezing of Bread

Deep freezing can be used to prevent the process of staling and inhibit microbiological activity in bread. However, the freezing of bread is not simply a matter of storing loaves in a standard freezer. Such storage will certainly stop the process of starch re-crystallization, but if the maximum benefit is to be achieved it must be taken into account that after thawing, staling resumes and appears to proceed at a faster rate than in unfrozen loaves. In addition, during the freezing and thawing process the bread passes through the temperature range at which it stales fastest.

Avrami's theory of phase change kinetics (1940, 1941) can be represented by an equation which was used by Cornford et al. (1964) to study the relationship between the elastic modulus of bread crumb and its storage time and temperature. They demonstrated that the relative rate of increase in the limiting modulus became greater as storage temperatures were lowered towards the freezing point.

Bread stales more rapidly in the refrigerator than at ambient temperatures, but can be stored indefinitely in the deep freeze. The fact that the staling process gets faster as the temperature is reduced can be explained by polymer theories of recrystallization which apply to partially crystalline polymers such as starch. This is due to the fact that there are actually two separate events that go to make up the process of crystallization in polymer systems such as gelatinized starch systems. The first of these two events is nucleation and the second crystal growth or propagation. These polymer theories suggest that crystal nucleation is favoured at lower temperatures but that crystal growth is favoured when the polymer chains are more mobile to form double helixes (Figure 10.5).

Nucleation is the coming together of two or more polymer chains in the right arrangement to form a nucleus for subsequent crystal growth. Because the polymer chains are always moving, not all encounters lead to proper nucleation; the chains rebound or unwind again. In general, the number of successful nucleations increases as the temperature decreases. However, for crystal growth or propagation to occur, the polymers must be sufficiently mobile for the double helixes to aggregate or pack together into the crystallites. This mobility will increase with increasing temperature, so the rate of the propagation process also increases with increasing temperature.

So we can conclude that the rates at which optimum nucleation and growth occurs are different. The net result is that as the temperature of the product is reduced towards its glass transition temperature the rate of the overall crystallization process increases to a maximum, then falls. At storage temperatures below 0°C (32°F), the most important single parameter which determines the storage stability potential of a food product is its glass transition temperature.

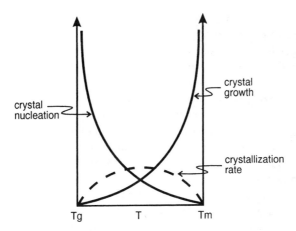

FIGURE 10.5. Crystallization kinetics of partially crystalline polymers (source: Morris, 1990).

It was not until the 1980s that food scientists recognized the importance of the glass transition temperature in food systems and used it to predict product properties, quality and stability. Levine and Slade (1988) introduced the idea that foods could achieve mechanical stability if the freeze-concentrated, unfrozen water fraction is in a glassy state. For maximum storage stability, the product must be stored at a temperature that is below its $T_g$ (Levine and Slade, 1990). When products are stored below their $T_g$, they are known to have achieved the glassy state form. Once in the glassy state it is known that polymers, such as starch, are immobilized and that reactions and movement of solutes, such as sugar, and plasticizers, such as water, within their matrix are extremely slow. Bread stored below its $T_g$ does not stale. Levine and Slade (1990) indicated that frozen storage stability is controlled by the temperature difference between the freezer temperature and the glass transition temperature.

The changes in the molecular motion of polymer chains from a glassy (rigid) state to a rubbery (flexible) state also depend on the amount of water in the system (Slade and Levine, 1991). Breads with moisture contents of around 40% after baking have their starch and proteins in the rubbery state above $T_g$ and at ambient temperatures. In bread, the $T_g$ is at $-7$ to $-9\,^{\circ}C$ (20–16 $^{\circ}F$), below which the staling process is inhibited. For temperatures above $T_g$ and below the crystal melting temperature of starch, the material is 'rubbery' and sufficient motion of the polymer occurs, allowing crystallization. Under these circumstances, bread stored at, say, 21 $^{\circ}C$ (70 $^{\circ}F$) is considerably above the glass transition temperature of the system, and would be expected to stale.

The freezing of bread can be achieved in a variety of ways. Cold air blown over the product is the commonest way. The air temperature and the air velocity will affect the rate of heat removal or freezing rates. The effect of freezing rates on bread crumb quality is related mainly to ice crystal formation. With respect to crystal formation upon freezing, slow freezing favours the formation of larger ice crystals, while fast freezing favours the formation of small ice crystals. Crystal growth is one of the factors which could affect the storage life of frozen bread, since ice crystals grow in size during storage and cause crumb weakening by damaging cell walls and disrupting internal structures to the point where the thawed product is quite unlike the original in texture and eating quality. More importantly, freezing rates control the amount of unfrozen water in the matrix. With rapid freezing, there is an increased frequency of ice crystal nucleation. This has a significant effect on the glass transition temperature of the product. Faster removal of heat enables the product to achieve a glassy state at a higher temperature, thus improving its frozen storage stability (George, 1993).

It is important, therefore, to reduce the product temperature below its $T_g$ as quickly as possible. With small-sized fermented goods, such as bread rolls, this can be achieved even in a common storage freezer unit with its limited air movement, provided the unit is not overloaded. With pan bread, such as 400 and 800 g pan loaves, blast freezing is necessary in order to achieve an adequately fast freezing rate and to prevent excessive staling during the initial freezing. Dehydration of the loaves is likely to occur with prolonged frozen storage time. This arises from the low humidity of the frozen air and can be avoided by adequate wrapping of the products in low-moisture permeability film and close packing of the frozen loaves in the freezer. During thawing, however, loaves should be well spaced out to promote maximum air movement around them and maximum rate of heat transfer into the cold product.

Partial defrosting and re-freezing of baked breads should be avoided since they may give rise to the phenomenon referred to as 'freezer burn' – the formation of white opaque patches in the crumb or on the crust of baked products (Cauvain and Young, 2000).

## References

Avrami, M. (1940) Kinetics of phase change, II. *Journal of Chemical Physics*, **8**, 212.

Avrami, M. (1941) Kinetics of phase change, III. *Journal of Chemical Physics*, **9**, 177.

Baik, M.-Y. and Chinachoti, P. (2001) Effects of glycerol and moisture gradient on thermomecahnical properties of white bread. *Journal of Agriculture, Food & Chemistry*, **49**, 4031–8.

Bechtel, W.G., Meisner, D.F. and Bradley, W.B. (1953) The effect of the crust on the staling of bread. *Cereal Chemistry*, **39**, 160–8.

Boussinggault, J.G. (1852) Experiments to determine the transformation of fresh bread to stale bread. *Annales de Chimie Physique*, **36**, 490.

Beuchat, L.R. and Golden, D.A. (1989) Antimicrobials occurring naturally in foods. *Food Technology*, **43**(1), 134–41.

Cairnes, P., Miles, M.J. and Morris, V.J. (1991) Studies of the effect of the sugars ribose, xylose and fructose on the retrogradation of wheat starch gels by X-ray diffraction. *Carbohydrate Polymers*, **16**, 355–65.

Cauvain, S.P. and Young, L.S. (2000) *Bakery Food Manufacture & Quality: Water Control & Effects*, Blackwell Science, Oxford, UK.

Chinachoti, P. (2003) Preventing bread staling. In *Breadmaking: Improving Quality* (Ed. S.P. Cauvain), Woodhead Publishing Ltd, Cambridge, UK, pp. 562–74.

Clark, H.A. (1946) Rope and mould. *Bakers' Digest*, June, 49–51.

Cluskey, J.E., Taylor, N.W. and Senti, F.R. (1959) Relation of the rigidity of flour, starch and gluten gels to bread staling. *Cereal Chemistry*, **36**, 236–46.

Cornford, S.J., Axford, D.W.E. and Elton, G.A.H. (1964) The elastic modulus of bread crumb in linear compression in relation to staling. *Cereal Chemistry*, **41**, 216–29.

Czuchajowska, Z. and Pomeranz, Y. (1989) Differential scanning calorimetry, water activity and moisture contents in crumb centre and near-crust zones of bread during storage. *Cereal Chemistry*, **66**, 305–309.

Doerry, W.T. (1990) Water activity and safety of baked products. *American Institute of Baking Research Department Technical Bulletin*, **12**(6), 6 pp.

Dragsdorf, R.D. and Varriano-Marston, E. (1980) Bread staling: X-ray diffraction studies on bread supplemented with alpha-amylase from different sources. *Cereal Chemistry*, **57**, 310–14.

Eliasson, A.C. (1983) Differential scanning calorimetry studies on wheat starch–gluten mixtures. II. Effect of gluten and sodium stearoyl lactylate on starch crystallization during ageing of wheat starch gels. *Journal of Cereal Science*, **1**, 207–13.

Eliasson, A.C. (1985) Retrogradation of starch as measured by differential scanning calorimetry, in *New Approaches to Research on Cereal Carbohydrates* (eds R.D. Hill and L. Munk), Elsevier Science, Amsterdam, pp. 93–8.

Erlander, S.R. and Erlander, L.G. (1969) Explanation of ionic sequences in various phenomena. X. Protein–carbohydrate interactions and the mechanism of for the staling of bread. *Die Starke*, **21**, 305.

George, R.M. (1993) Freezing processes used in the food industry. *Trends in Food Science and Technology*, **4**, 134–8.

Ghiasi, K., Hoseney, R.C., Zeleznak, K. and Rogers, D.E. (1984) Effect of waxy barley starch and reheating on firmness of bread crumb. *Cereal Chemistry*, **61**, 281–5.

Graybosch, R.A. Waxy wheats: Origin, properties and prospects. *Trends in Food Science & Technology*, **9**, 135–42.

Guy, R.C.E, Hodge, D.G. and Robb, J. (1983) An examination of the phenomena associated with cake staling. *FMBRA Report No. 107*, November, CCFRA, Chipping Campden, UK.

He, H. and Hoseney, R.C. (1991) Differences in gas retention, protein solubility and rheological properties between flours of different baking quality. *Cereal Chemistry*, **68**, 526–30.

Herz, K.O. (1965) Staling of bread – a review. *Food Technology*, **19**, 90–103.

Hoseney, R.C., Lineback, D.R. and Seib, P.A. (1978) Role of starch in baked foods. *Bakers' Digest*, **52**(4), 11, 14, 16, 18, 40.

I'Anson, K.J., Miles, M.J., Morris, V.J. et al. (1990) The effects of added sugars on the retrogradation of wheat starch gels. *Journal of Cereal Science*, **11**, 243–8.

Inns, R. (1987) Modified atmosphere packaging, in *Modern Processing, Packaging and Distribution Systems for Foods* (ed. F.A. Paine), Van Nostrand Reinhold, New York, pp. 36–51.

Jankiewicz, M. and Michniewicz, J. (1987) The effect of soluble pentosans isolated from rye grain on staling of bread. *Food Chemistry*, **25**, 241–6.

Jay, J.M. (1992) *Modern Food Microbiology*, AVI Publishing, Westport, CT, pp. 3–12.

Katz, J.R. (1928) Gelatinization and retrogradation of starch in relation to the problem of bread staling, in *Comprehensive Survey of Starch Chemistry*, Vol. 1 (ed. R.P. Walton), Chemical Catalog Co., New York.

Kim, S.K. and D'Appolonia, B.L. (1977a) Bread staling studies. I. Effect of protein content on staling rate and bread crumb content on staling rate and bread crumb pasting properties. *Cereal Chemistry*, **54**, 207–15.

Kim, S.K. and D'Appolonia, B.L. (1977b) Bread staling studies. II. Effect of protein content and storage temperature on the role of starch. *Cereal Chemistry*, **54**, 216–24.

Kim, S.K. and D'Appolonia, B.L. (1977c) Bread staling studies. III. Effect of pentosans on dough, bread and bread staling rate. *Cereal Chemistry*, **54**, 225–29.

Kim, S.K. and D'Appolonia, B.L. (1977d) Bread staling studies. IV. Effect of pentosans on the retrogradation of wheat starch gels. *Cereal Chemistry*, **54**, 150–60.

Knightly, W.H. (1977) The staling of bread. *Bakers' Digest*, **51**(5), 52–6, 144–50.

Knightly, W.H. (1988) Surfactants in baked foods: Current practice and future trends. *Cereal Foods World*, **33**, 405–12.

Krog, N., Olesen, S.K., Toernaes, H. and Joensson, T. (1989) Retrogradation of the starch fraction in wheat bread. *Cereal Foods World*, **34**, 281–5.

Kulp, K. and Ponte, J.G. (1981) Staling of white pan bread: Fundamental causes. *CRC Critical Reviews of Food Science and Nutrition*, **15**, 1–48.

Legan, J.D. (1993) Mould spoilage of bread: The problem and some solutions. *International Biodeterioration and Biodegradation*, **32**, 33–53.

Legan, J.D. (1994) Spoilage of bakery products and confectionery. *PHLS Microbiology Digest*, **11**(2), 114–7.

Legan, J.D. and Voysey, P.A. (1991) Yeast spoilage of bakery products and ingredients. *Journal of Applied Bacteriology*, **70**, 361–71.

Levine, H. and Slade, L. (1988) Principles of 'cryostabilization' technology from structure/property relationships of carbohydrate/water systems – a review. *Cryo-letters*, **9**, 21–63.

Levine, H. and Slade, L. (1990) Influences of the glassy and rubbery states on the thermal, mechanical, and structural properties of doughs and baked products, in *Dough Rheology and Baked Product Texture* (eds H. Faridi and J.M. Faubion), Van Nostrand Reinhold, pp. 157–330.

Leung, H.K., Magnuson, J.A. and Bruinsma, B.L. (1983) Water binding of wheat flour doughs and breads as studied by deutero relaxation. *Journal of Food Science*, **48**, 95–9.

Lineback, D.R. (1984) The role of starch in bread staling, in *International Symposium on Advances in Baking Science and Technology*, Department of Grain Science, Kansas State University, Manhattan, Kansas.

Maga, J.A. (1975) Bread staling. *CRC Critical Reviews of Food Technology*, **29**, 443–486.

Martin, M.L. and Hoseney, R.C. (1991) A mechanism of bread firming. II. Role of starch hydrolyzing enzymes. *Cereal Chemistry*, **68**, 503–507.

Martin, M.L, Zeleznak, K.J. and Hoseney, R.C. (1991) A mechanism of bread firming. I. Role of starch firming. *Cereal Chemistry*, **68**, 498–503.

Miller, B.S., Johnson, J.A. and Palmer, D.L. (1953) A comparison of cereal, fungal and bacterial alpha-amylase as supplements for bread making. *Food Technology*, **7**, 38–42.

Morris, V.J. (1990) Starch gelation and retrogradation. *Trends in Food Science and Technology*, July, 2–6.

Ooraikul, B. (1982) Gas packaging for a bakery product. *Canadian Institute of Food Science and Technology*, **15**(4), 313–15.

Osborne, B.G. (1980) The occurrence of ochratoxin A in mouldy bread and flour. *Food and Cosmetics Toxicology*, **18**, 615–17.

Ponte, J.G. and Tsen, C.C. (1978) Bakery products, in *Food and Beverage Mycology* (ed. L.R. Beuchat), AVI Publishing Co., Westport, CT, pp. 191–223.

Potter, N. (1986) *Food Science*, 4th edn, AVI Publishing Co., Westport, CT, p. 463.

Rao, P.A., Nussinovitch, A. and Chinachotti, P. (1992) Effects of selected surfactants on amylopectin recrystallization and on recoverability of bread crumb. *Cereal Chemistry*, **69**, 613–18.

Ring, S.G., Colonna, P., I'Anson, K.J et al. (1987) The gelation and crystallization of amylopectin. *Carbohydrate Research*, **162**, 277–93.

Rogers, D.E., Zeleznak, K.J., Lai, C.S. and Hoseney, R.C. (1988) Effect of native lipids, shortening and bread moisture on bread firming. *Cereal Chemistry*, **65**, 398–401.

Russell, P.L. (1982) Recent work on bread staling. *FMBRA Bulletin No. 2*, April, CCFRA, Chipping Campden, UK, pp. 69–80.

Russell, P.L. (1983a) A kinetic study of bread staling by differential scanning calorimetry. *Starch/Starke*, **35**, 277–81.

Russell, P.L. (1983b) A kinetic study of bread staling by differential scanning calorimetry and compressibility measurements. The effects of added monoglycerides. *Journal of Cereal Science*, **1**, 297–303.

Russell, P.L. (1987) Aging of gels from starches of different amylose/amylopectin content – a DSC study. *Journal of Cereal Science*, **6**, 147–58.

Schoch, T.J. (1945) The fractionation of starch, in *Advances in Carbohydrate Chemistry*, Vol. 1. (eds W.W. Pigman and M.L. Wolfrom), Academic Press, New York.

Schoch, T.J. and French, D. (1947) Studies on bread staling. 1. Role of starch. *Cereal Chemistry*, **24**, 231–49.

Seiler, D.A.L. (1968) Prolonging the shelf-life of cakes. *The British Baker*, 7 June, 25–6, 60–2.

Seiler, D.A.L. (1983) Preservation of bakery products. *FMBRA Bulletin No. 4*, August, CCFRA, Chipping Campden, UK, pp. 166–77.

Seiler, D.A.L. (1984) Controlled atmosphere packaging for preserving bakery products. *FMBRA Bulletin No. 2*, April, CCFRA, Chipping Campden, UK, pp. 48–60.

Seiler, D.A.L. (1989) Modified atmosphere packaging of bakery products, in *Controlled/Modified Atmosphere/Vacuum Packaging of Foods* (ed. A.L. Brody), Food and Nutrition Press, Trumbell, CT, pp. 119–33.

Seiler, D. (1992) Basic guide to product spoilage and hygiene. Part 1: Mould growth on bread. *Chorleywood Digest No. 121*, CCFRA, Chipping Campden, UK, pp. 117–9.

Seiler, D. (1993) Basic guide to product spoilage and hygiene. Part 2: Yeast spoilage of bread. *Chorleywood Digest No. 124*, CCFRA, Chipping Campden, UK, p. 23.

Silliker, J.H. (1980) Cereals and cereal products, in *Microbial Ecology of Foods* (ed. J.H. Silliker), Academic Press, Orlando, FL, pp. 669–730.

Slade, L. and Levine, H. (1987) Recent advances in starch retrogradation, in *Industrial Polysaccharides: The Impact of Biotechnology and Advanced Methodologies* (eds S.S. Stivala, V. Crescenzi and I.C.M. Dea), Gordon and Breach, New York, p. 387.

Slade, L. and Levine, H. (1991) Beyond water activity: Recent advances based on an alternative approach to the assessment of food quality and safety. *Critical Reviews of Food Science and Nutrition*, **30**, 115–360.

Tamstorf, S., Joenssen, T. and Krog, N. (1986) The role of fats and emulsifiers in baked products, in *Chemistry and Physics of Baking* (eds J. Blanshard, P. Frazier and T. Galliand), Royal Society of Chemistry, London, pp. 75–88.

Wagner, M.K. and Moberg, L.J. (1989) Present and future use of traditional antimicrobials. *Food Technology*, **43**(4), 143–7, 155.

Willhoft, E.M.A. (1973a) Mechanism and theory of staling of bread and baked goods and associated changes in textural properties. *Journal of Textural Studies*, **4**, 292–322.

Willhoft, E.M.A. (1973b) Recent developments on the bread staling problem. *Bakers' Digest*, **47**(6), 14–21.

Wynne-Jones, S. and Blanshard, J.M.V. (1986) Hydration studies of wheat starch, amylopectin, amylose gels and bread by proton magnetic resonance. *Carbohydrate Polymers*, **6**, 289–306.

# 11
# Principles of Dough Formation

Clyde E. Stauffer

## 11.1. Introduction

The first, basic step in breadmaking is combining water with wheat flour and kneading (imparting mechanical energy to) the mixture to form an elastic dough (Bushuk, 1985; Hoseney, 1985). Flour from wheat, rather than from other cereal grains, is used because wheat storage protein has unique properties; no other cereal storage protein possesses the ability to form a visco-elastic dough when wetted and kneaded. A full explanation at the molecular level for this uniqueness still eludes researchers.

The events that occur when gluten proteins are hydrated and worked are also elusive. Part of the obscurity is due to the complexity of the system. The basic properties of dough are established by the characteristics of the storage (gluten) proteins in the flour. These characteristics, however, are modified by other flour components, both soluble and insoluble, as well as the additional ingredients added to dough. In studying dough formation we are limited to observing physical events on a macroscale, at the supra-molecular level. Numerous techniques that study molecular properties have been applied to dough: X-ray analysis, nuclear magnetic resonance (NMR), differential scanning calorimetry (DSC), electron spin resonance spectrophotometry (ESR) and scanning electron microscopy (SEM), to name just a few. The interpretation of the results, however, is always complicated by the complexity of the system. X-ray analysis, for example, led to a model of the dough matrix (Grosskreutz, 1961) that included gluten proteins, phospholipids and solid (starch) contributions, but there is no way to confirm independently the accuracy of that model. While these techniques each help us clarify certain aspects of dough structure, the concepts that will be set forth in this chapter are of necessity highly speculative. This fact must be kept firmly in mind while reading this or any other publication on dough formation.

The macro-properties of dough change with time. At the end of the mixing process (Chapter 4) the dough has certain visco-elastic characteristics that are considered optimum for subsequent processing. The resting period (floor-time) changes these properties and makes the dough more pliable (relaxed). Dividing and rounding reverses this to some extent and the dough appears more elastic (less relaxed). An intermediate proof period decreases the elasticity, allowing good molding into the shape of a loaf. During proofing the characteristics are further modified, not only by relaxation but also by changes in matrix composition from the products of fermentation (ethanol, carbon dioxide), by the action of additives (oxidants and enzymes) and possibly by the action of native flour proteases. Again, our understanding of the molecular alterations resulting in these modifications in dough properties is rudimentary, at best.

Governing all our discussions about dough formation (and the breadmaking process) is the fact that the ultimate criterion of 'good' or 'poor' structures and processes is the final product – a good loaf of bread. The two main contributors to bread quality are volume (stability in the prover or proof-box and good oven spring) and a fine, silky crumb. These desirable outcomes depend, obviously, on certain optimum properties in the dough matrix. Two characteristics define 'good' dough:

1. the ability to retain gas (carbon dioxide), generated during fermentation (proofing), in the form of numerous small gas cells;
2. a proper balance of viscous flow and elastic strength so that the loaf can expand adequately during proofing and the early stages of baking, yet retain its rounded form.

Gluten (hydrated wheat storage protein) is the component of dough that determines how well these requirements are met. While other flour components affect gluten functionality, and mechanical energy input during mixing is crucial to developing the proper characteristics, it is still the physicochemical nature of gluten proteins with which we will be mainly concerned in this discussion of dough formation.

## 11.2. Flour and Dough Components

Wheat flour components (dry basis) can be classified into six groups:

1. starch;
2. storage (gluten) proteins;
3. non-starch polysaccharides (pentosans);
4. lipids;
5. water-soluble proteins;
6. inorganic compounds (ash).

Starch is relatively inert during dough mixing, but plays a role as a 'filler' that contributes to increased dough visco-elasticity. (Starch, of course, has a critical influence during the baking process, when it gelatinizes, and during subsequent storage, when retrogradation accounts for the major part of bread staling (see Chapter 10).) Endogenous inorganic materials are relatively unimportant in dough formation, although added salt strongly influences dough properties. The other four component groups listed are actively involved in dough formation during mixing and subsequent processing.

## 11.2.1.  Starch

Starch represents by far the largest portion of flour, making up about 65% of ordinary flour (14% moisture basis). Wheat starch comprises about 23% amylose and 73% amylopectin (thus the two species represent 15% and 50% of the flour weight, respectively). Amylose is a linear chain of α-1,4 linked glucose units, with a molecular weight in the range of 100,000 Da, while amylopectin is a highly branched structure, with an estimated molecular weight in the range of 20,000,000 Da (Figure 11.1). Native starch exists as granules and has a high degree of crystallinity, evidenced by birefringence (the 'Maltese cross' seen when it is examined with a polarizing microscope). These granules are relatively inert during mixing but influence dough elasticity by their presence in the total matrix. In hard wheat flour, as much as 15% of the starch granules (10% of the flour weight) are 'damaged', that is they have been deformed during milling and contain cracks and fissures. Damaged starch granules absorb about four times as much water as intact granules, and increase dough water absorption (see below and Chapter 12). Also, damaged starch is much more susceptible to the action of α-amylase than is intact starch, a fact that enters into dough property modification during the proofing stage of processing.

## 11.2.2.  Gluten

The storage protein found in flour is not, strictly speaking, gluten; that term designates the hydrated glutelins (glutenins) and prolamines (gliadins) formed when a dough is mixed. However, for convenience in this discussion the anhydrous storage protein of wheat endosperm will be called gluten (as the term is commonly used). Of the total protein in wheat flour, about one-sixth is soluble protein (albumins and globulins), falling into group 5 in the list above.

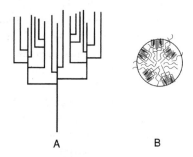

A                 B

FIGURE 11.1. (A) Amylopectin; (B) a starch granule.

Thus a flour having 12% protein contains only about 10% gluten. The molecular characteristics of gluten proteins will be examined in more detail below.

## 11.2.3. Pentosans

Non-starch polysaccharides represent only about 2–2.5% of flour (Michniewicz et al., 1990), but have a disproportionate influence on dough properties. They are sometimes called hemicelluloses because they constitute part of the cell wall materials (formed in conjunction with cellulose) in the seed, made by the plant as the wheat berry is synthesized and ripening. More often the name pentosan is used because approximately 80% of the sugars present are the pentoses D-xylose and D-arabinose. The pentosans are a heterogeneous group of macromolecules, but the preponderant backbone structure is a xylan, a chain of β-1,4 linked D-xylose units. Various other sugars are attached to this chain by α-1,3 linkages; the major side chain sugar is arabinose, but small amounts of glucose, fructose and mannose are also found. Besides the xylans, a significant amount of arabinogalactan (polygalactose chain with arabinose side chains) is present in the water-soluble portion.

About 65% of wheat flour pentosans are water-insoluble (WI) pentosans (Michniewicz et al., 1990); these are almost exclusively xylans. The water-soluble (WS) pentosans are approximately half arabinoxylans and half arabino-galactans. Pentosans are gums; they absorb several times their weight in water and form highly viscous solutions. This is especially important in rye flour (Chapter 13), where the pentosan content may be as high as 10%; the viscosity that allows rye flour to form a dough is due almost exclusively to the pentosans. The water-absorbing property of pentosans is influential in wheat flour doughs (see below) and the viscosity due to the WS pentosans influences the visco-elastic behavior of dough.

Flour pentosans form gels when treated with certain oxidants (Geissman and Neukom, 1973). The mechanism involves ferulic acid, an α, β-unsaturated aromatic carboxylic acid that is esterified to arabinoxylans (Figure 11.2). Oxidants that generate free radicals, for example hydrogen peroxide, promote cross-linking between ferulic acid residues on adjacent polymer chains and lead to gelation of suspensions of both WI and WS pentosans. The formation of covalent linkages with gluten proteins has been postulated (Neukom and Markwalder, 1978), via reaction between ferulic acid and tyrosine side chains of the protein (Figure 11.3). Sulfhydryl compounds react with α, β-unsaturated aromatic acids. [14]C-tagged cysteine binds to WS pentosans (Sidhu et al., 1980b), presumably through addition of the —SH group across the activated double bond. Cysteine inhibits the oxidative gelation of pentosans (Hoseney and Faubion, 1981). Carboxylic acids with an activated double bond (e.g. fumaric acid and cinnamic acids) drastically reduce the mixing stability of doughs (Hoseney and Faubion, 1981). These facts have been interpreted to mean that during normal dough mixing, feruloyl moieties attached to pentosans are attached to gluten

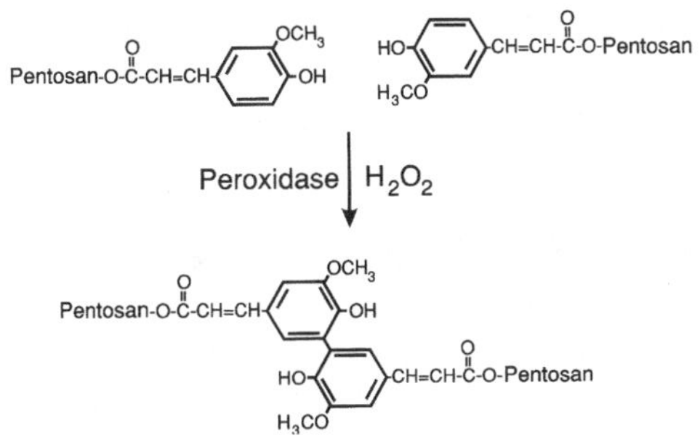

FIGURE 11.2. Oxidative crosslinking of ferulic acid in flour pentosans.

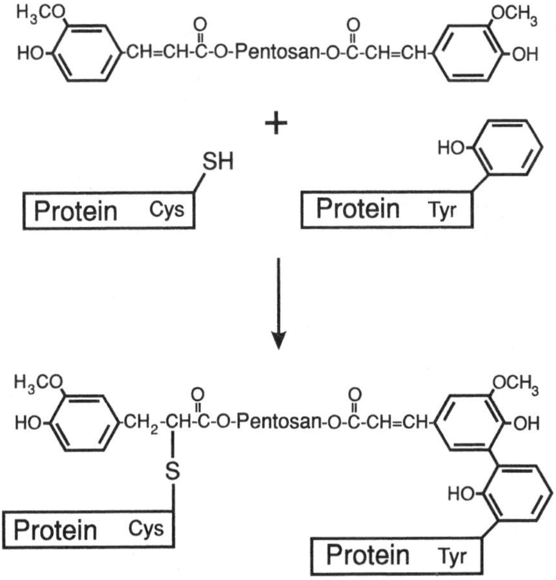

FIGURE 11.3. Reactions of water-soluble pentosans with dough proteins. Left: addition of –SH to ferulic acid. Right: ferulic acid linking to tyrosine.

proteins via addition of sulfhydryl groups across the activated double bond, generating cross-links and enhancing dough elasticity (Figure 11.3).

## 11.2.4.  Lipids

Wheat flour contains about 2.5% lipids. Of this, about 1.0% is non-polar lipids (triglycerides, diglycerides, free fatty acids and sterol esters). The two main groups of polar lipids are galactosyl glycerides (0.6%) and phospholipids (0.9%). During mixing, both classes of lipids are complexed with gluten and become relatively un-extractable with any of the usual solvents (Daniels et al., 1976;

Wootton, 1976; DeStefanis et al., 1977). Grosskreutz (1961) interpreted his X-ray spacings to indicate the presence of bimolecular layers of polar lipids in the gluten complex, and polar lipids have been proposed as adhesive agents between starch granules and gluten protein in several other models (Chung, 1986).

Flour lipids appear to have little effect on mixing requirements; the mixogram is identical for control and defatted flours (Schroeder and Hoseney, 1978). On the other hand, the addition of anionic surfactants such as sodium dodecyl sulfate strengthens the dough and increases mixing time (Danno and Hoseney, 1982). While these are not flour components, the effect offers further insight into the nature of the gluten complex (and how mechanical mixing modifies it) and will be discussed in more detail below.

Lipids have a major influence on baking performance of bread, especially with respect to oven spring (loaf volume) and the keeping quality of the finished product. However, this aspect of lipids in baking is outside the scope of this chapter.

## 11.2.5.  Water-soluble Proteins

The water-soluble fraction of flour (approximately 2–3% of total flour weight) contains albumins and globulins as well as WS pentosans. The proteins include enzymes, enzyme inhibitors, lipoproteins, lectins and globulins of unknown function. It is reported that two-dimensional electrophoresis shows over 300 components. These undoubtedly have some biological function in the seed, relating to its primary role as the progenitor of the next wheat plant. The only clearly identified role played by any of these compounds in baking is the action of β-amylase on starch, generating maltose which serves as a fermentable sugar for yeast during proofing of lean doughs.

Reconstitution studies with fractionated flour components show that the water-soluble fraction plays a role in the breakdown of over-mixed dough (Schroeder and Hoseney, 1978). A mixogram of only the gluten and starch fractions showed a long mixing time with extremely long mixing tolerance; when the water-soluble fraction was also included the mixogram resembled that of the control (unfractionated) flour. The fraction was dialyzed (removing low molecular weight materials) and then heated (denaturing proteins). The remaining soluble material (presumably mainly WS pentosans), when added back to the gluten plus starch fractions, gave a mixogram similar to that of the control, that is mixing to a peak followed by relatively rapid breakdown.

## 11.2.6.  Ash

Wheat flour contains about 0.5% ash (higher if a longer extraction rate is used by the miller; see Chapter 12). This inorganic material has little influence on dough formation. The addition of salt, however, increases the resistance of dough to mechanical mixing and decreases water absorption, presumably due to enhanced gluten aggregation. The mechanism of this effect of salt (sodium chloride) as well as other inorganic salts will be discussed in more detail below.

## 11.3. Flour Components and Water Absorption

An important factor in commercial bread dough production is the proper water-to-flour ratio. In common usage this ratio is called 'absorption', and expressed as a percentage of the flour mass. The adjective 'proper' differs depending upon the kind of dough being made (absorption is much lower for a bagel dough than for a white pan bread dough) and the method used in its measurement (Farinograph absorption can be 2–4% lower than operational absorption). Operational (or baking) absorption, of course, means the water-to-flour ratio that results in a dough having the handling (machinability), proofing and baking (loaf volume) and finished product (appearance, eating quality) characteristics necessary to give the desired baked food (bread). As stated earlier in this chapter and discussed in Chapter 1, the criterion of 'good' or 'poor' is determined by the final product and the consumer. The contribution of various flour components to absorption has usually been made using some sort of instrumental measurement, with the instrument defining 'correct' absorption. This is most often the Brabender Farinograph (Chapter 12), which records mechanical resistance as a simple mixture of flour and water is kneaded. Farinograph absorption is the water-to-flour ratio that results in a recorder trace which, at its maximum, is centered on the 500 (or 600 in the UK) Brabender units line. This is generally lower than baking absorption – the water-to-flour ratio, determined by an experienced mixer operator, that gives optimum dough handling and final product qualities. The absorption numbers that are presented here must be understood as representing the relative water uptake by the various components, and not as an attempt to allow precise calculation of baking absorption for a given flour based upon analytical data.

Four flour components absorb water: protein, native starch, damaged starch and pentosans. The relative absorptions (in grams of water per gram of component) are given in Table 11.1. Using analytical data (typical for a hard red spring wheat flour) shown in column 3, an absorption of 68.4% is calculated (total of column 4), which is reasonable for such a flour.

TABLE 11.1. Influence of flour components on absorption.

| Component | Water per g component (g) | Amount per 100 g flour (g) | Absorption per 100 g flour |
|---|---|---|---|
| Protein | 1.3 | 12 | 15.6 |
| Intact starch | 0.4 | 57 | 22.8 |
| Damaged starch | 2.0 | 8 | 16.0 |
| Pentosans | 7 | 2 | 14.0 |

Both soluble and insoluble (gluten) proteins absorb water (Greer and Steward, 1959; Bushuk, 1966). We might expect that absorption by the insoluble gluten proteins would have more effect on dough rheology than solution of the soluble proteins, but the specific evidence for such a conclusion is somewhat indirect. The greater influence of gluten proteins on baking absorption is highlighted in a study by Tipples et al. (1978). They measured total protein, wet gluten, gluten 'quality' (wet gluten quantity divided by total protein), damaged starch and pentosan contents of flours from a number of milling streams of Canadian hard red spring wheat. They also measured Farinograph absorption and baking absorption (for several different baking protocols) of the flours. The most important predictor of Farinograph absorption was damaged starch content; inclusion of total protein improved the prediction equation significantly, but including the other factors produced no further improvement (i.e. no increase in $r^2$). Of the five baking tests they used, the one most like commercial (North American) production methods was the remix test that included 0.3% malt. This test is similar to the standard American Association of Cereal Chemists (AACC) baking test, with a long (nearly 3 h) bulk fermentation, but the dough is then remixed before being molded and panned. The most important single factor in predicting baking absorption was gluten quality. Adding simple protein level as a prediction factor significantly increased reliability ($r^2$), but the other factors (damaged starch and pentosan content) were not statistically significant.

Native starch granules are relatively impermeable to water. This may be due in part to the lipids and protein found on the surface of the granules, probably derived from the cell wall of the amyloplasts present in the ripening wheat berry (Greenwood, 1976). While native starch is the largest single contributor to absorption, this is due to its preponderance in flour. During baking, of course, when these granules swell and gelatinize, the contents become readily hydratable and are probably the main water-binding species in baked bread.

Most damaged starch is formed during milling (though amylase action in sprouted grain can also cause starch damage). During the process of reducing chunks of endosperm from the break rolls to flour on the reduction rolls, the particles are subjected to extreme pressure. The granules are somewhat elastic and return to their original shape after the pressure is relieved, but some granules are left with cracks and fissures. These represent spots where water can readily penetrate to the interior of the granule and interact with the amorphous regions found there. More pressure at the reduction rolls is needed to break up hard wheat endosperm than that of soft wheat; hence hard wheat flours typically have a higher damaged starch content (6–12%) than soft wheat flours (2–4%). These cracks also represent points of susceptibility to amylase action, in contrast to intact starch granules which are resistant to amylolytic attack under ordinary conditions. Digestion by amylases in dough releases maltose, which can be fermented by yeast, as discussed in Chapter 3. Also, during proofing digestion of damaged starch decreases its water-holding capability, releasing more water into the dough matrix and increasing pan flow. Significant amylolytic activity requires some period of time, and is not a factor during the relatively short time involved in dough formation.

Studies on the water-absorbing capabilities of pentosans give rather varied results. Kim and D'Appolonia (1977) added isolated pentosans to flour and measured the change in Farinograph absorption. The addition of 1% WS pentosan increased absorption by 4.4%, while 1% WI pentosan increased absorption by 9.9%. Michniewicz et al. (1990) added WI pentosan to various hard wheat flours at different levels, and measured the changes in Farinograph absorption. They found increases in absorption ranging from 3.2 to 5.6 g of water per gram of pentosan; the increment was smaller when the intrinsic baking quality of the test flour was better. Patil et al. (1976) used flour fractionation and reconstitution studies to explore the effect of flour water solubles and WS pentosan fractions on absorption, mixing time and loaf volume. They found essentially no effect of WS pentosan on baking absorption (and a small, variable effect on mixing times). In recent studies examining the effect of added xylanases on dough consistency, it was reported that allowing time for the enzymes to degrade pentosans has approximately the same effect as increasing absorption by 10% (H. Moonen and H. Levine, unpublished). Based on all these reports, a median value of 7 g of water absorbed per gram of flour pentosans was chosen for inclusion in Table 11.1.

## 11.4.   Wheat Gluten Proteins

Wheat proteins have occupied a central position in protein studies since early times. Gluten was first recognized as the rubbery component of wheat flour in 1729 (Bailey, 1941), although at that early stage it was not called protein (the term had not yet been coined). The common method of characterizing proteins based on their solubility was developed using wheat proteins (Osborne, 1907). According to Osborne's scheme, proteins were divided into four groups:

1. albumins, soluble in distilled water;
2. globulins, soluble in dilute salt solutions;
3. prolamines, soluble in 70% aqueous ethanol;
4. glutelins, soluble in dilute acid.

Gluten proteins are members of the latter two groups.

### 11.4.1.   Amino Acid Composition

Wheat gluten proteins are anomalous, even compared to other cereal storage proteins, in their amino acid make-up (Kasarda, 1989). About one-third of the residues are glutamyl residues, which are almost entirely in the form of glutamine (the amide of the side chain carboxyl group). The amide, a non-ionizing group, readily forms hydrogen bonds with electron donors (other amides and water molecules). The content of basic amino acid residues (arginine, lysine and histidine) is relatively low, and the amount of carboxylic acid residues (aspartic and glutamic acid) is even lower. As a result the proteins have a rather low surface charge density, even at pH values somewhat removed from the isoelectric point. Since the charge repulsion between molecules is low the protein chains can approach each other and interact (form hydrogen bonds) in the aqueous dough matrix. The addition of sodium chloride further suppresses charge repulsion, increasing molecular interaction.

Gluten also contains a higher level (about 14% of the residues) of proline than that is usual in proteins. This amino acid favors formation of β-sheets and similar structures that are thought to be responsible for some of the elastic characteristics of gluten (see discussion below).

While the content of hydrophobic amino acids is not unusual, the lack of ionic character makes hydrophobic interactions between protein chains possible. The hydrophobicity of gluten proteins has been demonstrated experimentally by chromatography of acid-solubilized gluten on hydrophobic gel media such as phenyl-sepharose (Chung and Pomeranz, 1979). These authors examined gluten from two flours having different baking properties, and found that glutenin from the good-baking flour was more strongly absorbed to the gel than the glutenin from the poor-baking flour. Surprisingly, the relationship was reversed for gliadin; that from the poor-baking flour was bound somewhat more strongly than gliadin from the good-baking flour. Kaczkowski et al. (1990), on the other hand, found gliadin from good-baking wheat to be slightly more hydrophobic than gliadin from wheat of medium-baking quality. They used binding capacity for sodium dodecyl sulfate as their criterion of hydrophobicity, a difference in technique that might account for the discordant results.

## 11.4.2.  Gliadin

Actually a heterogeneous group of prolamines, more than 70 different gliadin species have been identified, using chromatography and electrophoresis. They are rather hydrophobic, hence their insolubility in water or salt solutions, but can be divided into groups based upon their degree of hydrophobicity. More hydrophobic gliadins (the γ-gliadins) increase bread loaf volume, while gliadins from the more hydrophilic end of the spectrum (Θ–gliadins) decrease loaf volume (Weegels et al., 1990; van Lonkhuijsen et al., 1992). Gliadin proteins are relatively small, with molecular weights in the range 30,000–100,000 Da. They are single-chain proteins (i.e. no cross-links between chains) and such disulfide bonds that occur are all intra-molecular (Figure 11.4A). Concentrated solutions of isolated gliadin are highly viscous, with little measurable elasticity.

## 11.4.3.  Glutenin

Glutenin is the type example of Osborne's glutelins. Like gliadin it is quite hydrophobic (its amino acid composition is similar to that of gliadin) but it has a very different molecular structure; glutenin is a polymeric protein. The average

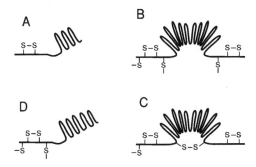

FIGURE 11.4. Schematic depiction of gluten proteins. (A) Gliadin; (B) HMW glutenin subunit, showing possible action of the β spirals as 'molecular spring'; (C) HMW glutenin subunit, showing disulfide bond preventing extension of the β spirals; (D) LMW glutenin subunit.

molecular weight of native glutenin is stated to be about $3 \times 10^6$ Da, a number that is highly approximate and serves only to characterize the wide molecular weight distribution of glutenin (Kasarda, 1989). Polymerization takes place via intermolecular disulfide bonds. Reduction of these bonds with a reagent such as dithiothreitol (DTT) frees the basic glutenin subunits, which can be separated using SDS-PAGE (electrophoresis in a polyacrylamide gel in the presence of a high concentration of sodium dodecyl sulfate, a technique that separates proteins on the basis of their molecular weights). Two groups of subunits are identified. High molecular weight glutenin subunits (HMW-GS) have apparent molecular weights in the range 80,000–120,000 Da, while the molecular weights of low molecular weight glutenin subunits (LMW-GS) are about 40,000–55,000 Da. The molar ratio of LMW-GS to HMW-GS is 2:1 or higher; the amounts of the two kinds of subunits are roughly equal on a weight basis.

The molecular architecture of glutenin subunits is unusual (Figure 11.4). In HMW subunits cysteine is concentrated in the regions near each end of the chain, with a long stretch of other amino acids between these two ends. The cysteine residues are involved in both intra- and inter-molecular disulfide bond formations. LMW subunits have a similar concentration of cysteine residue, but at only one end of the chain (Figure 11.4D). Thus, both ends of the HMW protein can enter into polymerization reactions, while only one end of the LMW protein can react this way. The interior regions of both species, but the HMW subunits in particular, are postulated to form β-turn spirals, which in turn can fold into a helical sheet structure that can possibly be likened to a coil spring (Figure 11.4B). This is only a hypothesis, but it is an attractive one that could account for the elastic nature of glutenin. An intra-molecular disulfide bond can restrain this 'spring' (Figure 11.4C); this bond can be broken during mixing (see below) to 'develop' the gluten structure. Isolated glutenin, when re-hydrated, forms an elastic, rubbery mass that has almost no viscous flow characteristics.

Numerous proposals have been put forward for the structure of glutenin polymers in dough (Figure 11.5). Unfortunately, because glutenin is such an

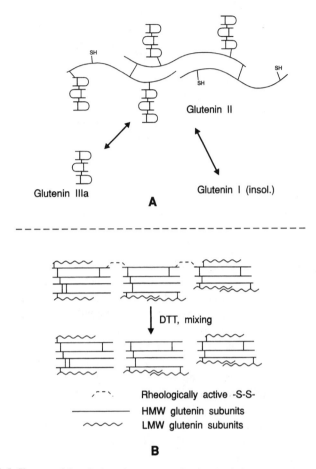

FIGURE 11.5. Two models of glutenin structure in dough. (A) Adapted from Graveland et al. (1985); (B) adapted from Gao et al. (1992).

intractable protein to examine, these must of necessity be considered highly speculative. Graveland et al. (1985) postulated a basic 'building block' of three glutenin subunits linked through disulfide bonds (glutenin IIIa) and a tetramer of this basic structure (glutenin IIIb). These react with linear proteins having two or more reactive sulfhydryl sites, to form a larger molecule called glutenin II. Glutenin I is a highly polymerized, insoluble protein which is thought to be the glutenin protein present in wheat flour. It is partially depolymerized during mixing and reforms during the resting stage of dough processing. Gao et al. (1992) examined the effects of small amounts of DTT on dough consistency in the Farinograph, and arrived at a slightly different model. They also postulated a subunit structure similar to Graveland's, but specified both HMW and LMW subunits in their 'building block'. In their work, DTT reduction of a small portion of disulfide bonds produced the maximum decrease in consistency;

they called these 'rheologically effective —S—S—'. Jones et al. (1974) used a similar approach (a change in Farinograph consistency in the presence of small amounts of DTT) to conclude that only about 3% of the disulfide groups in flour affect the rate of dough development, and about 12% of the disulfide groups are involved in mixing resistance. The best current picture of glutenin is that it is a large linear polymer, and interactions between glutenin chains in dough is through non-covalent forces, namely hydrogen and hydrophobic bonds (Ewart, 1977).

## 11.5.   Stages in Dough Formation

The word 'dough' connotes a semi-solid mass that resists mixing. In a recording mixer such as the Mixograph, a dough in the mixing bowl gives enhanced (and variable) resistance as the mixing head rotates, while a batter (a semi-liquid mass) causes low resistance with little variability (narrow band width). The progression from a simple mixture of water and flour to a dough correlates with an increased resistance to mixing, however that may be recorded. A Mixograph is just as much a mixer as any large commercial dough mixing machine, and the trace can be easily translated into consistency changes during dough mixing. In Figure 11.6 a typical Mixograph trace shows the various stages of dough

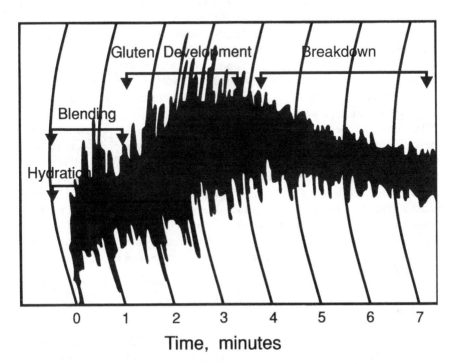

FIGURE 11.6. A Mixograph trace, indicating the four main stages of dough formation.

formation: hydration, blending, gluten development and breakdown. These same stages can be observed in a commercial horizontal mixer (Pyler, 1988). Bakers in the USA refer to them as the 'pickup stage' (Pyler, 1988, Figure 14.3), 'cleanup stage' (Figure 14.4), 'development' (Figures 14.5–14.7) and 'letdown' and 'breakdown' (Figures 14.8 and 14.9), respectively. Similar changes can be observed in most dough mixing and development systems in use around the world.

## 11.5.1.  Hydration

In flour most of the protein exists as a flinty material. An analogy for the hydration of this protein is hydration of a bar of soap. If the soap is simply immersed in a bowl of water, the water slowly penetrates the outer layer of the bar. Rubbing the soap wipes off this soft, hydrated layer, and water proceeds to penetrate further into the soap. In the same way, the initial action of the mixer hastens the conversion of the flinty protein bodies into a soft, hydrated (but not truly dissolved) protein dispersion that is further modified during gluten development. Simultaneously the WI pentosans and damaged starch granules are absorbing water, and the water-soluble flour components (and added water-soluble ingredients such as salt and sugar) are dissolving.

The soap analogy is not strictly accurate. When water is brought into contact with flour particles and the process is observed under a microscope, the particles seem to explode; strands of protein are rapidly expelled into the aqueous phase (Bernardin and Kasarda, 1973). Movement of the cover glass stretches the protein strands, indicating their extensibility (Amend and Belitz, 1990). The rapid extrusion of protein fibers appears in part to be due to surface tension at the air–water–protein interface. Amend and Belitz (1990) submerged flour particles in acetone, which was then replaced by water. The particles swelled but no fiber formation was evident.

The input of mechanical energy is crucial to dough formation. A simple exercise demonstrates this fact. Blend cold wheat flour with powdered ice (in a 100:65 ratio) and then allow it to warm to room temperature. The result is a thick slurry that has no dough-like properties. When this slurry is stirred it rapidly increases in consistency, forming a soft (undeveloped) dough. Hydration alone is not sufficient to make a dough. Tkachuk and Hlynka (1968) substituted $D_2O$ for water to show the importance of the formation of hydrogen bonds in dough. The mixing energy required to develop a dough using $D_2O$ was much greater than that when water was used indicating that the hydrogen bonds formed with $D_2O$ were significantly stronger.

## 11.5.2.  Blending

Flour particles are agglomerates of starch granules embedded in a network of protein (Figure 11.7). As the protein network is softened by hydration and agitated by mixing, the starch granules become less firmly attached to the protein,

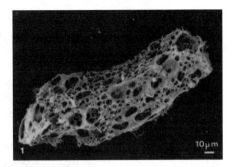

FIGURE 11.7. Hydrated flour particle after starch is removed by enzymatic digestion, showing the protein framework (from Amend, 1995, by permission).

but nevertheless remain associated with the protein fibers (Figure 11.8). Most of the starch can be removed by washing and kneading the dough (the basis for isolating wet gluten) but it cannot be totally removed. SEM photos of optimally mixed dough indicate that most of the starch is readily removable, but a small number of granules appear to have protein fibrils strongly attached to them (Amend and Belitz, 1990, Figure 20). The actual strength of the starch–protein interactions has not, of course, been measured, but only inferred from observations such as those described.

During this early stage of mixing, all the ingredients of dough are being blended, to give a dough mass that is, at least at the millimeter scale, homogeneous. Lipids (flour and added lipids) are uniformly distributed and brought into contact with the protein fibers, and soluble materials are fully dissolved and distributed in the aqueous matrix.

## 11.5.3.  Gluten Development

The pivotal step in forming a wheat-flour dough is the increase in consistency (increased resistance to mixing) that is generally called 'dough development'.

FIGURE 11.8. Protein film and associated starch granules in a hydrated, stretched flour particle (from Amend, 1995, by permission).

During this stage of mixing, the flour–water mixture is converted from a thick, viscous slurry to a smooth visco-elastic mass, characterized by a dry, silky appearance (and feel) and the ability to be extended into a thin continuous membrane. The most important practical parameter is mixing time (which can be equated to total energy input; however, in this context the impact of mixing speed on the degree of dough development discussed in Chapter 2 should be noted), the time required to reach the peak consistency (maximum resistance) of the dough. Dough mixed to this point gives the maximum loaf volume, as compared with dough that is under-mixed or significantly over-mixed. It should be noted that in commercial practice mixing is usually extended slightly beyond the peak, giving a dough with better machinability in the subsequent molding step, and one less likely to exhibit a 'wild shred' during baking. Data published by Millar (2003) has also shown that maximum bread volume is achieved after mixing beyond an 'NIR optimum' for dough and that the finest cell structure that could be achieved in bread made by the CBP occurred some time before the volume optimum. Such observations support the practical experience of bakers for a need to mix beyond peak dough resistance.

The previous paragraphs describe gluten (dough) development on the macroscopic scale. Scanning electron microphotographs of gluten at various stages of development have been published. One such series is shown in Figure 11.9 (Amend, 1995). At early stages (corresponding to the 'hydration' segment of

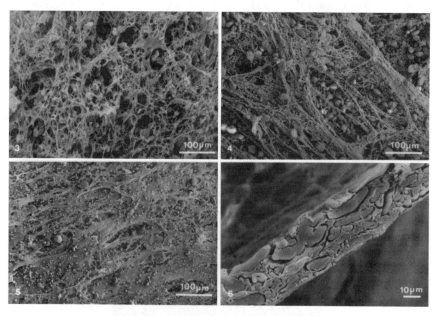

FIGURE 11.9. Gluten network in dough at various stages of mixing. (a) Early in the process (at about the middle of the hydration stage); (b) partially mixed dough; (c) dough at maximum development stage (from Amend, 1995, by permission).

Figure 11.6) the fibrils of hydrated protein adhere to each other, forming a rather coarse, random network of large strands (Figure 11.9a). The action of the mixer stretches these strands, thinning them but also orienting them along the direction of the stretching action, allowing them to interact with each other (Figure 11.9b). At the peak of consistency (Figure 11.9c) the protein fibrils have been significantly reduced in diameter, and they appear to interact two-dimensionally, rather than just along the individual strand axes. In other words, at this stage the gluten appears able to form the continuous film, or gluten sheet, that is used by a baker (by hand-stretching a piece of dough) to evaluate completeness of mixing.

The crucial question, and one that continues to generate much research, is 'What happens on the molecular level during dough mixing?' The research is complicated by the complexity of the dough system, and the fact that the main species involved (glutenin) is a high molecular weight polymeric protein that is, to a large extent, insoluble. Nevertheless, progress toward the answer is being made.

It seems clear that mixing breaks the high molecular weight glutenin into smaller units, which then reform to some extent. Graveland et al. (1985 and references therein), for example, found that during short high-energy mixing the amount of glutenin having a lower molecular weight (less than 1 MDa) increased sharply, but then decreased again when the dough was allowed to rest. This is the basis for the model shown in Figure 11.4A; the insoluble glutenin I (assumed to be the form in the dry endosperm) is depolymerized, perhaps down as far as the glutenin III subunits, but also repolymerizes to glutenin II during a subsequent resting stage. Ewart (1977) came to a similar conclusion in his consideration of mixing action that glutenin macromolecules are broken during mixing. The points of scission are thought to be at the disulfide bonds, forming thiyl radicals ($-S-S- \rightarrow 2 -S^*$). The presence of free radicals in flour was shown by Redman et al. (1966; see also Dronzek and Bushuk, 1968). Sidhu et al. (1980a) showed that fumaric acid formed an adduct with cysteine residues of glutenin, a reaction that probably proceeds via addition of the thiyl radical to the $\alpha, \beta$ double bond of fumaric acid.

The importance of stress-mediated scission of disulfide bonds in developing gluten is consonant with several lines of evidence. It is well known that increasing the absorption in a dough in the bakery increases the mixing time. Tipples and Kilborn (1977) found that the critical speed of mixing (the minimum mixing rpm necessary to achieve good loaf volume) increased as absorption increased. They could make a well-developed dough at 100% absorption (using a high-quality Canadian wheat flour) by running the mixer at a high speed. As the dough consistency decreased, a higher rate of energy input was required to achieve the necessary stress to break disulfide bonds. The mixing time (at fixed rpm) required to develop doughs is highly correlated with the amount of glutenin in the flour (Orth and Bushuk, 1972; Singh et al., 1990); more glutenin requires more energy input to be broken down and rearranged.

As disulfide bonds are broken, they reform between adjacent molecules that have been aligned along the lines of stress in the dough. Several different combinations can be envisioned:

$$2 -S^* \rightarrow -S-S-$$
$$-S^* + -SH \rightarrow -S-S- + H^*$$
$$-S^* + -S_1-S_2- \rightarrow -S-S_1- + -S_2^*$$

The end result of these rearrangements is the linear glutenin polymers envisioned in Figure 11.5.

A common picture of this process is simple thiol–disulfide interchange, as proposed by Goldstein (1957). This is unlikely, however, because such a reaction proceeds via nucleophilic attack of the thiolate anion on the disulfide. At dough pH (approximately 5), less than 0.1% of the thiol groups would be ionized ($pK_a$ of —SH is approximately 8.5). These interchanges are more likely to involve a free radical mechanism, as described here.

While rearrangement of glutenin is the major consumer of mixing energy, it is not the only process occurring. The protein also incorporates lipids from the flour and any added emulsifiers and shortening. Grosskreutz (1961) used X-ray studies to conclude that developed gluten has a lamellar structure, with lipid bi-layers interleaved with protein layers. Other researchers (e.g. Chung, 1986) have proposed different models for the protein–lipid interaction. All that can be confidently stated is that most polar lipids and a significant fraction of non-polar lipids become tightly associated with the gluten protein (DeStefanis et al., 1977; Chung et al., 1978). The precise role played by these included lipids in dough properties (and final loaf volume) is still not fully clarified (Pomeranz, 1985).

The final result of development is thought to be an alignment of extended, nearly linear polypeptide chains, interacting through ionic and hydrophobic forces (Ewart, 1977). This will be discussed in detail below.

## 11.5.4.  The Formation of Other Bonds

The dominance of the di-sulphide bond in dough formation is undisputed but other bonds are formed during mixing which contribute to dough development. The formation of hydrogen bonds has already been introduced. Disruption of the hydrogen bonds, for example with urea (Wrigley et al., 1998) weakens the dough while for metal chloride ions (e.g. sodium chloride) gluten strength is increased (Eliasson and Larsson, 1993) because higher charge densities result in more hydrogen bonding in the structure.

The recent application of spectroscopic techniques led Belton (1999) to develop the so called 'loop and train' model for the interaction of glutenin subunits in dough. In his model, Belton postulates that individual glutenin subunits interact with one another by disulphide bonds at the ends of the subunits and hydrogen bonds along repeat regions. The 'loops' formed at repeat regions are where the water is bound, and when extension is applied to the system, such as during mixing, the loops disappear and the 'trains' are formed. If the extension force is removed and the polymer relaxes, then loops may be re-formed.

More recently a hypothesis has been developed (Tilley et al., 2001) for the formation of dityrosine cross-links in dough as a contribution to dough development. Tilley et al. postulated that the addition of a free tyrosine source prevents the over-formation of tyrosine cross-links and enhances dough stability. Millar et al. (2005) examined the effect of adding free tyrosine and concluded that the effect of tyrosine addition varied with flour type and in one case a soft milling variety with weak gluten characteristics recorded an improvement in dough rheological properties as assessed with the DoCorder. The role of enzymic activity in the modification of tyrosine cross-links has also been reported (Tilley and Tilley, 2005).

## 11.5.5.  Breakdown

If mixing continues after peak development is reached the dough becomes softer, less resistant to mixing action and loses its ability to retain gases during proofing. SEM photographs indicate that the protein strands become shorter and thicker compared with those in optimally mixed dough (Amend and Belitz, 1990). The viscosity of dough proteins extracted into 1% sodium dodecyl sulfate solutions were lower in over-mixed doughs compared to optimally mixed doughs, indicating a smaller average molecular weight (Danno and Hoseney, 1982).

Several $\alpha, \beta$-unsaturated carbonyl compounds, such as fumaric acid, maleic acid, sorbic acid, ferulic acid and $N$-ethylmaleimide all increase the rate of dough breakdown during mixing (Schroeder and Hoseney, 1978). $^{14}$C-Fumaric acid reacts with cysteinyl groups in gluten proteins during mixing, forming $S$-succinyl adducts (Sidhu et al., 1980a). It did not react with cysteine in soluble proteins or with added —SH compounds, leading the authors to conclude that it was combining with thiyl radicals on the gluten proteins. Flour water solubles also contribute to the breakdown phenomenon (Schroeder and Hoseney, 1978; see above). Presumably it is the ferulic acid present in the WS pentosans which causes this effect. Fumaric acid and sorbic acid have been suggested as agents for reducing mixing time; at normal levels of use, and in practical situations, their effect may be too powerful, and the practice has not been widely accepted.

To summarize, dough breakdown appears to be simply a continuation of the process by which flour glutenin I is converted to (relatively) medium weight protein polymers that impart the desired rheological properties to dough.

## 11.5.6.  Unmixing

Tipples and Kilborn (1975) reported an unusual phenomenon, the reversible decrease of resistance of a fully developed dough when it is mixed at a much lower rpm. When mixer speed is returned to that used for original development, dough consistency (and loaf volume potential) rapidly returns to that originally achieved. They termed this 'unmixing'. It is not the same thing as allowing a dough to rest (no mixing action). If a nearly developed dough is allowed to rest, when the mixer is restarted, the consistency first drops to the level that would be the case if it were mixed at low speed, then rises to full consistency. An explanation that has been made (Ewart, 1977) is that with low-speed mixing the gluten molecules are no longer being constrained to extended parallel alignment by shear forces. They tend toward more random configurations, and the low-shear mixing allows these molecules (presumably somewhat more globular in shape) to form interactions that stabilize the less extended configurations.

## 11.5.7. Air Incorporation

In the early 1940s, Baker and Mize (1941) showed that achieving a fine crumb grain depended, in part, on incorporating air into the dough and subdividing the air bubbles into small cells. These serve as nuclei for expansion of the gases formed during fermentation and baking. Junge et al. (1981) determined the course of air incorporation during mixing in a Mixograph (Figure 11.10). Little air is incorporated during the hydration and blending stages of mixing. Entrapment begins only after the dough begins to develop resistance to mixing and some internal structure that can envelop the air bubbles. An interesting point is that incorporation continues well past the mixing peak, into the breakdown portion of the mixogram. Thus it is not simply the elasticity of the dough that is responsible for entraining air; viscosity also seems to play a role (and perhaps also the ability of dough proteins to stabilize foams).

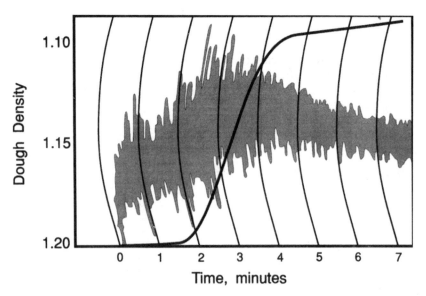

FIGURE 11.10. Incorporation of air in dough mixed in a Mixograph (adapted from Junge et al., 1981).

Chamberlain and Collins (1979) found an interesting corollary to the observation of Baker and Mize: the entrapped gas must contain some nitrogen. They mixed doughs under a pure oxygen atmosphere. The final bread had an extremely coarse grain, with only a few large cells. Their conclusion was that yeast consumed all the oxygen during early stages of fermentation, leaving relatively few gas bubble nuclei for expansion of fermentation gases during proofing and baking, resulting in large voids in the bread (see also Chapters 2 and 4).

## 11.6.  The Gluten Matrix

The final product of dough mixing is a visco-elastic mass that, after appropriate proofing and baking, produces an aerated solid called bread. Bread has a sponge-like structure (the voids are interconnected) with the structural elements being primarily gelatinized starch and denatured protein. The rheological characteristics of dough are primarily responsible for achieving the desired result. Dough rheology, however, is (or should be) traceable to the nature of the matrix elements which are, in this case, gluten proteins. A great deal of dough research has to do with measuring its rheological characteristics, correlating them with bread characteristics (the effects of additives such as oxidants, reductants and surfactants, proofing behavior, loaf volume and crumb grain), and attempting to connect those measurements with such physical characteristics of gluten as can be determined. Much of this research has been presented and reviewed. Some excellent sources (but by no means the only ones) are Bloksma and Bushuk (1988), Bloksma (1990a,b), Faridi and Faubion (1990), Hoseney and Rogers (1990), Eliasson and Larsson (1993) and Cauvain (2004).

While much more is known about dough now than, say, in the 1960s, the current situation might be summarized as follows:

- dough is an extremely complicated system that cannot be fully described in simple rheological terms (springs and dashpots);
- many practical instruments make measurements that are difficult to interpret in fundamental rheological terms, and may or may not be applicable to events during proofing and baking;
- statements about the structure of gluten protein polymers are still largely hypothetical;
- there is great scope for further fundamental research in this area.

I intend to sound realistic, not negative, in this summary. The following discussion is not a detailed consideration of either rheology or protein structure. Rather, salient points from many publications will be chosen, to highlight certain significant findings that relate to the overall breadmaking process. The intention is to formulate a hypothetical framework, hopefully raising useful questions for guiding further research.

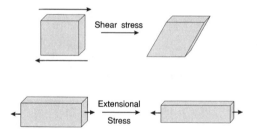

FIGURE 11.11. Diagram of shear and extensional deformation.

## 11.6.1.  Dough Rheology

Numerous discussions of dough rheology are available. The chapter by Menjivar (1993) clearly presents basic rheological concepts, while Bloksma and Bushuk (1988) apply them more specifically to dough. An important point is that two types of stress are involved: shear and extensional (Figure 11.11).

In shear stress opposing forces are applied parallel to each other, in opposite directions to the matrix element (Figure 11.11a). A strain is set up at right angles to the two surfaces. If it remains constant (and the element returns to its original shape when stress is relieved) the deformation is elastic, and elastic modulus is defined as: $E =$ stress/strain. Intuitively, $E$ is larger for more 'solid' materials; a cube of hard rubber has much higher value of $E$ than a cube of sponge rubber. If the strain decreases as a function of time, then when stress is relieved the element does not return to its original shape, and the deformation is viscous. For a simple (Newtonian) fluid, viscosity is defined as $\mu =$ stress/strain rate. In a dough mixer, Mixograph or Farinograph shear stress is the dominant mode.

In extensional stress the opposing forces are applied in opposite directions, but at the opposite faces of the matrix element (Figure 11.11b). The definitions of elastic modulus and viscosity are the same as in shear stress, but the dimensional effects on the element are different. Whereas in shear the element maintains the same cross-section, in extension the cross section decreases as the element lengthens (the volume remains the same in both cases). Extensional stress is applied to a dough by the Extensograph or Alveograph, and also during fermentation (proofing) and baking (oven spring).

Dough is visco-elastic, that is it has both viscous and elastic characteristics. The simplest mechanical model that can be used to interpret rheological studies on dough is the Burgers body (Figure 11.12). When stress is applied to dough the immediate response is elastic deformation (element A), followed by a delayed elastic response due to stretching of element B as element C undergoes viscous flow. Viscous flow by element D relaxes the instantaneous elastic strain on A. When the stress is relieved, any remaining elastic deformation of A is immediately removed. The removal of strain on element B is relieved only as C undergoes viscous flow (in the opposite direction). There is no force to reverse the flow that has occurred in D, so that amount of dough deformation remains when final equilibrium is reached.

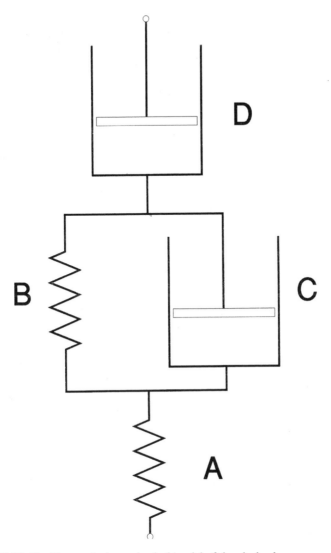

FIGURE 11.12. The Burgers body mechanical model of dough rheology.

A typical creep and recovery curve is shown in Figure 11.13. The strain (deformation) continues as long as stress is maintained on the dough piece. The contributions of the various elements of the Burgers body can be identified on the curves, based on the previous discussion. However, it should be noted that each element is a composite of many elastic and viscous elements in the dough, so that element A (for example, Figure 11.12) actually represents a spectrum of elastic moduli and D comprises a range of viscosities. By collating the results of many such creep and recovery studies an equation relating apparent dough viscosity (element D) to shear stress (Bloksma and Bushuk, 1988, Figures 6 and 7, was

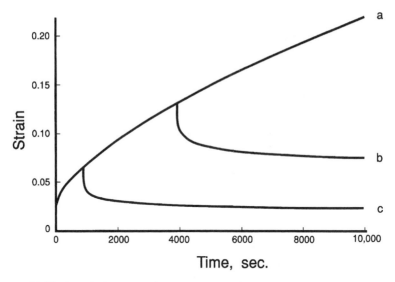

FIGURE 11.13. A typical creep and recovery curve for dough under extensional stress. (a) Stress maintained throughout; (b) stress removed after 4000 s, (c) stress removed after 1000 s (adapted from Hibbered and Parker, 1979).

developed. Dough is a shear-thinning material, and its viscosity was calculated as $1.6 \times 10^5$ Pa s at a shear rate of $10^{-3}$/s and $1.1 \times 10^2$ Pa s at a shear rate of $10^2$/s.

Bloksma (1990a) presents some figures relating shear rates in various laboratory instruments to the situation in dough. They are as follows:

- dough mixers, 10–100/s;
- Farinograph, Mixograph, 10/s;
- Extensograph, Alveograph, 0.1–1/s;
- proofing, $10^{-4}$–$10^{-3}$/s;
- baking (oven spring), $10^{-3}$/s.

The problem with relating test results to actual dough function thus becomes apparent. Extensograph and Alveograph testing involves dough with a viscosity of $(2–8) \times 10^3$ Pa s (calculated according to Eq. 1 of Bloksma and Bushuk, 1988), some two orders of magnitude lower than viscosity in proofing dough. While results from such testing may correlate with dough properties (and qualities such as loaf volume), these should not be taken as 'explanations' of what is actually occurring in the dough.

A typical extensogram is shown in Figure 11.14. The parameters of interest are $R$ (the height of the curve at 5 cm extension), $E$ (the length of curve until the dough piece breaks) and $A$ (the area under the curve). A dough having large values of $R$ but small $E$ is extremely 'bucky', while one with small values of $R$ and large $E$ is very soft and pliable. Extensogram curves have been

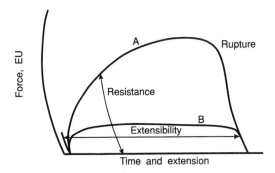

FIGURE 11.14. An extensogram. (A) Dough having good elasticity and good breadmaking properties; (B) dough with poor elasticity and poor breadmaking properties.

transformed into stress–strain diagrams (Rasper, 1975) but little use has been made of this work. More often, one or more of these measurements is correlated with dough properties. One example is the report by Singh et al. (1990), where they found that $E$, $R$ and $A$ for a series of 15 flours were all highly correlated with final loaf volume (the three parameters were strongly inter-correlated, as might be expected, so there was really only one Extensograph test of loaf volume potential).

Bloksma (1990b) expresses the opinion that:

The only rheological properties required for good breadmaking performance appear to be extensibility and a sufficiently large viscosity. Extensibility can be translated into structure; a large quantity of high-molecular-mass glutenins enhances extensibility. The latter of these two conditions, a large viscosity, is met by virtually all doughs; it has no discriminating power.

Before considering the meaning of the term 'extensibility', we must think about the structure to which Bloksma refers.

## 11.6.2.  Gluten Structure

Meredith (1964) proposed that in developed dough the gluten consists virtually of one giant molecule, comprising glutenin extensively cross-linked by disulfide and other bonds. The comparison was made (by other authors) to vulcanized rubber. Bloksma (1990a and references therein) pointed out that this comparison was invalid; temperature changes had opposite effects on the elastic and viscous moduli for dough and for rubber. Ewart (1968, 1977) pointed out that there were several other lines of evidence that substantiated rejection of the 'giant molecule' hypothesis, and proposed that gluten structure was due to interactions between long, linear glutenin polymers. He made the analogy between dough 'strength' and the strength of a rope; while rope fibers are not physically cross-linked, the longitudinal forces between fibers give it a high elastic modulus (resistance

A. Linear Glutenin

B. Glutenin plus Gliadin

FIGURE 11.15. Models of the structure of gluten. (A) The linear glutenin hypothesis (Ewart, 1977); (B) inclusion of gliadin in the structure.

to deformation in extensional shear). The glutenin 'fibers' impart elasticity to dough by virtue of the (non-covalent) bonds between them.

The linear glutenin hypothesis of Ewart envisions numerous long glutenin molecules, aligned somewhat as shown in Figure 11.15A. The contribution of gliadin to dough properties cannot be ignored (Weegels et al., 1990; van Lonkhuijsen et al., 1992). The scheme in Figure 11.15B includes gliadin molecules, which contribute to the interactions between glutenin chains. Some of the possible consequences for dough rheology due to these models are discussed below.

## 11.6.3.  Bonding between Protein Chains

There are three possible types of non-covalent bonds in dough: ionic, hydrogen and hydrophobic bonds (Wehrli and Pomeranz, 1969). Glutenin has a low density of ionizable (acidic and basic) amino acids, so that such bonds would appear to be relatively unimportant in dough. At low pH (for example, in a sponge subjected to long fermentation with consequent formation of much acetic and lactic acid) protonation of the few carboxylic side chains present leads to a significant net positive charge on gluten proteins, weakening inter-chain interactions by ionic repulsion.

The high percentage of amide (glutamine) side chains contributes to extensive hydrogen bonding between chains. The importance of this interaction to gluten elasticity was clearly demonstrated by Beckwith et al. (1963). They treated gluten with methanolic hydrochloric acid, converting amide groups to esters. Conversion increased solubility, decreased intrinsic viscosity of protein solutions and decreased cohesion of the hydrated gluten. Individual hydrogen bonds are relatively weak (about 4.2–6.3 J/mol or 1–1.5 kcal/mol) but the presence of large numbers of them lends overall strength to the inter-chain interactions. The fact that the resistance of dough to elastic deformation decreases with increasing temperature (Bloksma and Nieman, 1975) emphasizes the importance of hydrogen bonds in the proteins. Besides inter-chain bonding, hydrogen bonds also stabilize the β-turn spirals in the central portions of glutenin molecules. These play a role in the interpretation of elasticity presented below.

The relative importance of hydrophobic bonding in dough is difficult to assess accurately. When a dough is mixed in deuterium oxide ($D_2O$) rather than ordinary water ($H_2O$) it is much more elastic (Hoseney, 1976). Both hydrogen (deuterium) bonds and hydrophobic bonds are stronger in the presence of $D_2O$, so this test is not decisive. The addition of various salts, however, does discriminate between the two types of bonds. The Hofmeister (lyotropic) series arranges ions according to their ability to 'salt in' (increase hydration of) proteins as well as other hydrophobic materials. This is interpreted as being primarily an effect on water structure. Lyotropic salts (e.g. magnesium thiocyanate) decrease water structure and increase solubility of ('salt in') hydrophobic chains. Non-lyotropic salts (e.g. sodium chloride and sodium phosphate) enhance water structure and decrease solubility of ('salt out') hydrophobic chains (Tanford, 1973). (The term 'chaotropic' is used synonymously with lyotropic. The advantage is its mnemonic nature; a chaotrope is an ion or molecule that increases the 'chaos' in water structure.) The effect of salts on dough elasticity, absorption and mixing tolerance has been studied by numerous authors (Salovaara, 1982; Kinsella and Hale, 1984; Holmes and Hoseney, 1987). Lyotropic salts (e.g. sodium thiocyanate) increased water absorption by the protein (enhanced its solubility in water), while non-lyotropic salts (e.g. sodium fluoride) decreased absorption. The reported results have been interpreted in terms of protein hydration (Stauffer, 1990), but they can equally well point to the role of hydrophobic bonds in gluten structure.

Glutenin and gliadin are rather hydrophobic proteins, as shown by numerous studies using gel chromatography on hydrophobic media such as phenylsepharose (Chung and Pomeranz, 1979; Weegels et al., 1990). Chung and Pomeranz (1979) found that acid-soluble glutenin extracted from a good-quality flour was more hydrophobic than that from a poor-quality flour. Hydrophobic gliadins increase bread loaf volume, while hydrophilic gliadins decrease loaf volume (Weegels et al., 1990; van Lonkhuijsen et al., 1992). Flour lipids (Daniels et al., 1976; Wootton, 1976) and added emulsifiers (DeStefanis et al., 1977) are bound to gluten during dough mixing, which must be in large part due to hydrophobic interactions. Hydrophobic interactions are weaker than hydrogen bonds (approximately 2500 J or 600 cal per $CH_2$ group) but again, because of the rather large number of available interaction sites, the overall contribution to gluten structure is significant.

The relative contributions of ionic, hydrogen and hydrophobic bonds to aggregation of glutenin proteins were estimated to be 17.3, 56.3 and 26.4%, respectively, in a good-quality gluten, and 12.8, 80.1 and 7.1%, respectively, in a poor-quality gluten (Vakar and Kopakova, 1976). In discussing chain inter-actions in gluten, then, hydrogen bonding (glutamine side chains) is of primary importance, and hydrophobic interactions play a lesser, but not negligible, role, particularly when lipids are involved.

## 11.6.4.   Gluten Elasticity

What is the source of gluten elasticity? A reasonable hypothesis (Tatham et al., 1985) is that β-turn spirals, and the hydrogen bonding between them (which connects them into β sheet structures), can be slightly extended and act as springs (Figure 11.16A). Under stress, hydrogen bonds can be slightly extended. While each such extension might amount to only a fraction of a nanometre, summed over many thousands (or even millions) of such deformations, the total dough deformation can amount to several percent, as indicated in Figure 11.13. While Tatham et al. (1985) proposed an analogy with elastin, Bloksma 1990b) pointed out that elasticity in the two proteins has opposite temperature dependence.

A second source of elasticity could be entropic. Ewart (1977) considered the individual glutenin molecules to be roughly spherical in shape; several such spheres are concatenated to form the 'linear glutenin' of his hypothesis. Under stress each glutenin molecule could be extended (Figure 11.16B) into a less-favorable configuration. Relieving stress allows the protein molecule to recoil to its preferred (lower-energy) state. A similar picture has been suggested by Amend (1995), based upon SEM pictures of extended (stretched) gluten membranes.

## 11.6.5.   Gluten Viscosity

For dough to undergo viscous flow the glutenin molecules must move relative to each other. Several mechanisms have been proposed by which this might

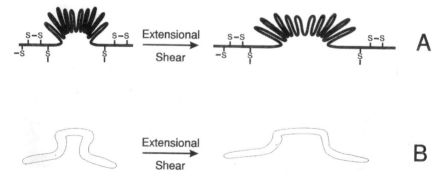

FIGURE 11.16. A proposal for the source of glutenin elasticity. (A) Extension of the β-turn spirals and sheets; (B) deformation of compact glutenin molecules into a more linear configuration.

occur. At the time Goldstein (1957) suggested sulfhydryl–disulfide interchange as such a mechanism, disulfide cross-links were considered to be the most important feature of gluten structure. Today that picture of the 'giant molecule' seems unlikely, for reasons given above. In freshly mixed dough, however, thiyl free radicals appear to be present. These disappear during a 10-min resting period (Graveland et al., 1985), probably through interaction with disulfide bonds. Thus during this relaxation period (bulk fermentation or floor-time in common parlance) the dough undergoes viscous flow (releasing elastic stress) via thiyl–disulfide interchange.

Movement of glutenin molecules is more likely to occur via hydrogen bond and hydrophobic bond interchange. Some of this may happen as a result of molecular motion and Brownian movement (Figure 11.17A). A certain fraction of the hydrogen bonds between chains can be disrupted, and one chain move relative to another, before re-establishing hydrogen bonds. This happens more readily at higher temperatures, and dough viscosity decreases by a factor of five over the range 26–60 °C (79–140 °F) (Bloksma, 1990a).

Gliadin may also play a role as a mobile, small-molecule intermediary of these interchanges (Figure 11.17B). While gliadin can certainly interact via hydrogen bonding because it contains a higher percentage of glutamine than glutenin, the fact that its hydrophobicity contributes to bread quality indicates involvement of this aspect of its nature, as emphasized in Figure 11.17B. By facilitating the movement of adjacent glutenin molecules, gliadin may be characterized as 'molecular ball bearings'.

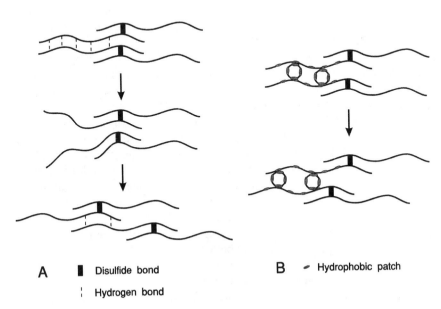

A    ▌ Disulfide bond          B    ◞ Hydrophobic patch

    ┆ Hydrogen bond

FIGURE 11.17. Viscous flow and dough relaxation. (A) Brownian motion and chain realignments; (B) involvement of gliadin in glutenin realignment.

A portion of gluten viscosity may be simply due to a high concentration of macromolecules in the aqueous phase. The viscosity of gum solutions increases tenfold for each 1% increase in concentration. The concentration of glutenin can be estimated at 15% as a lower boundary. Solutions of non-gelling gums at this concentration show a viscosity of the order of $10^6$ Pa s or more. Viscosity is also strongly dependent on the average molecular weight of the protein, and glutenin molecular weights are of the order of $10^6$ Da. Even if there were no interactions between glutenin chains, one would intuitively predict a high viscosity for a suspension such as that found in dough.

The postulated formation of links between gluten proteins and WS pentosans (Hoseney and Faubion, 1981; Figure 11.3) must not be overlooked. To the extent that this happens in dough, the glycoprotein (pentosan-glutenin) would have an even higher molecular weight than the glutenin complex alone, thus increasing viscosity. It would also hinder relaxation (Figure 11.17), increasing dough elasticity.

## 11.6.6. Extensibility

Extensibility is difficult to define in precise rheological terms, although it is easy to find familiar instances; bubble gum, taffy (toffee) and bread dough are three common examples. Under extensional stress, the material thins out to form a membrane. At the limit of extensibility, holes appear in the membrane and expand as extension continues. In the Extensograph this corresponds to the distance the center of the dough piece (of defined initial dimensions) can be stretched before the dough ruptures. To some degree this distance depends on the rate of extension; at a lower rate the dough will extend further before rupturing. Thus, viscous flow is involved to some degree. Elasticity is involved in defining the amount of stress that can be applied before the dough piece breaks (tensile strength). With a less elastic dough the amount of extension may be the same as for a more elastic one, but the actual stress at rupture will be higher for the more elastic dough (Figure 11.14).

Slade et al. (1989) show polarized light photomicrographs of stretched films of a synthetic chewing gum base (polyisobutylene elastomer) and of gluten. The similarities between the two photographs are striking (their Figures 11.17 and 18). These films were stretched in one direction, and the authors point out that film strength is maximum along that axis and minimum at right angles. Holes begin to form when the fibrils separate laterally, and the holes expand perpendicularly to the direction of stretching. In bread dough during proofing and expansion of gas cells, and in the Alveograph, extension of dough is biaxial with the gluten film being stretched along both dimensions. (In the third dimension, perpendicular to the film surface, the membrane grows thinner). This results in the maximum strength for gluten membranes.

Bloksma (1990a) reviews at some length the various components of extensibility (viscosity, elasticity and tensile strength) that influence the overall performance of dough during proofing and baking. While these stages of breadmaking are beyond the scope of this chapter, his discussion emphasizes that all those factors of importance to the baker are developed during dough mixing. In other words, he confirms the experience of every bakery technical service person: if a bakery is having trouble producing good bread, one of the first places to look is at the mixer. If the mixing is right, the rest of the process should be relatively trouble-free.

## References

Amend, T. (1995) Der Mechanismus der Teigbildung: Vorstoß in den molekularen Strukturbereich. *Getreide Mehl und Brot*, **49**, 359–62.

Amend, T. and Belitz, H.-D. (1990) The formation of dough and gluten – a study by scanning electron microscopy. *Zeitschrift für Lebensmittel Untersuchung Forschungen*, **190**, 401–9.

Bailey, C.H. (1941) A translation of Beccari's lecture 'concerning grain' (1729). *Cereal Chemistry*, **18**, 555–61.

Baker, J.C. and Mize, M.D. (1941) The origin of the gas cell in bread dough. *Cereal Chemistry*, **18**, 19–34.

Beckwith, A.C., Wall, J.S. and Dimler, R.J. (1963) Amide groups as interaction sites in wheat gluten proteins: Effects of amide–ester conversion. *Archives of Biochemistry and Biophysics*, **103**, 319–30.

Belton, P. (1999) On the elasticity of gluten. *Journal of Cereal Science*, **29**, 103–107.

Bernardin, J.E. and Kasarda, D.D. (1973) Hydrated protein fibrils from wheat endosperm. *Cereal Chemistry*, **50**, 529–36.

Bloksma, A.H. (1990a) Rheology of the breadmaking process. *Cereal Foods World*, **35**, 228–36.

Bloksma, A.H. (1990b) Dough structure, dough rheology, and baking quality. *Cereal Foods World*, **35**, 237–44.

Bloksma, A.H. and Bushuk, W. (1988) Rheology and chemistry of dough, in *Wheat Chemistry and Technology*, 3rd edn (ed. Y. Pomeranz), American Association of Cereal Chemists, St Paul, MN, pp. 131–217.

Bloksma, A.H. and Nieman, W. (1975) The effect of temperature on some rheological properties of wheat flour doughs. *Journal of Texture Studies*, **6**, 343–61.

Bushuk, W. (1966) Distribution of water in dough and bread. *Bakers' Digest*, **40**(5), 38–40.

Bushuk, W. (1985) Flour proteins: Structure and functionality in dough and bread. *Cereal Foods World*, **30**, 447–51.

Cauvain, S.P. (2004) *Bread making: Improving quality*. Woodhead Publishing Ltd., Cambridge, UK.

Chamberlain, N. and Collins, T.H. (1979) The Chorleywood Bread Process: the role of oxygen and nitrogen. *Bakers' Digest*, **53**(1), 18–24.

Chung, O.K. (1986) Lipid–protein interactions in wheat flour, dough, gluten, and protein fractions. *Cereal Foods World*, **31**, 242–56.

Chung, O.K. and Pomeranz, Y. (1979) Acid-soluble proteins of wheat flours. II. Binding to hydrophobic gels. *Cereal Chemistry*, **56**, 196–201.

Chung, O.K., Pomeranz, Y. and Finney, K.F. (1978) Wheat flour lipids in breadmaking. *Cereal Chemistry*, **55**, 598–618.

Daniels, N.W.R., Wendy-Richmond, J., Russell-Eggitt, P.W. and Coppock, J.B.M. (1976) Studies on the lipids of flour. III. Lipid binding in breadmaking. *Journal of Science of Food and Agriculture*, **17**, 20–9.

Danno, G. and Hoseney, R.C. (1982) Effects of dough mixing and rheologically active compounds on relative viscosity of wheat proteins. *Cereal Chemistry*, **59**, 196–8.

DeStefanis, V.A., Ponte, J.G. Jr, Chung, F.H. and Ruzza, N.A. (1977) Binding of crumb softeners and dough strengtheners during breadmaking. *Cereal Chemistry*, **54**, 13–24.

Dronzek, B. and Bushuk, W. (1968) A note on the formation of free radicals in dough during mixing. *Cereal Chemistry*, **45**, 286.

Eliasson, A.-C. and Larsson, K. (1993) *Cereals in Breadmaking – a Molecular Colloidal Approach*, Marcel Dekker, New York.

Ewart, J.A.D. (1968) A hypothesis for the structure and rheology of glutenin. *Journal of Science of Food and Agriculture*, **19**, 617–23.

Ewart, J.A.D. (1977) Re-examination of the linear gluten hypothesis. *Journal of Science of Food and Agriculture*, **28**, 191–9.

Faridi, H. and Faubion, J.M. (eds) (1990) *Dough Rheology and Baked Product Texture*, Van Nostrand Reinhold, New York.

Gao, L., Ng, P.K.W. and Bushuk, W. (1992) Structure of glutenin based on farinograph and electrophoretic results. *Cereal Chemistry*, **69**, 452–5.

Geissman, T. and Neukom, H. (1973) On the composition of the water-soluble wheat flour pentosans and their oxidative gelation. *Lebensmittel Wissenschaft und Technologie*, **6**, 59–62.

Goldstein, S. (1957) Sulfhydryl und Disulfidgruppen der Klebereiweisse und ihre Bezeihung zur Backfähigkeit der Brotmehle. *Mitteilungen Gebiete Lebensmittel und Hygiene* (Bern), **48**, 87–93.

Graveland, A., Bosveld, P., Lichtendonk, W.J. et al. (1985) A model for the molecular structure of the glutenin of wheat flour. *Journal of Cereal Science*, **3**, 1–16.

Greenwood, C.T. (1976) Starch, in *Advances in Cereal Science and Technology*, Vol. I (ed. Y. Pomeranz), American Association of Cereal Chemists, St Paul, MN, pp.119–57.

Greer, E.N. and Steward, B.A. (1959) The water absorption of wheat flour: Relative effects of protein and starch. *Journal of Science of Food and Agriculture*, **10**, 248–52.

Grosskreutz, J.C. (1961) A lipoprotein model of wheat gluten structure. *Cereal Chemistry*, **38**, 336–49.

Holmes, J.T. and Hoseney, R.C. (1987) Chemical leavening: Effect of pH and certain ions on bread-making properties. *Cereal Chemistry*, **64**, 343–8.

Hoseney, R.C. (1976) Dough forming properties. *Journal of the American Oil Chemists Society*, **56**, 78A–81A.

Hoseney, R.C. (1985) The mixing phenomenon. *Cereal Foods World*, **30**, 453–7.

Hoseney, R.C. and Faubion, J.M. (1981) A mechanism for the oxidative gelation of wheat flour water-soluble pentosans. *Cereal Chemistry*, **58**, 421–4.

Hoseney, R.C. and Rogers, D.E. (1990) The formation and properties of wheat flour doughs. *Critical Reviews of Food Science and Nutrition*, **29**, 73–93.

Jones, I.K., Phillips, J.W. and Hird, F.J.R. (1974) The estimation of rheologically important thiol and disulphide groups in dough. *Journal of Science of Food and Agriculture*, **25**, 1–10.

Junge, R.C., Hoseney, R.C. and Varriano-Marston, E. (1981) Effect of surfactants on air incorporation in dough and the crumb grain of bread. *Cereal Chemistry*, **58**, 338–42.

Kaczkowski, J., Kos, S. and Pior, H. (1990) Gliadin hydrophobicity and breadmaking potential, in *Gluten Proteins 1990* (eds W. Bushuk and R. Tkachuk), American Association of Cereal Chemists, St Paul, MN, pp. 66–70.

Kasarda, D.D. (1989) Glutenin structure in relation to wheat quality, in *Wheat is Unique* (ed. Y. Pomeranz), American Association of Cereal Chemists, St Paul, MN, pp.277–302.

Kim, S.K. and D'Appolonia, B.L. (1977) Bread staling studies. III. Effect of pentosans on dough, bread, and bread staling rate. *Cereal Chemistry*, **54**, 225–9.

Kinsella, J.E. and Hale, M.L. (1984) Hydrophobic associations and gluten consistency: Effect of specific anions. *Journal of Agricultural and Food Chemistry*, **32**, 1054–6.

Menjivar, J.A. (1993) Fundamental aspects of dough rheology, in *Dough Rheology and Baked Product Texture* (eds H. Faridi and J.M. Faubion), Van Nostrand Reinhold, New York.

Meredith, P. (1964) A theory of gluten structure. *Cereal Science Today*, **9**, 33–4, 54.

Michniewicz, J., Biliaderis, C.G. and Bushuk, W. (1990) Water-insoluble pentosans of wheat: composition and some physical properties. *Cereal Chemistry*, **67**, 434–9.

Millar, S.J. (2003) Controlling dough development, in *Bread Making: Improving Quality* (ed. S.P. Cauvain), Woodehead Publishing Ltd, Cambridge, pp. 401–23.

Millar, S.J., Bar L'Helgouac'h, C., Massin, C. et al. (2005) Flour quality and dough development interactions – the critical first steps in bread production, in *Using Cereals Science and Technology for the Benefit of Consumers* (eds S.P. Cauvin, S.E. Salmon and L.S. Young), Woodhead publishing Ltd, Cambridge, pp. 132–6.

Neukom, H. and Markwalder, H.U. (1978) Oxidative gelation of wheat flour pentosans: A new way of cross-linking polymers. *Cereal Foods World*, **23**, 374–6.

Orth, R.A. and Bushuk, W. (1972) A comparative study of the proteins of wheats of diverse baking qualities. *Cereal Chemistry*, **49**, 268–75.

Osborne, T.B. (1907) *The Proteins of the Wheat Kernel*, Carnegie Institute of Washington, Washington, DC.

Patil, S.K., Finney, K.F., Shogren, M.D. and Tsen, C.C. (1976) Water-soluble pentosans of wheat flour. III. Effect of water-soluble pentosans on loaf volume of reconstituted gluten and starch doughs. *Cereal Chemistry*, **53**, 347–54.

Pomeranz, Y. (1985) Wheat flour lipids – what they can and cannot do in bread. *Cereal Foods World*, **30**, 443–6.

Pyler, E.J. (1988) *Baking Science and Technology*, Sosland Publishing Co., Kansas City, MO, Chapter 14.

Rasper, V.F. (1975) Dough rheology at large deformations in simple tensile mode. *Cereal Chemistry*, **52**, 24r–41r.

Redman, D.G., Axford, D.W.E. and Elton, G.A.H. (1966) Mechanically produced radicals in flour. *Chemistry and Industry*, 1298–9.

Salovaara, H. (1982) Effect of partial sodium chloride replacement by potassium chloride or some other salts on wheat dough rheology and breadmaking. *Cereal Chemistry*, **59**, 422–6.

Schroeder, L.F. and Hoseney, R.C. (1978) Mixograph studies. II. Effect of activated double-bond compounds on dough-mixing properties. *Cereal Chemistry*, **55**, 348–60.

Sidhu, J.S., Nordin, P. and Hoseney, R.C. (1980a) Mixograph studies. III. Reaction of fumaric acid with gluten proteins during dough mixing. *Cereal Chemistry*, **57**, 159–63.

Sidhu, J.S., Hoseney, R.C., Faubion, J. and Nordin, P. (1980b) Reaction of [14]C-cysteine with wheat flour water solubles under ultraviolet light. *Cereal Chemistry*, **57**, 380–2.

Singh, N.K., Donovan, R. and MacRitchie, F. (1990) Use of sonication and size-exclusion high-performance liquid chromatography in the study of wheat flour proteins. II. Relative quantity of glutenin as a measure of breadmaking quality. *Cereal Chemistry*, **67**, 161–70.

Slade, L., Levine, H. and Finley, J.W. (1989) Protein-water interactions: Water as a plasticizer of gluten and other protein polymers, in *Protein Quality and the Effects of Processing* (eds R.D. Phillips and J.W. Finley), Marcel Dekker, New York and Basel, pp. 9–124.

Stauffer, C.E. (1990) *Functional Additives for Bakery Foods*, Van Nostrand Reinhold, New York.

Tanford, C. (1973) *The Hydrophobic Effect*, John Wiley & Sons, New York.

Tatham, A.S., Miflin, B.J. and Shewry, P.R. (1985) The beta-turn conformation in wheat gluten proteins: Relationship to gluten elasticity. *Cereal Chemistry*, **62**, 405–12.

Tilley, K.A., Benjamin, R.E., Bagorogoza, Moses Okot-Kotber, B., Prakash, O. and Kwen, H. (2001) Tyrosine crosslinks: Molecular basis of gluten structure and function. *Journal od Agricultural and Food Chemistry*, **49**, 2627–32.

Tilley, M. and Tilley, K.A. (2005) Modifying tyrosine crosslink formation in wheat dough by controlling innate Enzymic activity, in *Using Cerealsscience and Technology for the Benefit of Consumers* (eds S.P. Cauavin, S.E. Salmon and L.S.Young), Woodhead publishing Ltd, Cambridge, pp. 142–6.

Tipples, K.H. and Kilborn, R.H. (1975) 'Unmixing' – the disorientation of developed bread doughs by slow speed mixing. *Cereal Chemistry*, **52**, 248–62.

Tipples, K.H. and Kilborn, R.H. (1977) Factors affecting mechanical dough development. V. Influence of rest period on mixing and 'unmixing' characteristics of dough. *Cereal Chemistry*, **54**, 92–109.

Tipples, K.H., Meredith, J.O. and Holas, J. (1978) Factors affecting farinograph and baking absorption. II. Relative influence of flour components. *Cereal Chemistry*, **55**, 652–60.

Tkachuk, R. and Hylynka, I. (1968) Some properties of dough and gluten in $D_2O$. *Cereal Chemistry*, **45**, 80–7.

Vakar, A.B. and Kopakova, V.V. (1976) Solubility of the glutenin fraction of gluten. *Vestn. Skh. Nauk* (Moscow), p. 45 (quoted in He and Hoseney, 1990.)

van Lonkhuijsen, H.J., Hamer, R.J. and Schreuder, C. (1992) Influence of specific gliadins on the breadmaking quality of wheat. *Cereal Chemistry*, **69**, 174–7.

Weegels, P.L., Marseille, J.P., de Jasger, A.M. and Hamer, R.J. (1990) Structure–function relationships of gluten proteins, in *Gluten Proteins 1990* (eds W. Bushuk and R. Tkachuk), American Association of Cereal Chemists, St Paul, MN, pp. 98–111.

Wehrli, H.P. and Pomeranz, Y. (1969) The role of chemical bonds in dough. *Bakers' Digest*, **43**(6), 22–6.

Wootton, M. (1976) Binding and extractability of wheat flour lipid after dough formation. *Journal of Science of Food and Agriculture*, **17**, 297–301.

Wrigley, C.W., Andrews, J.L., Bekes, F., Gras, P.W., Gupta, R.B., Macrithies, F. and Skerrit, J.H. (1998) Protein-protein interactions – essential to dough rheology, in *Interactions; Keys to Cereal Quality* (eds R.J.Hamer and R.C. Hoseney), American Association of Cereal Quality, St Paul, Minnesota, pp. 17–46.

# 12
# Flour Milling

Paul Catterall and Stanley P. Cauvain

## 12.1. Introduction

Question – is flour milling an art or a science? Answer – it is neither. It is a technology, the marrying of food science with the art of the practical miller, both of which have evolved over many years. Some may say that science will eventually overcome the art of milling and consign the practical miller to the flour bin of history, but with the variety of baking products still expanding daily and new wheat varieties being developed that event should still be quite a long time off. This chapter aims to help define the science and the art of a process which produces one of the most versatile of bakery raw materials and aims to provide a background to the link between wheat, the milling process and the properties of the final flour.

Wheat has been a major food source for thousands of years. The unique properties of its proteins when hydrated have given it a flexibility which has made it ideal for a multitude of different bread products from the flat breads of the Mediterranean and equatorial countries, for example chapatti, pizza and ciabatta, through to the sandwich breads of Europe, America and Australia.

We must remember that the wheat grain is a seed that is designed to protect the embryonic plant from the rigours of the outside world until conditions are right for its germination and subsequent growth. A representation of the structure of the wheat grain is given in Figure 12.1. The outer bran coat with its unique physical structure which folds the seed in on itself to form the characteristic crease protects the seed. As a result of this complex shape, milling engineers have spent many a sleepless night trying to find ways of breaking through these protective layers to extract the endosperm with its maximum food value.

## 12.2. In the Beginning

Flour milling can trace it origins back to prehistory, but the modern systems known as gradual reduction flour mills have only been developed over the last 200–300 years. Early humanity used pestles and mortars to grind wheat,

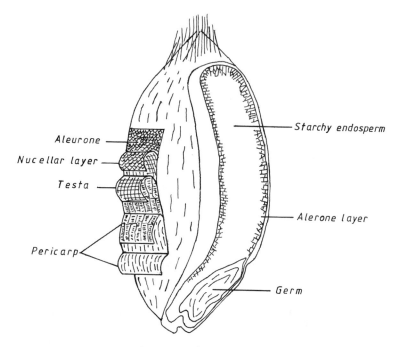

Aleurone

Nucellar layer

Testa

Pericarp

Starchy endosperm

Alerone layer

Germ

FIGURE 12.1. The wheat grain (not to scale).

pounding it to make a crude wholemeal flour. This basic milling technique developed into saddle rubbing stones and eventually to rotary grinding stones, with evidence for use of the latter dating back 7500 years. While initially the latter would have been hand operated, as communities developed, larger stones with increased capacity were used, powered by animals and later water. New power sources were developed; wind power has been used since the twelfth century and steam was introduced in the eighteenth century, but whatever the power source the basic principle of grinding wheat grains between two stones to produce flour remained unaltered. In fact, even in today's highly technological society, wheat ground to flour between two stones is perceived by some to be a superior product.

This basic one-pass system does have its drawbacks in that the bran skins and germ are ground as finely as the endosperm and the separation of these parts can be difficult. To overcome this, a system was developed by French and Hungarian millers in the seventeenth and eighteenth centuries which linked together several sets of stone mills, the gap between each pair of stones being set slightly closer than the one before. After each pass the resultant flour was sieved, and some bran separated before the remaining material was sent to the next set of stones. As a result flour colour improved and a 'white' flour was produced. The development of the gradual reduction flour-milling process had begun.

## 12.3.  The Modern Flour-milling Process

The principles of modern milling are still very much the same as described for stone grinding above, except that now the wheat is ground between pairs of cast-iron rolls before the stocks are sieved and then reground. At each grinding stage a little flour will be produced and up to four or five other sieve separations are made. Because the particle sizes of the various sieve fractions will differ, they will all be treated separately on different rolls. At first sight the system in a modern flour mill appears to be quite complicated, but it will be attempted, over the next few pages, to shed some light on the whole process to help make it easier for the student to follow. We will start with the very first stage, when the wheat arrives at the mill.

### 12.3.1.  Delivery of the Wheat

Flour mills have been and still tend to be built to be near their major source of wheat supply. At the beginning of the twentieth century they were often built at ports to take advantage of imported grain. Those built inland were built adjacent to canals, rivers and railways to minimize the cost of transport. Large 20,000-tonne bulk grain carriers are still used to transport wheat around the world by sea, and small 1500-tonne coasters are used in more local waters. A discussion of the full extent to which grain is transported by sea can be found elsewhere (Sewell, 2003).

Since raw material and transport costs have been and continue to be an important factor in all industries, many countries prefer to use more of their home-grown wheats. This desire prompted a response by wheat breeders which has led to the development of wheat varieties with improved breadmaking performance, and imports can therefore be reduced. In consequence the dependence on port mills has declined. Mills can now be built closer to the wheat-growing areas and the raw material transported by road.

When wheat arrives on the mill site it will need to be sampled so that tests can be carried out to determine quality against the buying specification. Sampling has to be done efficiently so that the results obtained from the tests are representative of the whole load.

Generally sampling is done in one of two ways:

1. Using a manual spear. This method requires the operative to climb on top of the load and force the spear into the grain. Turning the handles takes a range of samples at various points up the spear. Whilst this system is simple and cheap, it is prone to abuse. Pushing the spear vertically into a load is very difficult, and the temptation will always be to push it in at a shallower angle, with the result that wheat in the bottom of the lorry will not be sampled. Walking on the load to sample it is an unacceptable practice for a foodstuff and as a result many more mills are changing to the pneumatic sampler.

2. Using a pneumatic sampler. This type of sampling spear is far more reliable. It can either be steered manually or pre-programmed to follow a defined sampling pattern, and is driven vertically down into the wheat to take samples at all depths. The intake operative no longer has to walk on the wheat, so that the risk of contamination is reduced although the lack of personal contact does have a slight disadvantage in that it puts a greater emphasis on the laboratory staff to pick up problems, such as contamination and taints, which previously could have been more readily spotted in the lorry, truck or railcar.

## 12.3.2.  Wheat Testing

On arrival in the mill laboratory the wheat sample is thoroughly blended to ensure that the results which will be obtained are representative of the whole delivery. It is important that tests are carried out on the wheat prior to tipping the lorry for the following reasons:

- to ensure the wheat is of the quality required;
- to make sure it is not contaminated with foreign bodies or infested;
- to ensure it will be stored with wheats of similar quality.

The tests carried out will vary according to the type of mill, and the flour being produced, but will generally include some or all of the following.

### 12.3.2.1.   Appearance, Off-odours and Taints

Initially the wheat is inspected by trained laboratory staff who will be looking for any unusual odours, but they will mostly be looking for mustiness from damp, mouldy wheat or evidence of contamination from the transporting vehicle, either a previous load or even from the fuel used.

### 12.3.2.2.   Screenings (impurities)

'Screenings' is a general term applied to all impurities in a parcel of wheat. They can be divided into two main types, intrinsic and extrinsic.

Intrinsic impurities are those which are reasonably associated with the wheat itself, but for obvious reasons are not required in the finished flour, for example shrivelled or diseased grain, straw, the seeds of weeds or crops growing in the vicinity of the wheat (cockle, millet bindweed, etc.) and ergot. The latter is a fungus associated with wheat but more commonly with rye. It appears as dark purple structures which replace the individual ears of grain and can contain ergotoxin, a poison that may lead to abortion in both humans and livestock.

Extrinsic impurities are contaminants of wheat which should not reasonably be present, for example string, paper, nails, wire, wood, or evidence of contamination from rats and mice. Such materials can come from contamination in storage, or could have been picked up during combining.

All of the above examples of impurities, and more, have been found in wheat, and a test has been developed for use in laboratories. It is a simple sieving test based on using two slotted screens with holes of different sizes. A sample of wheat is placed on the top deck and the apparatus is shaken either mechanically or manually. All impurities larger than wheat are retained on the top deck (3.5 mm), the wheat itself is retained on the middle deck, and any fine impurities pass through to the bottom container. These fine impurities can be further inspected for signs of infestation before being weighed together with the coarse impurities to give a total figure which is expressed as a percentage of the original wheat sample. If this figure exceeds the specification agreed with the wheat merchant, then the wheat can be rejected. High levels of screenings will contribute to poor extraction rates, black specks in the flour (especially wholemeals) and possibly taints. In relatively small amounts they will not be a problem, and should all be removed when the wheat is cleaned in the part of the mill known as the screenroom.

### 12.3.2.3.   Wheat Density

This property is commonly referred to as hectolitre weight or bushel weight. Various manufacturers produce equipment for its measurement and the most common apparatus all work in a similar manner. A cylinder of known volume is filled using a standard method and weighed. The figure obtained is converted to kilograms per hectolitre (kg/hl) and is one of the first tests carried out to determine the 'quality' of the grain. Hard breadmaking wheats will have a higher figure, in excess of 80 kg/hl, compared with softer biscuit types at around 70 kg/hl. A poor harvest will give low hectolitre weights because of the presence of small shrivelled and sprouted grains. The milling industry around the world has to maintain the optimum amount of flour from the wheat, which is referred to as the 'extraction rate'. Low hectolitre weights will give a poor extraction rate, can cause too many wheat grains to be removed in the screenroom, will produce a dirty 'specky' flour and are associated with low Hagberg Falling Number wheats.

Having assessed the whole grain, and assuming the wheat has passed these initial tests, a sample of the wheat can be ground in a laboratory mill and a set of further tests carried out. The choice of which particular tests to be carried out varies, according to the particular needs of the miller and location. There is no common consensus view as to which tests are important though commonly they will include the measurement of protein content and associated gluten-forming properties.

### 12.3.2.4.   Protein Content

Generally, the value for wheat protein content is accepted as nitrogen $\times 5.7$. The classic method for assessing protein content uses the Kjeldahl apparatus (CCFRA FTP 0009, 1991) though this has now been largely superseded by the Dumas method. Both methods take far too long for use at the point of wheat intake.

More rapid methods for protein determination have been developed, the most common one being based on near-infrared (NIR) (CCFRA FTP 0014, 1991) which can produce a result in approximately 25 s.

Protein content and quality are of vital importance in flour milling. They are the characters which make wheat unique and are the main properties on which wheat is traded, with higher protein wheats generally commanding a higher value. Protein contents can vary widely from one lorry load to another and so accurate measurements are required to be able to segregate the wheat into suitable protein bands which can then be blended to give consistent grist formulations. The effects on protein of product character will be discussed further when flour testing methods are considered.

### 12.3.2.5.    Gluten Content

For this test a sample of flour is prepared from the ground wheat by sieving. The gluten quality is tested using a small mixing machine which kneads the dough for a set time, before washing the starch away. A salt solution is used to help keep the gluten firm (CCFRA FTP 0013, 1991). The gluten sample can then be weighed and manually assessed for its vitality and strength. Alternatively a small dough can be prepared manually in a beaker and the starch washed away under a running tap.

Wheat which has been dried incorrectly may have damaged protein so that its gluten will appear as very short and 'bitty' in character, and in extreme cases the sample may not even form a gluten. Generally a smooth gluten with a good light grey colour indicates a good breadmaking wheat.

### 12.3.2.6.    Moisture

The moisture content of wheat is important to both farmer and miller because wheat has to be stored for up to 12 months before use and therefore has to be low enough to prevent spoilage. As a general rule it should not exceed 15%, but if the miller intends using the wheat reasonably quickly then slightly higher figures might be accepted.

There are several ways to determine moisture content in the wheat laboratory. NIR is often used and also heat balances. In the latter case, an infrared or incandescent heat source is mounted over an electronic balance and the sample is heated until a constant weight is achieved. Conductive methods are also available but the key factors are that they have to be quick and accurate. Oven drying methods, such as 130 °C for 1.5 h (BS, 1987), take far to much time for them to be used at wheat intake but are standards against which the rapid methods are calibrated.

### 12.3.2.7.    Hagberg Falling Number

This is a measure of the cereal *alpha*-amylase in wheat, and is a critical parameter for large-scale bakeries (Chapter 3). To carry out the measurement a suspension of flour from the laboratory grinder and water is blended in a test tube then heated in a boiling water bath and stirred continually for 60 s. At the end of this period the stirrer is brought to the top of the tube and released. The *alpha*-amylase in the sample slowly breaks the solution down converting the starch to dextrins, and the stirrer gradually sinks under its own weight. When the stirrer reaches the bottom of the tube the test is stopped. The resulting time in seconds is the Hagberg Falling Number, and is proportional to the amount of cereal *alpha*-amylase present, the higher the figure the lower the amylase (CCFRA FTP 0006, 1991). The Hagberg figure includes the first 60 s mixing, so 60 would be the lowest figure possible. Figures over 350 s are unreliable because the contents of the test tube will be at the same temperature as the water bath, and the amylase will have been denatured. This would normally happen at temperatures above 75 °C (167 °F).

### 12.3.2.8.  Hardness

This is a measure of the wheat endosperm texture. As a general rule hard wheats are used in breadmaking and softer wheats are used for biscuits and cakes. However, this simple classification cannot be relied upon as there are many hard wheats that do not make bread and could only be described as 'feed' wheat, and there are also some so-called 'breadmaking wheat varieties' that can be classified as soft.

Wheat hardness testing can be carried out using a specially calibrated laboratory mill (Hook, 1982), where the time taken to grind a standard volume of material is measured. NIR may be used (AACC Method 39–70A, 1995) and has the advantage of reducing the number of tests carried out by the laboratory technician and therefore speeding up the whole of the sample testing process. The Perten single-grain kernel characterization system has been developed to deliver an automated and objective measure of wheat hardness (Gaines et al., 1996). The force-deformation characteristics of wheat grains are determined by crushing a number of individual grains, commonly around 300. The same equipment also provides information on grain weight, diameter and moisture.

### 12.3.2.9.  Electrophoresis

This test is carried out to determine the specific variety of the wheat (BS 4317: Part 30, 1994). In essence the test splits the protein fraction into individual amino acids and records them as a unique pattern which can then be used to compare an unknown pattern with those from known samples. Electrophoresis is not a common test at the wheat intake; however, it is very valuable for ensuring authenticity when specific varieties are being purchased for specialized applications and there is increasing interest in the development of rapid and reliable tests.

Wheat varietal identification is becoming increasingly important both in the context of ensuring that the wheat has the appropriate qualities and for purposes of traceabilty. Such issues have placed greater emphasis on the need for rapid testing so that more recent developments in electorphoresis have been focused on the 'lab-on-a-chip' technology. These are based on microfluidic devices which integrate all of the chemical and processing steps necessary for separating proteins using capillary electrophoresis (Lookhart et al., 2005). Another improvement to the electorphoretic techniques is the automation of varietal identification using pattern-matching software (Bhandari et al., 2005).

### 12.3.3.  Wheat Storage

All the laboratory tests should be carried out quickly preferably while the lorry, truck or railcar is at the mill, waiting to tip wheat in order to avoid the intake of undesirable wheat parcels. The results of the tests should be assessed and a decision made as to whether the wheat meets the requirements of the specification laid down, and if so, into which storage bin it should be put. After the wheat has been tipped but before it reaches the storage bin, it will pass over some preliminary wheat-cleaning equipment consisting of coarse screens and magnets designed to remove the larger impurities in the wheat which may otherwise damage the mill equipment and possibly block the bins charging or discharging.

The art of milling is to produce a consistent product from a varying raw material. This is achieved through blending, and the process starts in the storage silos where wheat can be segregated into different types. These segregations depend on the type of flour being milled; obviously bread and biscuit types will be kept separate. They may also be segregated by variety or Hagberg Falling Number and they will often be separated into different protein bands. By segregating the delivered wheat in this way and then eventually blending it back at controlled levels, a more consistent finished product will be achieved.

The quantity of storage space at the mill can vary widely and is dependent on many factors. In the UK, where mills are sited very near to the wheat-growing areas and road transport is good, mills need only have a very low storage capacity which could be as little as for 1 week's production. In some other countries storage on farms is very limited and the distances involved are much greater, so in these situations mills may contain many months' stocks of wheat. The care of wheat in storage will depend on the length of time it is expected to be there. If the mill has relatively small capacity and a high turnover of wheat, there should be no need for any special care, so long as the grain is regularly monitored. However, if the grain is to be kept for long-term storage, then it should be regularly turned bottom to top to help blend it and prevent any hot spots from occurring from localized microbiological activity.

### 12.3.4.  The Mill Screenroom

From the storage, silos wheat will be drawn off and passed through the screenroom to be cleaned. This is a dry cleaning process and is designed to remove a wide range of impurities typically found in wheat. The equipment used relies on five basic principles to achieve this separation:

1. size;
2. specific gravity;
3. shape;
4. magnetism;
5. air resistance.

### 12.3.4.1.  Size

This is the most basic type of separation, in which the machine generally uses a double deck system, the top deck being a coarse screen which removes coarse impurities such as string, straw and paper, and allows the wheat and finer impurities to fall through to the second deck. This would be a finer screen allowing small impurities such as sand and dust to pass through, and to let the wheat overtail. However, any contaminants which are the same size as wheat will not be removed and require a different technique to take them out.

### 12.3.4.2.  Specific Gravity

Some cleaning machines rely on the different densities of materials for separation of contaminants from wheat. The wheat will pass onto an inclined oscillating sieve bed and is fluidized by a controlled flow of air up through the sieve. Dense items such as stones fall through the screen onto a lower deck where as a result of the oscillations, they then travel up the deck and overtail into a container. To operate effectively, the machines require a delicate balance between the three variables: air flow, oscillation and angle of inclination.

### 12.3.4.3.  Shape

Machines that use this principle are commonly called cylinder or disc separators. They work in one of two ways. One type has discs with small pockets cut in them just the right size for small round seeds to fall in. As the discs rotate in the bulk of the wheat the seeds are lifted out and transferred to a separate conveyor. The second type of separating machines has pockets which are just big enough for wheat to fall into, and the wheat is lifted out leaving oats, barley and un-threshed grain behind. Usually these machines are used in tandem to remove both types of impurity.

### 12.3.4.4.  Magnetism

Plate magnets are situated at strategic points throughout the screenroom to collect any ferrous metal contamination. At this early stage in the process they are primarily used to protect the mill equipment, although their role in terms of product safety must not be overlooked. More recently, metal detectors have been appearing as part of the mill screenroom machinery. They have the added advantage of rejecting non-ferrous as well as ferrous materials.

12.3.4.5.   Air Resistance (Aspiration)

Light impurities, such as dust and fine dirt, and those with a large surface area, such as chaff, lend themselves to being removed by a controlled flow of air. The wheat is spread into a wide curtain to expose the maximum surface area and air is drawn through it. Aspiration machines are used in several places throughout the screenroom, as the actual handling of the grain creates more dust which needs to be controlled and removed.

## 12.3.5.   Conditioning

The final process which takes place in the screenroom is conditioning of the wheat. Water may be added to the grain to a predetermined level and then left to stand for up to 24 h. The amount of water used and the standing time vary according to the type of wheat. For example, Canadian wheat will be damped to between 16 and 17% moisture and left for a minimum of 12 h, while a soft English biscuit wheat would only be damped to about 15% for only 4–7 h. Increasingly there is a tendency for conditioning times for wheat to become shorter, not least because of the limited storage capacity at many flour mills.

Conditioning is a critical stage in the milling process. Modern damping systems are fully automated, with moisture levels being continually monitored to maintain a constant level of water addition. The purpose of the conditioning is to aid the removal of the bran layers from the endosperm. By keeping the bran layers damp they will detach more easily and stay in larger pieces which aids their removal from the endosperm through the rest of the milling machinery. Badly conditioned wheat will mean that the bran layers will break up into small pieces which are difficult to remove and give the resulting flour a specky appearance and contribute to an increase in grade colour figure or ash content, and possibly a lower extraction rate.

The moisture levels used during conditioning, especially for the harder wheats, are not maintained through to the finished flour where high flour moisture content would cause rapid deterioration. During the conditioning process a moisture gradient is developed within the cross section of individual grains, high on the outside and lower in the centre. The removal of the bran, together with the heat generated within the milling process, brings the moisture of a typical white flour back within the range of 14–14.5%. After being fully cleaned and conditioned, the wheat is now ready for the milling operation itself.

## 12.3.6.   The Mill

From the early days of the twentieth century, flour milling has developed into a complex network of machines all doing a very specific job; however, the basic principle of the process has changed very little in the last 150 years or so. In essence the wheat is ground between pairs of cast-iron rolls, and separated on a sieve into various sizes and quality fractions which are then ground again. This section will try to explain the various processes involved and their effects on final product performance. New mills are built using the latest technologies and engineering ideas and then evolve through their life with new techniques and products, so it must be remembered that there is no such thing as a standard flour mill. What we will try to do is describe a typical mill even though no such animal exists!

The modern flour-milling process can be separated into six very distinct areas (Figure 12.2):

1. the break system – the first grinding stages for the wheat;
2. scalping, grading, dusting – the separation of the ground materials after each of the break rolls;
3. the scratch system – the final removal of bran from the system, although sizing systems are more commonly used in modern plants;
4. purifiers – the cleaning up of semolina stocks (endosperm fragments) by grading and aspiration to remove bran fragments;
5. the reduction system – the reduction of semolina to flour;
6. flour dressing – the separation of flour from the other materials (mainly bran).

The flour finally produced for the baker does not necessarily pass through every one of these stages. After each section some flour is created and will be removed for blending, leaving the remaining stocks to continue through for further processing.

## 12.3.7. Break System

The rolls which go to make the break system are the first of the grinding operations, and consist of a series of three to five, but usually four pairs of fluted rolls designed to break the grain open and extract as much of the endosperm as

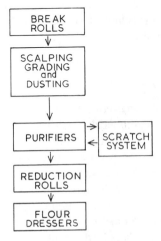

FIGURE 12.2. Schematic of a flour-milling process.

possible for the production of white flour. At this stage the endosperm will be in the form of coarse particles known as semolina, and must be separated from the individual grains with a minimum amount of disintegration of the bran. This is achieved by a combination of the flutes of the roll surfaces, which are cut on a slight spiral, and a speed differential between the rolls of the order of 2:1, which apply a scissors-type action to the wheat. The designs of the flutes have been determined by experience, and get finer as the intermediate products pass from the first break rolls to the fourth. After each set of rolls the products pass to a sifter for separation into a series of sub-fractions before going on to the next part of the process. Even at this early stage in the mill process some flour will be produced, either released from along fracture lines of the endosperm or due to attrition of the different particles. This will be sieved out and bypass the rest of the process. At this stage the amounts will be small since it is not the aim of the break system to produce finished flour.

## 12.3.8.  Scratch System and Bran Finishers

These are more finely fluted rolls than in the break system and are found in some mills. They are designed to remove the last fragments of endosperm from the smallest bran fragments. Stocks are usually transferred to the 'scratch' system from the purifiers where these bran-rich materials have been separated out. In some mills special bran finishers are installed after the third, fourth or even fifth break roll to 'dust off' the last remaining flour from these bran stocks. This scratch system helps to increase the extraction rate, but the flours produced are of a low grade and have poor functionality in breadmaking. A variation of the scratch system is called the 'sizing' system. This is predominantly used in more modern mills where throughputs are higher, causing large variations in the feed to the first part of the reduction rolls. To even this out the normal coarse semolina from the first and second breaks are fed through a finely fluted sizing roll. This process gauges the semolina to a more uniform particle size and also flattens the germ allowing it to be sieved out.

## 12.3.9.  Scalping, Grading and Dusting

These are the terms used to describe the separation of the stocks after the action of each of the break rolls, and which take place inside multi-sectioned oscillating sieves. Scalping is the separation of the coarse overtails, generally bran fragments with some endosperm attached, which can then be passed on to the next reduction roll in the system. Throughs of the scalping sieve can be divided into coarse and fine semolina in a process known as grading. These fractions will pass straight to the purification system, while the flour that has been produced will be removed and transferred to the flour collection system. Bran and the majority of the wheat-feed (mainly small particles of bran) will be removed from the system after the final break roll sifters.

## 12.3.10.  Purifiers

The main aim of the purifiers is to clean or 'purify' the semolina stocks coming from the break sifters. Typically, a purifier is a long oscillating sieve-bed, mounted at an incline above various receiving hoppers. Air is drawn up through the stocks being sieved which, together with the reciprocating action, stratifies the stocks and lifts out any fine bran present. The air flow can vary as the products pass down the length of the machine, leaving the purified product to fall into the hoppers beneath them to make their way into the reduction system. A well-set purifier is important in supplying cleaned semolina to the reduction system in order to improve the efficiency of the latter. White flour with bran specks will produce unattractive bakery products and have a lower protein quality resulting in a reduced flour performance in breadmaking.

## 12.3.11.  Reduction System

This is the final grinding stage in the flour production process. The cleaned semolina stocks are reduced down to the finished flour by a series of up to 12 pairs of roller mills. Reduction rolls are differentiated from the break rolls by being smooth rather than fluted and by having a reduced speed differential between individual rolls in the pairs.

The first section of reduction rolls will generally be dealing with the cleaner semolina stocks, mostly from the first and second break sifters, and will therefore be producing the whitest flours with the best functionality for breadmaking. The middle section deals with the tailings from the first section and also the poorer quality stocks from the later break purifiers. The final two or three rolls deal with the stocks overtailing the first two sections, and will be producing lower-quality flours.

The set-up and performance of the reduction system in a flour mill has a significant effect on one of the most important parameters for breadmaking, namely the water absorption. There are three main factors in a flour specification which can have a direct effect on the water absorption:

1. moisture content;
2. protein content;
3. level of starch damage.

These three factors, together with the pentosan content of the flour, enable the water absorption to be calculated very accurately. Given that the protein and moisture are predetermined by the specification and that the pentosan content is not a normal flour test, the major influence on water absorption is therefore the starch damage. In the milling process, damage to the starch granules present in the flour is achieved by grinding the flour on the reduction system which physically disrupts the starch grains and allows access to more sites for the water to bond.

Grinding hard can also have an effect on the particle size of flour. Some flours, specifically those used for dusting, need to be coarse and free flowing so that they do not clump and block the feeders. By reducing the degree of grinding, the mill can produce a flour which meets these particular requirements.

## 12.3.12.  Flour Dressing

As in the break system, each pair of reduction rolls is followed by a sifter making between three and five separations. Flour is removed and the remaining stocks are graded between the reduction rolls which follow. Towards the end of the system the flour removed is of poorer baking quality. The overtails pass out of the system into the wheatfeed bin.

The operation of this sifting system is critical to maintaining the efficiency of the flour mill and can be hindered by the action of the reduction rolls on the semolina stocks. As mentioned earlier, the reduction rolls are running at similar speeds, and so have a tendency to flatten the semolina particles and produce small flakes. If the flakes are passed to the sifters, the sieving action would be very inefficient and too much material would overtail the system. To prevent this happening, flake disrupters are installed above most reduction rolls. These machines are made up of two metal discs, held apart by short pins, mounted inside a case and driven by a high-speed motor at approximately 3000 rpm. Flour stocks pass into the centre of the discs and are thrown outwards through the pins by the centrifugal force, breaking up any flakes on their way through.

At the end of the flour dressing process all the flours from the various machines are brought together and blended to produce what is referred to as a 'straight run' flour. This is the normal white flour supplied to the majority of customers, and accounts for between 76 and 78% of the initial wheat mass. This figure is referred to as the 'extraction rate' and is an indication of the mill efficiency.

If required by the customer, the miller can be more selective with the streams of flour being produced and make a 'patent' flour. This will use only the high-grade streams with the whiter colour, generally from the first section of the reduction system. In this case the flour protein will be slightly lower due to the protein gradient in the wheat grain, that is the outer layers have a higher protein content, but the quality is not quite as good. A patent flour would have an extraction rate of approximately 60% and would therefore be more costly for the baker. The dark flour not used in the manufacture of a patent flour is referred to as 'low-grade'. While not suitable for normal bakery use, there are certain products where low-grade flours have a distinct advantage if they are available. These would include malt breads, speciality rye breads, and some types of boiled pie paste where the poor colour can actually be used to an advantage.

At the end of the milling process there will be a range of different products:

- straight run white flour (or patent and low-grade flours);
- coarse and fine wheat brans for use as an ingredient in many bakery products, health foods and breakfast cereals;
- wheatfeed used as an animal feed and containing the finest brans and possibly the wheatgerm, and low-grade flours if these cannot be sold separately.

Brown flours would be blends of bran and straight run flour and will have an extraction rate of about 90%, while wholemeal (wholewheat) flours must be made from 100% of the wheat.

### 12.3.13.  Storage and Packing

From the mill, flour will pass through a final redresser before bulk storage. The final redresser will be a fine sieve of about $300\,\mu$m mesh. The sieve here is used as a precaution should one of the many other sieves in the mill burst. The overtails will be monitored on a regular basis to check for this particular problem.

Bulk storage bins can be made from either wood, concrete or steel, depending on a balance between personal preference, materials available, product safety and price. Early bins were made of wood, which at the time was a good material and offered a degree of insulation which helped prevent condensation. Concrete can be used but such bins are liable to crack, are very heavy and require deep foundations. Both these materials have the disadvantage that pieces can break off and contaminate the flour, which helps to confirm steel bins as the most popular choice for modern flour mills. These are cheaper, do not crack like concrete, are lighter and easier to install, and if required can be dismantled and re-sited. Bin cross sections may be either square or round. Round bins are stronger but take up more space for a given capacity. Square bins need corrugated sides to give added strength, but have the advantage that there is no dead space between adjoining bins where infestation can build up.

From bulk storage the flour is transferred for either bulk delivery or packing. It is always good practice, from any form of storage, to pass the flour through a redresser to check that the flour has not been contaminated with any foreign material or hard lumps of flour. Packing covers a range of sizes depending on final usage, from 1 kg for domestic use up to 1 tonne tote bags for medium-sized bakeries. Flour used to be delivered in hessian bags which were returned to the miller; however, they were a serious infestation risk and these bags have now been generally replaced with paper sacks.

Bulk tankers can make deliveries from 5 tonnes to a maximum defined by local transport weight regulations. The smaller deliveries require specialized tankers which either have compartments for different flours or have one flour that can be metered accurately to different customers. These smaller bulk deliveries do not offer the same advantages in terms of cost benefit as those derived from a full tanker delivery, although they do significantly reduce the amount of packaging to be disposed of by the baker.

## 12.4.  Food Safety and Product Protection

The consumer's requirement for increasing confidence in safety and wholesomeness of the food that they eat has over the years worked its way through the supply chain from the multiple supermarkets, through to the food manufacturer, and then to the primary ingredient suppliers. The growers are now also understanding that they have a role to play in food safety, and the dialogue between the miller and the farmer, which at one time was only about price, is now becoming more intense on more serious food safety issues.

To succeed in any business you have to supply your customers with what they want. If you do not then someone else will. Flour milling is obviously no different and has had to respond by improving product safety measures. Millers have had to change their culture from being what could be described as an extension of the agricultural industry to being a food factory in their own right. Customer concerns about possible contaminants can be allayed by applying Hazard Analysis Critical Control Point (HACCP) principles to flour milling.

Contaminants in flour can be broken down into three basic categories:

1. foreign bodies;
2. chemical;
3. biological.

## 12.4.1.  Foreign Bodies

These are probably the most common problem with flour, but potentially the easiest to deal with. Typical contaminants would be those associated with the milling process and include pieces of hardened flour from blow-lines and bins, sifter wire or nylon, pieces from sieve cleaners and cotton fibres.

The benefit of the flour mill is that it has a collection of sieves at strategic points throughout the process. Some of these are process sieves to separate the different stocks within the mill. However, others referred to earlier, such as the final redresser, the last one in the milling process, and the redressers after bulk storage, are critical for monitoring and ensuring the safety of the finished product. Overtails of these sieves can be checked at relevant times, and the findings recorded. While small amounts of bran and hardened flour will always be found, if the level suddenly increases from the norm, this should be taken as an indication that a problem has occurred and that corrective action is needed. Likewise if any unusual object is found its source must be investigated to prevent any further problem. It must never be assumed that 'the sieve is doing its job by removing the object'. The redressers are there to indicate problems in the process, and any unusual findings should therefore prompt a response.

The same principles should be applied to metal detectors and magnets. Magnets have been used in flour milling for many years, but for more selfish reasons than for protecting the customer. They were initially used to protect milling equipment from damage or from being blocked. Their use has now been extended to cover product protection but they are obviously limited to just ferrous metals. Their value in terms of product protection has also been limited by the higher volumes of product passing over them in the newer high-speed mills. Any ferrous contaminant can be 'blown off' by the force of the flour; even the higher strength rare earth magnets are not foolproof. In terms of product protection for metal contaminants, the way forward is obviously the metal detector. Capable of reacting to ferrous and non-ferrous metals down to 1 mm in size, their positive action removes the object from the flour stream. They need to be situated as late as possible in the product flow and their action needs to be regularly checked to ensure that they not only see the contamination, but that they also reject it into a suitable container. This is usually done by introducing a 'test piece' into the product flow and recovering it from the reject bag.

## 12.4.2.  Chemical Contaminants

There are two main areas of concern under this heading: pesticide residues and taints. Reacting to consumer demand for cheaper foods, modern intensive farming relies to a significant extent on the chemical control of weeds and insects to maximize the yield from the growing crops. However, more recent concerns about the effect of these chemicals on the environment has resulted in tight controls being used to minimize their use.

All chemicals used on agricultural products are regulated by various laws, for example EU and CODEX, which define their maximum residue limits (MRLs). It is the farmer's responsibility to ensure that these levels are not exceeded; however, millers must police relevant agricultural practices by setting up regular checks to confirm the safety of the ingredients they are using. In the UK, a 'Pesticide Passport System' has recently been introduced to record what type and level of pesticides has been used on a particular load of wheat.

Taints are far more difficult to track down. The very nature of flour lends itself to absorbing taints, and as a general precaution it should never be stored near strong-smelling ingredients, such as spices. Strong cleaning chemicals and scented soaps should not be allowed in the mill, and again regular checks of the incoming raw material and outgoing products should be carried out. It must never be assumed that the technicians carrying out these tasks either know what they are looking for or can actually detect the taint at all. Phenolic taints, for example, often described as 'earthy' or 'musty' can be detected at very low levels by susceptible individuals, while others cannot detect them at all. It is always worthwhile investing in specialized training for the people who are going to carry out these tasks.

## 12.4.3.  Biological Contaminants

This is a general heading to cover both microbiological contamination, for example *Salmonella* and *E. coli*, and product infestation. The microbiology of wheat flour cannot be controlled to any significant degree by the normal processes of flour milling. The operation starts with a raw agricultural material, wheat, which is successively ground and sieved to produce a flour. Therefore, from an HACCP point of view, flour should always be treated by food producers as microbiologically dirty and should not be eaten raw or come into contact with finished bakery products. In many situations this is not a problem because most food products containing flour are baked before consumption. There are one or two examples of flour confectionery where this is not the case, and in these situations heat-treated flour or flour from steam-treated wheat can be used. However, all forms of heat treatment denature the gluten and this makes it unacceptable for the production of fermented products and pastries. What the miller can do is produce his flour using good manufacturing practice (GMP) to reduce any risks of contamination. By developing cleaning schedules for all areas, especially high-risk areas such as damping and conditioning equipment, any possible cross-contamination can be significantly reduced.

Another area of concern which has more recently become apparent is the potential presence of mycotoxins such as aflatoxin and ochratoxin from specific toxigenic fungi. It is well documented that most foods are prone to fungal growth at some stage during production, processing and storage (Frisvad and Samson, 1992). These fungi may well be killed or removed during processing; however, the toxins which have been produced will remain in the final product. The only form of control is to avoid using the conditions under which the moulds concerned will flourish. With levels as low as 4 ppb being quoted in specifications there is very little room for error, and with very limited historical information, this is going to be an area that will tax the miller and farmer in the future.

Infestation, on the other hand, is a problem with which the miller is well acquainted and has been since time began. It is not an area that the miller of 100 years ago worried about and one can still hear in some parts of the world the less professional miller using phrases like 'it's a flour mill, you expect to see flour beetles'. However, as we have already said, today's customers demand more in terms of quality and safety. Control of infestation within the mill takes several forms. Inspection of the raw material at intake can prevent infestation entering the building, and cleaning schedules and regular hygiene inspections can prevent the build-up of material which could form the nucleus for any infestation outbreak. Spot spraying of any problem areas can help keep the problem under control. Annual fumigation with methyl bromide is still a regular procedure in many parts of the world. Used in the late spring when insect activity is just beginning to build up, it helps control deep-seated activity which would be difficult to eliminate by other methods. It has, however, been confirmed as an ozone-depleter, and the effect of the Montreal Protocol (1993) means that alternatives are being actively sought and progressively implemented. Several options are still being considered, the application of heat (Dosland, 1995) being one of them, but they are more expensive and lack the penetrating effect of methyl bromide.

There is also a piece of equipment in the mill that is sometimes referred to as an 'infestation destroyer'. The in-line flake disrupters used after the reduction rolls have the effect of breaking up any insects and their eggs. However, these cannot be relied on as they leave the insect fragments in the flour, and this itself can cause problems to customers, especially in parts of the world where the filth test (AOAC, 1990), which evaluates the number of rodent hairs and insect fragments in a sample of flour is being used as a quality standard. With white flours having a potential shelf-life of up to 12 months, it is important that millers do all in their power to prevent contamination of their product.

## 12.5.    De-branning and Flour Milling

Flour milling has changed little since the transition from stone to metal roller mills. Milling systems have become more refined, controls have become improved and milling speeds have increased. Mill engineers have searched for alternatives that would in some way improve the finished product, or at least improve mill efficiencies and reduce manufacturing costs, but the refinements that have occurred show that it is very difficult to improve on the basic flour-milling process.

One technique that is being increasingly used in flour milling is the removal of the outer bran layers of the wheat before it enters the break system. Commonly referred to as 'de-branning' in flour milling, it is based on the 'pearling' process commonly used in the preparation of rice kernels. The presence of the crease in the wheat grain means that pearling is more difficult to achieve than is the case with rice. The de-branning process does not require the classic conditioning stage, instead the bran layers are carefully removed in a two-stage machine abrasive process followed by inter-particle friction (Satake, 1990). Bran removal is very tightly controlled to suit the ash content of the finished product. The endosperm is not disrupted during the process, so the problem of separating bran powder from the flour does not exist to the same degree as it does in the standard flour-milling process. The pearling stage is closely followed by a 'hydrating' section used to control the level of moisture in the finished flour. The hydration time is significantly reduced to between 30 min and 2 h, and is only necessary to replace moisture loss during the milling process. This shorter time means that this new process can quickly respond to the moisture content of the finished flour, and a control loop can be set up to maintain the level far more accurately than with current conditioning systems.

One of the advantages claimed for using de-branning is a reduction in the level of microbial contamination (Pandiella et al., 2005). This occurs because the microorganisms which naturally contaminate wheat are associated with the wheat hairs and bran layers. However, even de-branning cannot reach those microorganisms which are embedded in the crease of the wheat.

## 12.6.    Controlling Flour Quality and Specification

Let us start by defining quality. We often hear phrases like 'top quality' and when there is a problem 'poor quality'. Quality should never be defined from the manufacturer's perspective, it should always be defined by the customer and what they want. In many ways we can simply define 'quality' as being 'fitness for purpose' – the flour has to do what the customer wants it to do! Quality is also 'consistency' – it is no good delivering a flour to a customer and saying 'it's better than last week's'.

Flour is invariably the major raw material in bread and fermented goods and needs to be of the same quality all of the time, so that the bakers can, in turn, achieve high manufacturing efficiencies and provide their customers with a product of consistent quality. The problem for millers is that flour is used in a wide variety of applications from bakery to explosives and from foundries to face powder, and they have to find a way of controlling quality for each and every one of them, even though the basic raw material is variable. This is achieved using two main techniques:

1. by blending either the wheat, the flour or both;
2. by the addition of selected additives.

## 12.6.1.  Gristing Versus Blending

The start of the quality process is the blending of wheats or the recipe known to the miller as the 'wheat grist'. Based on the type of flour required and its specification, the miller will decide which wheats should be used in the grist. At intake the individual wheats parcels will have been segregated into a number of different types, such as biscuit, bread, high protein or low protein. The grist for a particular flour will have been predetermined by a variety of methods, but generally it will be based on the miller's experience of the character of different wheats and a knowledge of the customer's process (some customers have sufficient knowledge to determine the milling grist for themselves). In the UK some grists may contain high-protein wheats, such as Canadian Western Red Springs (CWRS), German or Australian. For bread flour the grist will be blended to a consistent protein level that will be up to 0.7% higher than that of the finished flour. This higher protein specification allows for the protein drop between wheat and flour due to the higher levels of protein in the bran layers than in the endosperm.

Where possible, the wheat protein can be gristed to a lower level to allow for the addition of dried vital wheat gluten. This material is very useful especially in seasons when wheat proteins are low because it can help maintain the flour protein at the required level (Cauvain, 2003). Another advantage of dried gluten is that with the ability to monitor protein continuously on-line using near-infrared (NIR) spectroscopy, a control loop can be set up to feed in the dried gluten thereby giving very tight control of protein levels (Maris et al., 1990). A problem with the use of dried gluten is that it does affect the hydration time of the flour and the rheological properties of the dough. Depending on the amount used, doughs can continue to hydrate after mixing, giving a tighter dough than is required, which causes problems including poor moulding. The gluten in the mixed dough will also appear more elastic and tougher, but this may help maintain the shape of oven-bottom breads. Some customers will consider the use of gluten an unnecessary ingredient in bread flour, and this was probably caused by high usage levels during the 1970s before its potential was fully understood. It is, however, a very useful ingredient, especially at a time when customers are requesting fewer and fewer 'additives' but it must be used at a consistent level – as mentioned earlier, the customer demands consistency.

The protein content of bread flours can vary widely according to the type of bread being manufactured and the breadmaking method being used and the desired characteristics in the final product. Typically, at the lower end of the range would be French flours for baguette from 9.5 to 11.0% (as is) using selected French-grown wheats, rising up through 10.5 to 12% for CBP sandwich-type breads, to 11.0 to 12% for general-purpose bakers' flours and, finally, 12.5% and above for competition and speciality breads (Chapter 2).

Once the grist is running through the mill, the miller can affect characteristics such as flour–water absorption by adjusting the amount of starch damage caused on the reduction system and flour colour by the amount of tail-end (low-grade, high ash) flours allowed into the final blend. Patent flours, that is flours with an exceptionally white colour, are particularly good for breadmaking. They give a good clean crumb colour and, although their protein content might be slightly lower than a normal straight run flour, their quality and baking performance are far superior. Unfortunately, the cost of these flours generally makes them less commercially attractive.

Blending wheats prior to milling is a very popular way of producing flours but it does have some disadvantages. Each wheat in a grist will have its own peculiar characteristics, depending on its variety or source, for example grain size, shape, hardness, moisture content, protein quantity and quality. They can all affect the way the grain behaves through the mill, so that if the grains are blended before milling, the mill settings will have to be a compromise between the various milling characteristics. However, if the wheats are milled individually, for example by variety or type, and then the flours produced are blended, one can be more accurate with the settings of the roller mills and the purifiers, and so maximize the quality from each wheat. Each individual flour milled can then be fully analysed before blending is carried out, with the result that a more consistent flour is produced.

The blending process can be as sophisticated or as simple as required. In its most basic form the flours can be metered together volumetrically using variable-speed discharges on the bottom of the bins. With more sophisticated batch blending using ribbon-type or similar mixers, the flour would be weighed into the mixer along with any other additions. A flour-blending system, while more expensive, gives better control of finished products and more accurate application of any additives.

## 12.6.2.  Additives

Additives are the final tool available to millers to help them improve the performance and consistency of their products. Over recent years the number of additives used have been slowly declining in part because of safety concerns about the use of chemicals in food. For example, potassium bromate, benzoyl peroxide and chlorine dioxide have now gone from the permitted list of additives for breadmaking in the UK (Potassium Bromate Regulations, 1990; Bread and Flour Regulations UK, 1995). In practice, millers are only left with ascorbic acid and various enzymes as methods of controlling the performance of their flours (Chapter 3). Other nutritional additions may be made to meet with legislative requirements for nutrition in various countries (e.g. Bread and Flour Regulations UK, 1995) or to meet the supply requirements for a particular market sector.

## 12.6.3. Ascorbic Acid

It has been known almost since the 1940s that ascorbic acid, chemically a reducing agent, can be used as an oxidizing bread improver (Melville and Shattock, 1938). During mixing, the atmospheric oxygen converts the ascorbic acid to dehydroascorbic acid, which is the oxidizing agent. Since the removal of potassium bromate in 1990, ascorbic acid has become the major additive for modifying flour performance in the UK and elsewhere. Its effect on the gluten is to reduce extensibility and increase elasticity, giving better shape and finer texture to the finished breads. It is added at low levels by flour millers to provide improved flour performance in those breadmaking situations where a separate improver addition will not be made in the bakery. If a separate improver addition is being made, then the additional contribution from the flour is too low to have a major effect on bread quality.

## 12.6.4. Enzymes

The addition of enzyme-active materials to breadmaking flours is now more widely practised. Their use is described in more detail in previous chapters. Enzyme-active materials have always been used by bakers to improve performance. The three types of enzymes most commonly used by the miller to supplement flour are as follows:

1. amylases which act on starch (amylose and amylopectin);
2. proteases which act on proteins;
3. hemicellulases which act on hemicellulose (also called pentosans).

### 12.6.4.1.  Amylases

One of the most common techniques used by bakers a few years ago was to include a small amount of malt flour in a bread mix. This would have contained large amounts of cereal *alpha*-amylase which would break the damaged starch down to dextrins, which in turn would be broken down by the *beta*-amylases into maltose, a useful yeast food. Malt flour was also used by the miller to improve flour performance, especially when the Hagberg Falling Number is particularly high. However, it was known that at excessive levels of addition, bread crumb becomes sticky and difficult to slice. This is caused by the *beta*-amylase being deactivated much earlier in the baking process than the *alpha*-amylase, which continues working until approximately 75 °C (167 °F) and produces an excess of sticky dextrins (Chamberlain et al., 1977).

In the early 1970s, fungal amylases were introduced, which were a much more controlled ingredient and had the added benefit that they are deactivated earlier in the baking process (Chapter 3). This meant they could be used at much higher levels without the stickiness problem, with additional benefits including a finer, whiter texture, better volume and an apparent improvement in the shelf-life of the bread (Cauvain and Chamberlain, 1988).

### 12.6.4.2.  Protease

The group of enzymes included under this heading are generally used where large quantities of hard wheats are included in the milling grist, for example North American. They help to reduce the strength of the dough and so improve handling and product texture, but they must be used with discretion to avoid a complete breakdown of the dough structure. A more common use of proteases would be in the production of biscuits and wafers where weaker proteins are more desirable and yet some degree of protein functionality is still required (Wade, 1995).

### 12.6.4.3.  Hemicellulases

The use of hemicellulases in breadmaking is a more recent introduction, only being allowed in the UK since 1996 (Bread and Flour Regulations UK, 1995). They break down in a controlled manner the pentosan component of the hemicellulose. This action continues through the mixing, fermentation and early baking stages, giving a soft but not sticky dough and which yields bread with improved volume and texture. They are useful in all flours, but of particular benefit in flours which contain high percentages of hemicellulose, such as brown, wholemeal (wholewheat) and rye flours and meals.

## 12.7.  Nutritional Additions

Bread has always been seen as a fundamental food, and as such has been well regulated to prevent adulteration and to protect the consumer. Fortification of flour because of its role as a major ingredient is seen as a method of ensuring that populations can receive extra minerals and vitamins that might be lacking in some diets (Rosell, 2003). Since the 1940s all flours in the UK, with the exception of wholemeal, have contained the following:

- calcium carbonate to provide calcium ions;
- a source of iron;
- thiamine;
- nicotinic acid.

Even the latest amendment of the UK Bread and Flour Regulations (1995) has retained this type of fortification even though diets in the UK are generally much improved. So rather than a reduction in nutritional additives, the future may actually bring an increase with a move towards what are termed 'functional foods', that is foods which can make a positive contribution to health through their consumption. As an example, folic acid, which helps prevent neural tube defects in unborn children (Alldrick, 1993, 1994), is already appearing in certain breads thereby offering consumers an element of choice. It is unlikely, however, that with a general move towards deregulation in the EU, any further nutritional additive will be required by legislation in the future.

## 12.8.    Other Flour Types

So far this chapter has discussed wheat flour from the perspective of white flour, since this accounts for approximately 90% of all flour used. Other flour types are available and the following are just a few examples.

### 12.8.1.    Brown

If white flour has an extraction rate in the range 76–78%, then brown flours have the equivalent of about 85–90%. They can be produced during the milling process by feeding back 10–15% of selected bran stocks into a white flour, or in a mixing plant by blending wholemeal and white flours in the ratio 50:50.

### 12.8.2.    Wholemeal (Wholewheat)

As the name suggests, this is the product of the whole grain. They can be produced on a standard roller mill, but they are perceived to be of better quality if made on a stone-ground plant. While we have seen several explanations as to why this is the case, we have not been convinced by them, and continue to believe that the only difference between stone-ground and roller-ground wholemeal is a very low level of sand which comes from the stones as they wear during grain processing!

Both brown and wholemeal flours tend to have higher protein content than white flours. Values up to 14.0% protein, with the flours often supplemented by dried gluten, are not uncommon because the dough has to hold the bran, which is effectively a non-functional ingredient in breadmaking. The particle size of the bran is also important to obtain the best performance (Cauvain, 1987). Coarse bran can give a good visual effect both in the bread crumb and on the crust, but if there is too much coarse bran present in the flour it can result in an open and unattractive crumb structure. At the other end of the scale, fine bran can have a deadening effect on the bread, resulting in a bland, small loaf with a dull grey crumb. The answer, as always, is a balance between the two and will be defined by customer requirements.

## 12.8.3.  Self-raising Flours

A combination of sodium bicarbonate together with a suitable acid ingredient will produce a flour for a variety of uses, including the manufacture of batters, cakes and scones. By varying the acid ingredient, the point in the process when the carbon dioxide is evolved can be varied. For example, if monocalcium phosphate is used, then 60% of the carbon dioxide will be generated at the mixing stage and 40% during baking. If this ingredient is changed to sodium aluminum phosphate, then this can be changed to 30% at the mixing stage and 70% during baking. The requirement for heat to be applied before the majority of the carbon dioxide is liberated can be useful if the product is required to stand before baking or if an extended shelf-life is required in the flour.

## 12.8.4.  Malted Grain Flours

The addition of malted grains, either kibbled or flaked, together with additional malt flours, either diastatic or non-diastatic (or both), can produce a very attractive bread with exceptional flavour characteristics.

Other types of flours available will include wheatgerm flour, soft grain flours and flour blends containing other cereals, for example rye, oats and maize, and seeds. Often such mixtures are prepared by millers for direct use by bakers under proprietary names, for example Hovis (germ-fortified flour) in the UK.

## 12.9.  Flour Testing Methods

The miller has a range of tests available to determine values for what are considered to be the most important performance specifications for a particular flour. Of these the measurement of protein quality presents the greatest challenge with different pieces of equipment being preferred in different countries. We have used the term 'protein quality', but having already defined quality earlier as 'fitness for purpose' we should perhaps therefore really refer to protein quality as 'protein rheology' because what is 'good quality' for one type of bread may well be 'poor quality' for another.

## 12.9.1.  Protein and Moisture Content

These are two of the most important parameters in flour, and are generally measured during production using an NIR analyser (CCFRA FTP 0014, 1991). This is by far the most common piece of testing equipment in flour mill laboratories and has already been mentioned under 'Wheat testing'. It is versatile and quick, conducting a range of tests, such as protein and moisture, but also starch damage, water absorption and even colour, in less than 30 s. However, with speed of measurement and the relative imprecision of the reference methods (e.g. colour) some accuracy is lost, and the NIR analyser is best kept for the main parameters of protein and moisture.

To maintain their accuracy, the instruments must be calibrated at regular intervals by comparing with national, or ideally international, standard methods. For protein content this is normally the Kjeldahl acid digestion method using sulphuric acid in the presence of a catalyst (CCFRA FTP 0009, 1991). The nitrogen figure produced is then converted to a protein value using the Kjeldahl factors as follows:

| | |
|---|---|
| Wheat bran | 6.31 |
| Other wheat products | 5.70 |
| Other foods | 6.25 |

The Kjeldahl method is not generally used during production because of the time it takes to provide a protein value, and because of the dangers arising from the strong reagents used. It has now largely been replaced by the Dumas method, a much safer system based on using combustion in the presence of oxygen (AACC Method 46–30, 1995). It is as accurate as the Kjeldahl method, and much quicker; in fact it is fast enough to be used as a quality assurance test during the milling operation, but requires more skill to operate than the NIR analyser. In some countries, including the UK, all proteins are declared on an 'as is' basis, that is as a percentage of all the constituents including the moisture. While this has certain advantages, it relies on the moisture content always being the same, otherwise it could appear that the protein content is varying. For example, a flour measured at three different moisture contents would give the following results:

| | Protein (%) |
|---|---|
| On a dry matter basis (dmb) | 10.0 |
| The same flour with a moisture of 8% | 9.3 |
| The same flour with a moisture of 15% | 8.7 |

One should always check, when comparing flour specifications from different countries, that the protein content is quoted to the same moisture content.

Moisture determination on the NIR analyser needs to be calibrated to an oven method, typically 130 °C (266 °F) for 90 min and 2 h for ground wheat (CCFRA FTP 0008, 1991). This test should be one of the simplest to carry out, but it can prove difficult to obtain consistent results especially when comparing results from different laboratories. This is not necessarily due to problems with the test method or the operatives; it is more likely that the pneumatic handling causes variations in the flour. Throughout the mill, during bulk delivery and also on customers' premises, flour is moved using either negative or positive air pressure. The air movement itself is sufficient to give a drying effect; however, when positive pressure is used, the increase in air temperature due to compression can enhance this effect. It is possible for flour to increase in temperature by up to 5 °C (9 °F) when blown from one bin to another. In this type of situation, flour moisture can easily drop by half a percentage point, which is sufficient to cause disputes between millers and customers who carry out their own checks.

## 12.9.2.  Flour Grade Colour

The flour grade colour (FGC) or sometimes grade colour figure (GCF) is measured using a Kent Jones Colour Grader (CCFRA FTP 0007/2, 1991). It is a measurement made of the reflection of light in the 530-nm wavelength region from the surface of a flour and water slurry contained in a glass cell. It gives a measure of bran contamination of white flours and varies with the amount of low-grade flours, from the tail end of the mill, which are allowed through to the final product. The measured value is also influenced by background endosperm colour which depends on the type of wheat in the grist, and the particular crop year, where the quality of the wheat can also influence the ultimate endosperm colour. Standardization is carried out using National Colour Standards (CCFRA FTP 0007/1, 1991), which ensure consistent results between laboratories, although there is an internal white tile which can be used on a day-to-day basis to prevent drift of data with time.

Grade colour is not a measure of the visual appearance of the flour. Two samples of flour that can look completely different may give the same grade colour value. Over the years, other methods have been considered to get round this problem and provide a figure that measures the actual appearance, but the colour grade figure is so well entrenched and understood by both millers and bakers in the UK and other countries that changing it would be like losing an old friend! Flour colour is closely associated with ash content that is used to measure flour purity in many parts of the world. Unfortunately, in countries such as the UK, where flour is fortified with minerals, the ash content appears unreasonably high and cannot be used as a direct measure of flour quality.

Flour colour or ash content can affect the flour performance in baking; generally the whiter the flour, the better the breadmaking properties (Cauvain et al., 1983, 1985). This fact is recognized in some countries, such as Italy, where the maximum ash content of soft wheat flours is defined in law.

There are three main categories, defined on a dry solids basis:

$$\text{Flour type OO} = 0.50\%$$

$$\text{Flour type O} = 0.65\%$$

$$\text{Flour type 1} = 0.80\%$$

'Specky' flour, that is flour contaminated with very small but visible pieces of bran, is not readily picked up using the flour-colour grader test, and in the past its detection has relied on a visual inspection by millers.

Using the latest advances in image analysis, a new method to test the bran content of flour has recently been introduced. The 'Branscan', as it is called, is designed for on- or off-line use. A sample of flour is compressed against a transparent window and a video camera with a PC-based image analysis system is used to calculate the number of bran specks. In the laboratory this is carried out by a technician but in the on-line system the process is completely automatic with the results being displayed on a graph with both the number and total area covered by the bran expressed as a percentage (Evers, 1993). This method is unaffected by variations in endosperm colour and composition, and can be set up to alarm when a specific figure has been exceeded.

## 12.9.3.  Water Absorption

Water absorption is a well-used term in flour technology, but unfortunately it will mean different things to different people. Bakers mixing batters will be adding more water to their products than when they make bread doughs, which in turn will be more than for biscuit doughs. Even within a particular product group, the recipe used will affect the amount of water added. Thus a water absorption test used has to eliminate these variations and deal with just flour and water; it cannot be complicated by a variety of products or ingredients. The dough must also have a predefined, 'optimum' viscosity so that it can at least provide a value that is reproducible.

The test which has become the standard uses the Brabender Farinograph (CCFRA FTP 0004, 1991). This machine measures and records the mixing characteristics of a dough made from just flour and water, and continues to record the properties as the dough develops to its maximum viscosity and until it starts to break down. The operator is required to add sufficient water to the flour to produce a dough with the maximum viscosity on the 600 line (this is the case for most breadmaking flours in the UK, although lower figures may be set on the Farinograph in other parts of the world). This may take two or three attempts, but once done, the machine is allowed to run to form the characteristic Farinograph curve. The water absorption is the amount of water added to the flour to achieve a given viscosity and is conveniently recorded in percentage terms directly on the burette. The Farinograph itself also gives valuable information on the rheology of the dough, as will be discussed later in this chapter.

The water absorption of a flour is influenced by four parameters:

1. Moisture content. A flour with a moisture content of 13% will have an apparent water absorption which is 1% higher than the same flour at 14%.
2. Protein content. Protein absorbs approximately its own weight in water, so that a higher-protein flour will naturally absorb more water than a lower-protein one.
3. Starch damage level. This is probably the major factor affecting the water absorption properties of flour. As discussed above, starch damage is achieved by grinding the flour hard on the reduction rolls. Excessive starch damage can

cause a greying of the crumb colour and an opening of the crumb structure (Cauvain and Young, 2006).

4. Pentosan (hemicellulose) level. These are present, at levels of 2–3% in white flours and up to 10% in wholemeal (wholewheat). These non-starch polysaccharides have a very high water-binding capacity and, although present in the dough at very low levels, can actually account for absorbing up to one-third of the water in the dough.

The water absorption capacity of flours will typically vary from 50–54% for biscuit flours up to 60–62% for standard UK bread flours.

## 12.9.4.  Hagberg Falling Number

As discussed under 'Wheat testing', this is a measure of the cereal *alpha*-amylase content of the flour. It is important that it is controlled at the wheat stage because it cannot be affected by the milling process. It can be lowered by the addition of malt flours, but while investigations have been carried out, no technique has been found to reduce the amylase and so increase the Hagberg Falling Number. The blending of flours with different Hagberg Falling Numbers can also be problematic because a simple arithmetic mean will not give a satisfactory approximation. Instead the values have to be converted to a liquefaction number using the following formula:

$$\text{Liquefaction Number (LN)} = 6000/(\text{Falling Number} - 50)$$

Rearranging the equation we can convert back from liquefaction number:

$$\text{Hagberg Falling Number} = \frac{6000}{\text{LN}} + 50$$

In the following example, two flours are blended 50:50; flour 1 has a Hagberg Falling number of 100 and flour 2 has a Hagberg Falling number of 300.

$$\text{Flour 1 LN} = 6000/(100 - 50) = 6000/50 = 120$$

$$\text{Flour 2 LN} = 6000/(300 - 50) = 6000/250 = 24$$

The liquefaction number of the blended flour will be

$$\text{LN (blended flour)} = (120 \times 50)/100 + (24 \times 50)/100 = 60 + 12 = 72$$

The Hagberg Falling Number (HFN) will be

$$\text{HFN} = (6000/72) + 72 = 83 + 50 = 133$$

The result is not the arithmetic mean of the two flours, which would have been 200, but is significantly biased towards the lower end.

## 12.9.5.   Flour Rheology

The rheological properties of flour are probably the most critical parameter in the whole of the flour specification. It is an indication of how a given dough will behave as it is being processed through the plant and through the oven and is related to the quality of the finished product. There are countless different bakery products and even more combinations of ingredients, and if you add to this the possible differences in the processing, then you have an infinite combination that would be impossible to define individually in terms of flour parameters. We therefore have some sympathy with the view that if you want to know whether a flour will make a pizza, you have to make a pizza. For this reason, test bakeries in flour mills are very common. Whether they are used to determine the suitability of wheat varieties in specific applications, or just as a final test in the quality assurance process, they are invaluable in helping millers understand the final process and in giving their customers that extra bit of confidence.

While loaves of bread can be measured for volume and assessed for colour and texture, some of the tests still carry an element of subjectivity as discussed in Chapter 1. More closely defined and more rapid tests than baking are required to help millers assess the suitability of their flours. The dough rheology assessment methods we use today, with all their imperfections, have stood the test of time and in some cases having been around for more than 50 years. While our scientific understanding of dough rheology has improved and breadmaking processes have changed, we still use three basic testing methods based on the Farinograph (CCFRA FTP 0004, 1991), the Extensograph (CCFRA FTP 0003, 1991) and the Alveograph (Faridi and Rasper, 1987).

## 12.9.6.   Farinograph

We have already mentioned the use of the Farinograph for measuring water absorption, but it is also very useful for measuring the mixing characteristics of flours and gives a good indication of flour performance in breadmaking.

There are three pieces of information that can be deduced from a farinogram (Figure 12.3):

1. Dough development time (A). This is the time taken from the start of mixing to the point of maximum viscosity just before the curve starts to weaken. It will be longer with strong flours and very short with biscuit flours.
2. Stability (B). This property is measured from the point when the top of the graph first crosses the 600 line (or other fixed point), to the point where it drops below it, that is the time the curve is above the line. It gives a measure of the tolerance of the flour to mixing.
3. Degree of softening (C). This is the difference in height, measured in Brabender units, between the centre of the graph at the maximum viscosity, and the centre of the graph at a point 12 min later.

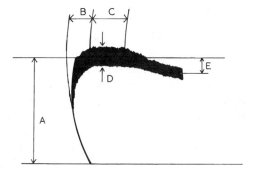

FIGURE 12.3. Typical farinogram. See text for key to symbols.

Some examples of the different farinograms obtained with different flours are given in Figure 12.4. The Farinograph test is probably the most rapid of the three being discussed under this heading; water absorption can be done in 10–15 min, the full curve probably taking another 10–15 min, depending on the flour. Because of these short testing times it is possible to use the Farinograph as a quality assurance tool. This is not the case with the next test.

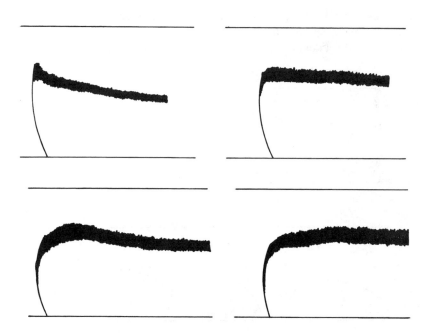

FIGURE 12.4. Examples of farinograms in which the flours become stronger, left to right and more elastic, top to bottom (based on data published by NABIM).

## 12.9.7.  Extensograph

In this test a flour–salt–water dough is prepared under standard conditions using the Farinograph. The salt is used at 2% (6 g in 300 g of flour), which is a level typically seen in bread, as it has a tightening effect on the dough which helps give a more realistic result. Various attachments on the Extensograph are then used to mould the dough to a standard shape before resting. After a set time of 45 min the dough is stretched, and the extensibility of the dough and its resistance are recorded. Immediately the doughs are re-moulded and allowed to stand for a further 45 min before being stretched once again. The dough pieces are once more re-moulded, and rested a further 45 min before the final stretch. This test is designed to give an indication of the baking performance of a dough over a time span of 135 min similar to that of a fermented dough. With modern short-time doughs the first stretch at 45 min is probably the most important one.

With modern, untreated bread flours the resistance will usually be around the middle of the graph (Figure 12.5), but with weaker biscuit types, the curve will usually be well below the 200 BU line. In these cases it may be necessary to use an amended method for the Extensograph that uses a higher level of salt (CCFRA FTP method 0016, 1991). This brings the curve nearer the centre of the graph and so gives a more reliable result.

## 12.9.8.  Alveograph

This testing equipment and method is very popular in France and elsewhere, and is used to great effect for defining the parameters for a good baguette-making flour. Essentially a dough is prepared using a set quantity of water and salt, and

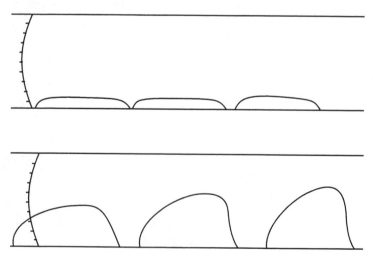

FIGURE 12.5. Typical extensographs for, a biscuit (top) and a bread flour (bottom), left to right, testing time 45, 90 and 135 min (based on data published by NABIM).

then extruded from the mixer and shaped following a standard method. After a resting period, the dough piece is clamped into a metal ring and inflated, while the pressure inside the bubble is measured against time and plotted on a graph (Figure 12.6). The characteristics of the dough can then be assessed using the shape and area of the curve thus obtained.

A modification of the Alveograph is the 'Consistograph', also supplied by Chopin. Both the Alveograph and the Consistograph provide information on the rheological properties of dough but the main difference between the two machines is that the latter is designed to test dough in which the water level is adjusted according to the water absorption capacity of the flour.

### 12.9.9.   Other Rheological Testing Equipment

The importance of dough rheology to dough processing and breadmaking in general has been appreciated for many years and has shaped the traditional testing methods commonly used today. The advent of improved computing capabilities for recording and interpreting data has resulted in a significant increase in the number of available testing methods. The evaluation of flour properties continues to evolve as equipment companies, flour millers and bakers continue to seek more appropriate ways of testing flour properties.

### 12.9.10.   The MixerLAB

This equipment can be used for the measurement of dough rheology in a manner similar to that seen with the Farinograph (Bason et al., 2005). Indeed one adaptation is possible to fit a standard Farinograph 50-g bowl to make appropriate measurements.

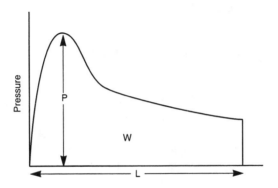

FIGURE 12.6. Typical alveograph (based on data published by NABIM).

## 12.9.11.   Amylograph and Rapid Visco Analyser (RVA)

While much attention is rightly focused on the rheological properties of the gluten it may be appropriate on occasions to make some measurements related to the largest of the flour components – the starch. This assessment is most commonly made by heating a mixture of starch and water at a pre-defined rate until gelatinization has been achieved. During the heating phase, changes in viscosity are recorded as the movement of the mixer blade is impeded by the starch–water mixture. Many variations of heating, holding and cooling may be applied to the starch–water mix according to the needs of the user. In the case of breadmaking, evaluation of the starch properties in the flour is limited.

The Amylograph and the RVA find use in the evaluation of rye flour (see Chapter 13) where they may be used to assess the level of enzymic activity (particularly amylase) and the degree of softening which may occur during gelatinization of the starch.

## 12.9.12.   Using Testing Equipment

None of the equipment and testing methods described above will predict the suitability of a flour for all bakery uses; with all testing methods, assumptions have to be made about the particular product, for instance:

- Using the Farinograph, the standard water absorption test is carried out at 30 °C (86 °F) in a low-speed, Z-blade mixer. We can reasonably ask the question 'What does this tell us about the effects on flour performance of harsher mixing regimes at cooler temperatures'?
- The Extensograph is a classic test but is only two-dimensional, that is a standard dough stretched at a standard rate. Dough consistency, resting times, temperatures, rates of stretch can all be changed but what effect will such changes have on the interpretation of the results and prediction of breadmaking potential?
- Using the Alveograph, well-documented tests exist for baguette-making flours, but the fixed level of water makes it unacceptable with stronger flours since dough consistency is not optimized for some breadmaking processes.

So what are the answers to our dilemma for assessing the breadmaking potential of flours from rheological testing methods? How can customers define the flours they require? To answer such questions we have to go back to our definition of quality – 'fitness for purpose'. Some products will be very tolerant to abuse or at least be used in breadmaking processes where small differences are less important. For these, flour can be defined by specifying a bare minimum of parameters; perhaps just protein content, moisture and colour are all that will be necessary. However, other products and processes are less forgiving, and so in these cases it is up to the baker and the miller to work together to find the right quality definitions. Generally it will be a combination of the tests available, or possibly it will include something developed specifically, for example viscosity using a viscometer, or particle size perhaps. Whatever the answer happens to be, the customer can be assured that millers will do everything in their power to get the flour right the first time.

## 12.10.   Glossary of Milling Terms Used in this Chapter

| | |
|---|---|
| **Aspiration** | A process using air to remove fine impurities from wheat. |
| **Break rolls** | Fluted rolls designed to open up grains and release the endosperm. |
| **Conditioning** | The process of bringing a wheat to the appropriate moisture content for milling. |
| **Extraction rate** | The amount of white flour extracted from the wheat, expressed as a percentage of the wheat entering the first break rolls. |
| **Flour dressing** | Sieving of the various flour stocks. |
| **Grist** | The blend of wheats for a particular flour. |
| **Hectolitre weight** | A measurement of wheat density, expressed in kilograms per hectoliter. |
| **Low-grade** | Flour of poor colour and poor functionality. |
| **Overtails** | Name given to materials (stocks) which are too large to pass through a given sieve size. |
| **Reduction rolls** | Smooth-surfaced rolls designed to reduce the size of semolina particles to flour. |
| **Semolina** | Coarse particles of endosperm. |
| **Screens** | Sieve meshes of a given size. |
| **Screenings** | Impurities removed from wheat as received at the mill. |
| **Screenroom** | Part of the milling process designed to clean grain. |
| **Silo** | A building for grain or flour storage. |
| **Stocks** | Mill feeds to the different machines used in the mill. |
| **Straight run** | Flour with approximately 76–78 per cent extraction rate obtained by blending individual machine flours. |
| **Throughs (thro's)** | Name given to the stocks which pass through a given sieve. |
| **Patent flour** | A flour produced from just top-quality mill streams with a very good colour (low ash content). |

## References

AACC (1995) *Approved Methods of the American Association of Cereal Chemists*, 9th edn, March, St Paul, Minnesota, USA: Method 39–70A, Wheat hardness as determined by near infrared reflectance; Method 46–30, Crude protein – combustion method.

AOAC (1990) *Official Methods of the Association of Official Analytical Chemists*, 15th edn, AOAC, Washington DC, USA. Method 972.37 a and b, Extraneous materials isolation.

Alldrick, A. (1993) Folic acid and neural-tube defects. *Chorleywood Digest No. 124*, February, CCFRA, Chipping Campden, UK.

Alldrick, A. (1994) Folic acid – revisited. *Chorleywood Digest, No. 133*, January, CCFRA, Chipping Campden, UK, pp. 9–10.

Bason, M.L., Dang, J.M.C. and Charroe, C. (2005) Comparison of the Dough LAB and Farinograph for testing flour quality, in *Using Cereal Science and Technology for the Benefit of Consumers* (eds S.P. Cauvain, S.E. Salmon and L.S. Young), Woodhead Publishing Ltd, Cambridge, pp. 276–82.

Bhandari, D.G., Church, S., Borthwick, A. and Jensen, M.A. (2005) Automated varietal identification using lab-on-a-chip technology, in *Using Cereal Science and Technology for the Benefit of Consumers* (eds S.P. Cauvain, S.E. Salmon and L.S. Young), Woodhead Publishing Ltd, Cambridge, pp. 529.

Bread and Flour Regulations (1995) UK SI 3202, HMSO, London.

BS 4317: Part 3 (1987) *Determination of Moisture Content of Cereals and Cereal Products (Routine Method)*, British Standards Institute, London.

BS 4317: Part 30 (1994) (ISO 8981: 1993) *Identification of Wheat Varieties by Electrophoresis*, British Standards Institute, London.

Cauvain, S.P. (1987) Effects of bran, germ, and low grade flours on CBP bread quality. *FMBRA Report No. 138*, December, CCFRA, Chipping Campden, UK.

Cauvain, S.P. (2003) Dried gluten in breadmaking, *CCFRA Review No. 39*, CCFRA, Chipping Campden, UK.

Cauvain S.P. and Chamberlain N. (1988) The bread improving effects of fungal amylase. *Journal of Cereal Science*, **8**, 239–48.

Cauvain, S.P., Davis, J.A. and Fearn, T. (1985) Flour characteristics and fungal amylase in the Chorleywood Bread Process. *FMBRA Report No. 121*, March, CCFRA, Chipping Campden, UK.

Cauvain, S.P., Chamberlain, N., Collins, T.H. and Davis, J.A. (1983) The distribution of dietary fibre and baking quality among mill fractions of CBP bread flour. *FMBRA Report No. 105*, July, CCFRA, Chipping Campden, UK.

Cauvain, S.P. and Young, L.S. (2006) The Chorleywood Bread Process. Woodhead Publishing Ltd., Cambridge, UK.

CCFRA FTP (1991) *CCFRA Flour Testing Panel Methods Handbook*, CCFRA, Chipping Campden, UK:

> Method no. 0003, Determination of rheological properties of doughs using a Brabender Extensograph.
>
> Method no. 0004, Determination of water absorption and rheological properties of doughs using a Brabender Farinograph.
>
> Method no. 0006, Determination of Falling number.
>
> Method no. 0007/1, Determination of Grade Colour using national standard: with particular reference to the use of an ultimate internal standard tile.Method no. 0007/2, Determination of Grade Colour for series 2 instruments.
>
> Method no. 0008, Determination of moisture content by oven drying.
>
> Method no. 0009, Determination of protein content with the Tecator Kjeltec 1030 auto system or equivalent analysis system.
>
> Method no. 0013, Determination of gluten content using the Falling Number Glutomatic.
>
> Method no. 0014, Determination of protein and moisture contents by near infrared reflectance.

Method no. 0016, Determination of rheological properties using a Brabender Extensograph (alternative method for use with weak, white flours).

Chamberlain, N., Collins, T.H. and McDermott, E.E. (1977) The Chorleywood Bread Process: The effects of alpha-amylase activity on commercial bread. *FMBRA Report No. 73*, June, CCFRA, Chipping Campden, UK.

Dosland, O. (1995) The Chester heat experiment. *Association of Operatives Millers – Bulletin*, September, 6615–8.

Evers, A.D. (1993) On-line quantification of bran particles in white flour. *Food Science and Technology Today*, **7**(1), 23–6.

Faridi, H. and Rasper, V.F. (1987) *The Alveograph Handbook*, AACC, St Paul, Minnesota.

Frisvad, J.C. and Samson, R.A. (1992) Filamentous fungi in foods and feeds: Ecology, spoilage, and mycotoxin production, in *Handbook of Applied Mycology*, Vol. 3, *Foods and Feeds* (eds D.K. Arora, K.G. Mukerji and E.H. Marth), Marcel Dekker, New York, pp. 31–68.

Gaines, C.S., Finney, P.F., Fleege, L.M. et al. (1996) Predicting a hardness measurement using the single kernel characterizations system. *Cereal Chemistry*, **73**, 278–83.

Hook, S.C.W. (1982) Determination of wheat hardness – an evaluation of this aspect of wheat specification. *FMBRA Bulletin No. 1*, February, CCFRA, Chipping Campden, UK, pp. 12–23.

Lookhart, G.L., Bean, S.R. and Culbertson, C. (2005) Wheat quality and wheat varietal identification, in *Using Cereal Science and Technology for the Benefit of Consumers* (eds S.P. Cauvain, S.E. Salmon and L.S. Young), Woodhead Publishing Ltd, Cambridge, pp. 293–7.

Maris, P.I., Fearn, T., Mason, M.J. et al. (1990) NIROS: Automatic control of gluten addition to flour. *FMBRA Report No. 143*, November, CCFRA, Chipping Campden, UK.

Melville, J. and Shattock, H.T. (1938) The action of ascorbic acid as a bread improver. *Cereal Chemistry*, **15**, 201–5.

Montreal Protocol (1993) On substances that Deplete the Ozone Layer, Montreal, 16 September 1987: Adoption of Annex D, Treaty Series No. 14, HMSO, London.

Pandiella, S.S., Mousia, Z., Laca, A. et al. (2005) Debranning technology to improve cereal-based foods, in *Using Cereal Science and Technology for the Benefit of Consumers* (eds S.P. Cauvain, S.E. Salmon and L.S. Young), Woodhead Publishing Ltd, Cambridge, pp. 241–4.

Potassium Bromate (Prohibition as a Flour Improver) Regulations (1990) UK SI 399, HMSO, London.

Rosell, C.M. (2003) The nutritional enhancement of wheat flour, in *Bread Making: Improving Quality* (ed S.P. Cauvain), Woodhead Publishing Ltd, Cambridge, pp. 253–69.

Satake, R.S. (1990) Debranning process is new approach to wheat milling. *World Grain*, **8**(b), 28, 30–2.

Sewell, T. (2003) *Grain: Carriage by Sea*. Protea Publishing Ltd, Cranleigh, Surrey.

Wade, P. (1995). *Biscuits, Cookies and Crackers*, Vol. 1, *The Principles of the Craft*, Blackie Academic & Professional, Glasgow, UK.

# 13
# Other Cereals in Breadmaking

Stanley P. Cauvain

## 13.1.  Introduction

The main thrust of the previous chapters has been accented towards the production of bread from 100% wheat flours but, although such products are universal, there are some bread products around the world which are based on or include a high proportion of non-wheat cereals. In breadmaking terms, rye is the closest of the cereals to wheat with similar protein contents but a distinctly limited ability to form gluten. The closeness of wheat and rye has led to their crossing and the first 'artificial' cereal – triticale – which may also be used for breadmaking (Gustafson et al., 1991).

In the past, maize (corn), barley, oats, sorghum, millet and rice have all found their way into bread products at some time, usually when wheat and rye have been in short supply. More recently, in countries where these other cereals are commonly grown, they have been utilized in breadmaking to reduce the proportion of wheat flour being used if economic conditions necessitate the reduction of wheat imports. In some cases it is possible to make a product which has some of the attributes of wheat breads with mixtures of some of these cereals by utilizing the gelatinization properties of their starches to form a bread-like, aerated structure. In these circumstances the lack of gas-holding capabilities of any proteins present in the flour must be compensated for by adding other bubble-stabilizing materials.

## 13.2.  Rye Bread

Of all the non-wheat breads, those based on rye are the most common with production in central and Eastern Europe and North America. Rye is also grown as a feed wheat in Asia and South America. Although probably of Mediterranean origin (Lorenz, 1991), cultivated rye (*Secale cereale*) is typically a plant of cool northern climates where it is better able to withstand the winter cold and grow farther north than any other cereal, hence its current favoured cultivation areas in northern Europe and North America. Rye is able to grow on a wider range of soil fertilities than wheat and is mostly sown in the autumn (fall) to over-winter.

Rye milling is a multi-stage process of similar complexity to that of wheat milling, with a number of break and reduction roll passages feeding sifters where the stocks are separated into various intermediate fractions and finished flours (Rozsa, 1976). The numbers of rye flours and meals available and their classification varies according to the country of production. Key quality attributes are ash contents and colour, with low-ash rye flours being referred to as 'white' and high-ash flours as 'dark'.

The high levels of pentosans present in rye flours largely inhibit their ability to form gluten, and so in rye bread doughs the proteins play a lesser role in the structure-forming process. The quantities of pentosans and other soluble materials are particularly important in relation to breadmaking properties, and rye flours contain large quantities of matter which is twice as soluble in water compared with wheat flours.

The starch in rye gelatinizes at a relatively low temperature, 55–70°C (130–160 °F), which coincides with the temperature range for maximum *alpha*-amylase activity. Rye has a low sprouting resistance and as a result particular attention is focused on enzymic activity in rye flour specifications, with measurement of Falling Number and Brabender Amylographs being common (Drews and Seibel, 1976). The overall enzymic activity in rye flours is high compared with that of wheat flours, together with the potential for significant cellulase and proteinase activities being present. Evaluations of the potential of rye flour for making a particular type of bread are usually based on the interpretation of Brabender amylograms.

Breads based on or containing rye are most commonly seen in the USA, and northern, central and eastern Europe. The methods of manufacture and forms that the breads take differ noticeably between these two main centres of production. The variety of rye breads is perhaps greatest in Germany where there are four main classes of bread, depending on the proportion of rye to wheat in the formula, which may use any of four raw materials to give a potential 16 bread combinations (Meuser et al., 1994). In the USA the range of rye breads tends to be less well defined and is generally fewer in number than in central Europe. Pumpernickel is a special rye bread produced on both sides of the Atlantic but in quite different forms.

In order to restrict amylolytic activity and breakdown of starch during baking, acidification of rye bread doughs has become common. Traditionally, lactic acid fermentation in a sour dough is preferred, although direct acidification can be achieved by the addition of acids such as citric or tartaric acid. Acidification of rye doughs improves their physical properties by making them more elastic and extensible and confers the acid flavour notes so characteristic of rye breads.

A preliminary heat treatment may be applied to some of the rye flour which will be used in the final mix. This process is commonly referred to as 'scalding' and is seen as a key element in the production of flavour in the final product (Petersen et al., 2005).

## 13.2.1.  Sour Dough Methods

The sour dough method begins with a 'starter' based on a pure sour culture prepared by inoculating sterile nutrient media with the appropriate bacteria.

Initially in the pure culture, some dark rye flour and water are mixed to form a soft dough at 27 °C (80 °F) which is blended with salt, more rye flour and water to form the sour dough. Commonly the starter is used at the rate of 20% of the final sour dough. Subsequent starters can take the form of a portion of the sour dough which has been stored at 5 °C (40 °F). The starter may be stored for a few days or even up to 6 months, though by this time the purity of the culture will have been lost and it is better to begin again with a new pure culture.

The activity of the sour dough needs to be controlled in order to inhibit the activity of unwanted microorganisms. This is usually achieved through the addition of about 2% salt (based on flour weight). A fairly slack dough consistency is required for the sour dough; typically water absorptions will range from 80 to 100% for rye meals and flours. The sour dough is formed at a temperature around 35 °C (95 °F) and allowed to cool to about 20 °C (68 °F) over a 24 h period. During this time the pH of the sour dough typically falls from around 5.8 to 3.5. The sour dough can be used for up to 9 days after preparation.

Dehydrated sours in dry powder form have become increasingly popular with bakers. Based on pre-gelatinized flours, organic acids or their salts and a dehydrated sour dough extract, they are more convenient to handle and ensure consistency of performance in doughmaking. Their rate of addition varies according to their source and individual preferences, but is typically about 2–6% based on total flour weight.

## 13.2.2.  Doughmaking

The sour dough is blended with the other ingredients using a low- or medium-speed mixer. The proportion of the sour to the other ingredients will vary according to the type of bread being made. A sample recipe for the production of rye wholemeal bread is given in Table 13.1.

TABLE 13.1. Example of rye wholemeal formulation.

|  | kg |
| --- | --- |
| Sour dough | 80.0 |
| Rye wholemeal | 60.0 |
| Salt | 1.2[a] |
| Yeast | 1.0 |
| Water | 25.0 |

[a] 0.8 kg salt contained in the sour dough.

Intensive mixing is not normally required for rye doughs because they are not able to form gluten. Mixing times between 5 and 30 min are used depending on the type of mixer and its speed. Coarse rye wholemeals and whole rye grain doughs require longer mixing times.

Dryness and crumbliness in rye bread can be avoided by presoaking or scalding part of the rye wholemeal before doughmaking. Between 10 and 20% of the meal is scaled with an equal weight of water 3 h before doughmaking and allowed to cool before adding it to the other ingredients.

Water-binding substances are sometimes used as 'improvers' in rye bread, typically up to 3% of pre-gelatinized potato, maize or rice starches, or a hydrocolloid or gum. A whole range of optional ingredients may be added to rye doughs for flavour or texture; they include buttermilk, soured milk, curd cheese, dried fruits and nuts.

## 13.2.3.  Baking

Rye breads may be baked in pans, as oven-bottom (hearth) or even as batch breads. A traditional practice for oven-bottom rye breads is to prove the dough in a dusted basket, in the past made of whicker but now more likely plastic, which leaves a 'ribbed' appearance on the dough surface when the dough pieces are tipped out onto the oven sole for baking. Oven-bottom rye breads are often given a deep cut from the surface to about half-way down the dough piece. This improves heat transfer to the rather dense dough and prevents ragged breads along the side of the dough. Cutting also adds variety of form to the dough surface.

Baking conditions vary with bread variety and oven type. The lighter varieties can be baked under similar conditions of time and temperature to wheat breads, while darker and wholemeal forms may require up to 55 min for an 880 g loaf. A common procedure is to bake it first for about 20 min at 260 °C (500 °F) using a considerable amount of steam and then complete the baking at a lower temperature, typically 200 °C (390 °F).

Rye wholemeal bread can be baked using a form of steam-pressure cooking to avoid the formation of a normal crust. The dough pieces are held in tightly closed pans and put into an autoclave-type steam chamber for 5–8 h at 200 °C or 16–24 h at 100 °C (312 °F). The advantages claimed for this type of process include better nutritional value and moisture retention in the finished bread. This baking process causes a darkening of the crumb and a bitter-sweet taste.

## 13.2.4.  American Rye Breads

The four basic types of rye bread produced in the USA were first described by Weberpals (1950):

- A light rye bread made with a blend of wheat and rye flours, typically in the proportion 60:40.

- A heavy rye bread made with a sour dough process.
- A light, sweet, pan rye bread, a mixture of wheat and rye flours and made with a straight dough process.
- Light and dark pumpernickel made with a sour dough process. This product is quite unlike the original German pumpernickel and often contains fat and molasses in the USA.

## 13.2.5.  Keeping Qualities of Rye Breads

The lower pH of rye breads, especially those made with sour doughs, inhibits microbial growth and confers a longer shelf-life than that commonly seen with wheat bread products. The shelf-life of rye bread may be further extended using a pasteurization or sterilization process. Such processing is usually carried out on the wrapped product so that the film used must be heat stable and have good barrier properties. The condensation developed within the wrapper is later reabsorbed into the product and has no adverse effect on product quality at the time of consumption. Some darkening of the products may occur but this has relatively little importance for rye breads, which tend to have naturally dark crumb colours. Conventional hot air ovens, steam chambers and microwave heating have all been used for the sterilization process. After treatment, the shelf-life of rye bread may extend for up to 24 months.

## 13.3.  Triticale

Interest in triticale stems in part from its nutritional properties, with higher protein levels and a more acceptable amino acid composition than wheat (Gustafson et al., 1991). Intensive work to produce high-yielding varieties with improved product performance has taken place in Poland (Achremowicz, 1993) and elsewhere for about 25 years, which might lead to its wider acceptance as a foodstuff.

Triticale grains are milled in a similar manner to wheat and rye and can yield straight run flours with relatively low ash contents. Wheat milling techniques tend to give higher extraction rates and are therefore preferred. Early strains of triticale gave flours with poor baking performance, in part because of higher *alpha*-amylase activity and lower paste viscosity than wheat flours (Lorenz, 1972) and in part because of weaker protein qualities, although a wide variation in the rheological properties in triticale flours has been noted (Macri et al., 1986). Later developments of triticale varieties and the use of bread improvers, such as sodium stearoyl lactylate (SSL) (Tsen et al., 1973), have brought the baking performance closer to that of wheat flours, although the 'quality gap' is still significant.

Achremowicz (1993) found that triticale bread baked with a three-stage method using a rye pre-ferment gave a loaf which rose well with the required shape, had an elastic crumb and pleasing odour and flavour. The bread remained fresh for a long period. The basic elements of the breadmaking process he used were as follows:

- a pre-ferment based on rye flour at 10% total flour fermented for 24 h at 28 °C (82 °F);
- a sour with 50% of the total triticale flour, 1–2% yeast and water to a total of 200%, fermented for 3 h at 32 °C (90 °F);
- a dough based on the sour, the balance of flour as triticale, 1.5% salt and water to a dough yield of about 160%, fermented for 30 min at 32 °C;
- unit pieces were proved for 25–45 min at 32 °C;
- baking at about 240 °C (464 °F).

In summary, we can see that triticale has some potential in breadmaking, although the performance of triticale flours is closer to that of rye rather than that of wheat. In areas where wheat cannot be grown successfully because the land or farming practices have marginal agricultural potential, triticale offers a viable alternative to wheat and greater potential than rye. Acceptable bread quality is achieved using rye bread manufacturing techniques rather than those used for wheat breads. However, given the improvements which it has undergone since its introduction, triticale has been successfully transformed from 'scientific curiosity to a viable crop' (Varughese et al., 1996).

# 13.4.   Other Grains and Seeds in Bread

The never-ending quest by marketing departments for new products to tempt consumers, and for ingenuity in product developers, has led to the development of many 'new' breads with special characters and in some cases specific nutritional properties. Levels of addition of non-wheat grains are usually quite modest and are made to a suitably strengthened wheat flour-based dough. Since the latter has to carry materials which are unlikely to contribute positively to the gas retention or dough rheology in the system, special attention must be focused on the qualities of the base wheat flour, improvers and the breadmaking process which is to be used.

A wide range of grains and seeds might be added to wheat flour doughs and so only a limited number of examples have been chosen to highlight some of the technical issues which may arise in bread production.

## *13.4.1.   Multi-grain Breads*

In multi-grain breads it is usual to add a portion of whole, cracked or kibbled grains to the wheat flour base. Ready-prepared premixes are available from millers and other suppliers which may include improvers, flavouring and colouring agents.

While cleaned grains can be added directly to the wheat flour dough, some problems can arise from this procedure. Like wheat, other grains are subjected to microbial contamination during growth and harvesting and this can be carried through to the final product. Of particular concern is the presence of the rope-forming bacteria *Bacillus subtilis* (Chapter 10) with its potential for causing product spoilage after baking. Thorough cleaning of the grain and some form of heat or surface treatment may be helpful in reducing the levels of microorganisms.

The other main problem with adding large particles of grain directly to bread dough is related to their hardness, which will be considerably greater than that of the surrounding bread crumb. This will certainly provide a contrast in textures but may also result in a rapid trip to the dentist! To obviate this problem, whole grains are often given a short period of soaking to reduce their hardness, a technique often used in treating grains for animal feed (Kent and Evers, 1994). Steam rolling and steam flaking may also be used, depending on the particular form of the grain required. Raising the moisture content of the grains is not without risk since the higher water levels will provide the potential for extra microbial growth.

## 13.4.2.  Modifying Nutritional Properties with Non-wheat Sources

Bread, especially wholemeal (wholewheat), is a naturally rich source of dietary fibre. The fibre largely derives from the bran skins of the wheat berry. There have been some product developments which have sought to raise the level of dietary fibre in breads by using fibre sources other than those occurring in wheat. In some cases the drive behind such developments has been to increase the nutritional value of white bread while retaining appeal to some sectors of consumers. While wholemeal and similar high-bran breads have the required nutrition, their appeal appears to be limited with younger consumers who seem to prefer the textural and flavour characteristics of white bread. The introduction of white-coloured fibre sources, such as pea and other fibres, enabled the development of bread products which met the requirements of both younger consumer groups and nutritionists. The definition of 'fibre' has been extended to include ingredients which have lower digestibility than wheat starch, for example Litesse (Danisco Sweeteners, www.danisco.com/sweeteners). As with bran in wholemeal flour, the non-wheat fibres generally play a negative role in promoting gas retention in the dough or in improving its rheological character, so that formulations must take into account the dilution effect of the fibre on the functionality of the system. Additions of extra protein as dried wheat gluten will be helpful.

Oats are another grain which has attracted attention for use in wheat breads. In this case it has been favoured because of the presence of *beta*-glucans and their potential beneficial effects in reducing cholesterol levels in the bloodstream. Oats also have a high lipid content for a cereal and so are prone to rancidity (McMullen, 1991) and long-term storage requires inactivation of enzymic activity to reduce the likelihood of rancidity. When added to wheat flour doughs, oat flours reduce the ability of the system to hold gas and the final loaf has a bland taste and firm texture. For these reasons oats have only become an established ingredient in bread products when its nutritional contribution can override its adverse effects on bread quality or where its unique flavour can provide positive benefits. An example of the latter is the use of oats in the manufacture of UK sandwich breads which have to be held at 5 °C prior to sale. In some cases the sandwiches may be eaten when they are still at refrigerated temperatures and at such temperatures white bread often appears to lack flavour. In these circumstances the addition of oats to white flour for the manufacture of sandwich bread helps to restore flavour in the final product.

## 13.4.3.  Malted Barley

Barley is grown in large quantities around the world and in many places rivals the production of wheat. Its main use with respect to human consumption is in the brewing industry. The malting of barley grains produces distinctive flavours which are carried through to the final product, and because of this the grain has found some use in the baking industry.

After the various stages of the malting process, a milled barley malt may be produced for use in breadmaking. This powdered form will have a high diastatic (enzymic) activity which can be used to supplement breadmaking flours. The enzymic activity is usually dominated by (cereal) *alpha*-amylase and proteolytic enzymes. The former can be beneficial in improving the gas retention abilities of flours, although such enzyme-active malt flours should be used with caution in breads which are later to be sliced, because of the high levels of dextrins which will be produced during dough processing and baking (Chapter 3). The proteolytic activity can also have both beneficial and adverse effects since they cause softening of the dough. Some softening can improve dough machinability, but too much may cause doughs to stick to equipment surfaces during moulding and intermediate proof.

In addition to using malt flours it is possible to use malt extracts and syrups. These too will have some enzymic activity but are more likely to be used by bakers for their contribution to flavour. They find use in some special 'malt' breads seen in some parts of the world, especially the UK. In such cases the sticky crumb resulting from the high enzyme activity is seen as a positive attribute. Gas retention in these specialist doughs can be so great and the crumb density at the centre so low that the loaves will collapse during conventional cooling. They therefore require the use of vacuum cooling to preserve the integrity of their structures.

## 13.5.   Wheatless Breads

While rye and triticale breads may fall within the category of 'wheatless' breads, this section heading is intended to cover those cereal products which are not normally considered for the production of bread because they lack proteins capable of forming gluten in even the smallest quantities. It is not proposed to discuss the use of such cereals in 'composite' flours neither where they have been blended with wheat to reduce the wheat component nor where dried gluten has been added. Such practices are common in countries not only wishing to make a bread product but also wishing to reduce the importation of wheats.

For the cereals which will be discussed under this heading the main structure-forming component in the flour will be based on the indigenous cereal starch, supplemented on occasions with other stabilizers to ensure stability of the gas bubbles in the 'dough'. Yeast and the generation of carbon dioxide gas by fermentation will be essential ingredients in the formulations and processing methods considered.

The technology of producing wheatless breads has much in common with the production of 'gluten-free' breads for sufferers of dietary disorders or allergies associated with gluten from wheat and rye flours. A range of 'gluten-free' products has been developed by a number of commercial companies based on wheat starch from which all traces of protein have been removed. The 'dough' which is formed from such starch-based formulations is much closer to the viscosity of a cake batter than a bread dough, and because of this similarity many of the techniques used to stabilize gas bubbles during processing and baking are derived from cake production. This being the case, it is appropriate to begin a discussion of these forms of wheatless breads by considering the mechanisms which operate in cake batters during processing and baking.

### 13.5.1.   Formation of Cake Batters

Cakes are produced by forming a complex emulsion and foam system – the batter – which is processed by being heat set. When eggs and sugar are whisked together during the mixing of the batter, large numbers of minute air bubbles are trapped in the batter by the surface-active proteins of the egg white and the lipoproteins, which form a protective film around the gas bubbles and prevents them coalescing. Other surface-active materials, such as distilled monoglycerides, may be used to augment the effect of the egg proteins (Cauvain and Cyster, 1996). This is especially true if oil or fat is added to the batter. If the level of added fat is increased sufficiently, the solid fat can take on the role of stabilizing the air bubbles. In this latter case the fat crystals align themselves around the air bubbles, thereby trapping and stabilizing them. This is possible because the oil fraction of the fat allows the crystals to move and at the same time stick to each other like links in a chain. The ratio of liquid to crystalline fat and the size of the individual crystals are very important in controlling cake quality (compare with the effects of solid fat in the CBP; Chapter 3).

As the temperature of the batter rises in the oven, the trapped gas bubbles expand and eventually, just as the mass of batter is setting, they burst into one another to form the porous structure of cake crumb. In the case of cake batters, carbon dioxide gas will be generated by the added baking powder and will add to the inflation of the initial air bubbles.

The similarities between bubble creation in cake batters and bread doughs are clear even if the mechanisms by which they are stabilized are quite different. There is no significant gluten formation in cake batters, in part due to the higher water levels, the presence of higher fat levels and the presence of sugar. During the early stages of heating in the oven, the viscosity of the batter slows down the movement of the gases in the system so that the batter expands. In the absence of gluten formation a significant contribution to the final product structure comes from the gelatinization of the starch in the flour and it is that property which must be exploited when making wheatless breads.

## 13.5.2.   Bread without Wheat

Since bread products do not contain the high levels of fat and sugar that occur in cakes, we have to seek alternative means for trapping and stabilizing gas bubbles in wheatless breads. Gums, stabilizers and pre-gelatinized starch can all be used to provide part or all of the gas occlusion and stabilizing mechanisms we require. For example, Satin (1988) reported on the replacement of dried gluten with xanthan gum in cassava and sorghum-based breads and then replacement of the gum with some pre-gelatinized starch. The addition of a protein source with some cereals is also helpful in heat setting the wheatless bread 'dough' to prevent product collapse. Satin (1988) suggested powdered egg whites or a legume flour.

Yeast can be used as the alternative source of carbon dioxide in the dough rather than baking powder, or they can be used together. The yeast will provide gas for bubble expansion in proof provided that sufficient substrate is available for fermentation to proceed. If insufficient substrate is available, then small additions of sugar may be needed to assist the yeast. Salt should be added for flavour.

Mixing methods for wheatless bread doughs can vary from the basic, for example by hand, to mechanical, planetary-style and even high-speed, continuous batter-type mixers.

Because wheatless bread doughs are more fluid than wheat doughs and closer in viscosity to cake batters, it is usual to bake the material in a rectangular or round bread pan. The pans should be well greased but they do not normally require any form of additional lining.

There are many potential recipes for wheatless breads and the following are a few examples by way of illustrating the potential for such developments.

## 13.5.3.  Cassava Bread (Satin, 1988)

|                   | % flour weight |
| ----------------- | -------------- |
| Cassava flour     | 100            |
| Sugar             | 4.9            |
| Egg white powder  | 2.3            |
| Salt              | 2.3            |
| Oil               | 5.7            |
| Yeast             | 5.7            |
| Water             | 120            |

- Take 15 parts of the flour and boil with the water for 4 min.
- Replace evaporated water, add the remaining ingredients and mix for 10 min.
- If powdered yeast is used rehydrate with a little sugar and water before use.
- Scale into a pan, leave to prove and bake.

## 13.5.4.  Rice or Maize Bread (Satin, 1989)

|                      | % flour weight |
| -------------------- | -------------- |
| Rice or maize starch | 100            |
| Sugar                | 4.4            |
| Salt                 | 1.7            |
| Oil                  | 1.7            |
| Yeast (fresh)        | 1.0            |
| Water                | 140            |

- Take 17 parts of the starch and boil with 92 parts water until translucent.
- Add the remaining ingredients and mix for 5 min.
- Scale into a pan, leave to prove and bake.

## 13.5.5.  Sorghum Breads

Sorghum (*sorghum bicolor*) is widely grown in areas of low rainfall and poor soil conditions and is cultivated throughout central and southern Africa, central Asia and the Indian sub-continent. It has many food uses (Dendy, 2001) and there has been some interest in its use for making 'bread-like' products (Taylor et al., 2005) as an alternative to the importation of wheat. There has also been interest in sorghum as a component of gluten-free products (Arendt et al, 2005).

An example of a sorghum-based bread product is illustrated in Figure 13.1 and is based on the following recipe:

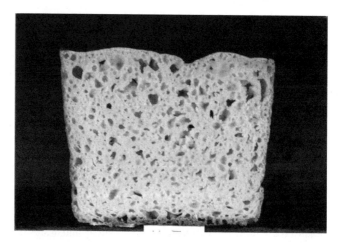

FIGURE 13.1. Sorghum bread.

|                                  | % flour weight |
|----------------------------------|----------------|
| Sorghum flour                    | 100            |
| Yeast (compressed)               | 4.2            |
| Salt                             | 2.1            |
| Skimmed milk powder              | 12.0           |
| Sodium carboxy methyl cellulose  | 1.0            |
| Water                            | 100            |
| Baking powder                    | 2.0            |
| Soya flour                       | 2.0            |

- Mix the ingredients for 6 min using a planetary mixer.
- Scale 460 g (1 lb) into rectangular bread pans.
- Prove until the dough reaches the top of the pan, then bake.

## 13.5.6.  Sorghum and Maize Bread Using a Continuous Mixer (Figure 13.2)

|                                  | % flour weight |
|----------------------------------|----------------|
| Sorghum flour                    | 50             |
| Maize starch                     | 50             |
| Yeast (compressed)               | 8.3            |
| Salt                             | 2.1            |
| Skimmed milk powder              | 12.0           |
| Sodium carboxy methyl cellulose  | 1.0            |
| Dried egg albumen                | 15             |
| Water                            | 80             |

FIGURE 13.2. Sorghum and maize bread using a continuous mixer.

- Blend the ingredients on a planetary mixer and then pass through a continuous cake batter-type mixer.
- Scale, prove and bake.

## 13.5.7.  Sorghum Flat Breads (Figure 13.3)

|  | % flour weight |
| --- | --- |
| Sorghum flour | 50 |
| Maize starch | 50 |
| Yeast (compressed) | 2.1 |
| Salt | 1.8 |
| Fat | 0.7 |
| Improver | 1.0 |
| Water | 80 |

- Mix for 5 min.
- Sheet to about 15 mm (0.5 in) thick and cut out suitable shapes.
- Bake on flat sheet or hotplate for approximately 15 min.

The success, or otherwise, of much wheatless bread production, as with any other bread type, lies in being able to obtain consistently a raw material of known characteristics. It is unfortunate that in many cases the attributes required from the flour, meal or starch are, as yet, not fully appreciated, so that baking results and final bread quality may be somewhat variable. Nevertheless sufficient work has been done to show the considerable potential for wheatless breads and the opportunities this can provide for bread production in countries which have little wheat available.

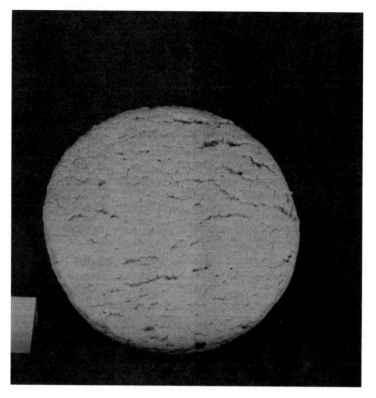

FIGURE 13.3. Sorghum flat bread.

## 13.6.   Unleavened Breads

Many of the unleavened breads seen in different parts of the world are based on wheat flour, although seldom in the same form as we would see for the production of 'Western-style' breads since they are not always based on 'white' flour. In many cases the wheat flours used can be described as 'high extraction', brown and occasionally wholemeal and so can be considered under the general heading of this chapter. A few unleavened products are based on maize and other non-wheat cereals.

The most usual form of unleavened breads is flat. They seldom contain yeast, though they may make use of sour or old doughs and in some cases sodium bicarbonate. In addition to their flatness, most of the unleavened breads will have a characteristic split or 'pocket' created during baking and which remains in the cooled product. This distinctive feature can be used to hold a filling because the bread will 'puff-up' when reheated.

There are a wide range of flat breads made around the world and some of these are given in Table 8.1. In Egypt the term 'balady' probably covers around 30 variations (Evans, 1983). To illustrate the range of flat breads a few examples are given below, but bear in mind that with traditional products such as these, recipes will vary widely (everyone has their own favourite!).

## 13.6.1.  Recipes for Unleavened Breads

## 13.6.2.  Balady

|  | % flour weight |
| --- | --- |
| Starter |  |
| Old dough | 11 |
| Flour | 100 |
| Water | 50 |
| Dough |  |
| Flour | 100 |
| Water | 75.0 |
| Salt | 0.6 |
| Starter | 10.0 |

- Use a flour of about 85% extraction (or a blend of 75% white and 25% wholemeal).
- Mix for about 20 min to produce a soft dough.
- Scale at 180 g (6 oz), mould round.
- After 15 min intermediate proof, flatten.
- Prove for 1 h and bake.

## 13.6.3.  Chapattis

|  | % flour weight |
| --- | --- |
| Flour | 100 |
| Water | 70 |
| Salt | Optional |
| Oil | Optional |

- Use a flour with about 85% extraction.
- Mix to form a soft dough, rest for 30 min.
- Scale around 50 g (2 oz), mould round.
- After a short rest roll out to about 2 mm (0.1 in) and bake with one or two turns.

## 13.6.4.  Naan

|  | % flour weight |
| --- | --- |
| White flour | 100 |
| Yeast | 3.8 |
| Salt | 0.6 |
| Baking powder | 1.3 |
| Caster sugar | 2.5 |

| | |
|---|---|
| Egg | 15.0 |
| Oil | 7.5 |
| Natural yoghurt | 15.0 |
| Milk | 43.8 |

- Mix to a dough and ferment for 1 h.
- Scale at 110 g (4 oz) and mould to an oval shape.
- Bake (traditionally on the sides of a Tandoor oven).

## 13.6.5.  Papadams

| | % flour weight |
|---|---|
| Blackgram flour | 100 |
| Water | 45 |
| Salt | 8 |
| Sodium bicarbonate | 1 |

- Blackgram flour comes from the pulse, *Phaseolus mungo*.
- Mix to a soft and sticky dough, rest for 30 min.
- Scale at 20 g (0.75 oz) and pin out.
- Dry to reduce moisture content to about 15% before frying.

## 13.6.6.  Tortillas

| | *% flourweight* |
|---|---|
| Whole maize flour | 100 |
| Water | 33 |
| Lime (calcium hydroxide) | 0.1 |

- Mix to form a 'masa' (dough).
- Divide and mould round before pinning out to 2 mm (0.1 in) thick.
- Bake on a hotplate.

## 13.6.7.  Flours for Unleavened Breads

Although unleavened and flat breads are products which are not expected to achieve and maintain high specific volumes (low densities), they do demand quite specific characteristics from many of the flours used in their production, especially wheat flours. The difficulty for the miller supplying bakers of flat breads is that the product requirements are subject to significant local variation and flour properties are often ill-defined.

There are two critical areas in the production of unleavened breads where wheat flour characteristics play major roles; one is in the doughmaking stage and the other is during baking. In dough preparation the quality of the protein in the wheat flour will significantly affect the way in which the dough piece deforms under sheeting. High resistance to deformation and significant elasticity will result in the dough pieces tearing when thinly sheeted, and loss of shape after forming as elastic regain in the dough takes effect. The flour must be capable of producing a dough with good extensibility, so that tearing during sheeting is minimized and that the dough is capable of considerable and rapid expansion during the short baking cycle. Treatment of wheat flours is not normally required for these products since most treatments tend to make the dough less extensible. Protein content is probably of lesser importance than protein quality.

## 13.7.  Conclusion

Although the majority of bread products are based on wheat, it is possible to use other cereals in bread production. In many cases the other cereals are seen as 'additions' to make wheat breads 'more interesting' (e.g. multi-grain breads), although many cereals are capable of making products with special characters of their own, for example rye bread.

In many parts of the world, unleavened or flat breads have evolved with the passage of time and they too have become products in their own right with specific characters and raw material requirements.

In many countries, for example in Africa, the growing of wheat presents some difficulties with our present skills and technology. In these countries the concept of bread is well known, having been carried there with trading and social links, but the purchase of wheat itself presents other challenges for the populations concerned. In such cases, alternative and locally available cereals may be used to make breads with many of the attributes of those made with wheat.

Breadmaking has always provided diversity of form, shape and flavour, even when the products have been based exclusively on wheat, but as we improve our understanding of the properties of our available raw materials the opportunities for adding other cereals increase dramatically. Customers are always seeking novel products and new taste sensations, and with the wide range of cereals available for breadmaking there are many more new products waiting to be developed.

## References

Achremowicz, B. (1993) Use of Triticale for bread production in Poland. *Chorleywood Digest No. 132*, November/December, CCFRA, Chipping Campden, UK, pp. 115–20.

Arendt, E.K., Schober, T.J., Messerschmidt, M. and Bean, S.R. (2005) Comparison of the breadmaking potential of different sorghum hybrids in, (eds S.P. Cauvain, S.E. Salmon and L.S. Young) *Using Cereal Science and Technology for the Benefit of Consumers*, Woodhead Publishing Ltd, Cambridge, pp. 62–7.

Cauvain, S.P. and Cyster, J. (1996) Sponge cake technology, *CCFRA Review No. 2*, January, CCFRA, Chipping Campden, UK.

Dendy, D.A.V. (2001) Sorghum and the millets in, *Cereals and Cereal Products: Chemistry and Technology* (eds D.A.V. Dendy and B.J. Dobraszczyk), Aspen Publishers Inc., Gaithersburg, MA, pp. 341–66.

Drews, E. and Seibel, W. (1976) Bread-baking and other uses around the world, in *Rye: Production, Chemistry and Technology* (ed. W. Bushuk), American Association of Cereal Chemists, St Paul, Minnesota, USA, pp. 127–78.

Evans, K. (1983) Ethnic breads. *FMBRA Bulletin No. 5*, October, CCFRA, Chipping Campden, UK, pp. 240–7.

Gustafson, J.P., Bushuck, W. and Dera, A.R. (1991) Triticale: Production and utilization, in *Handbook of Cereal Science and Technology* (eds K.J. Lorenz and K. Kulp), Marcel Dekker, New York, pp. 373–400.

Kent, N.L and Evers, A.D. (1994) *Kent's Technology of Cereals*, Elsevier Science Ltd, Oxford, UK.

Lorenz, K. (1972) Food uses of Triticale. Hybrids of wheat and rye can be used in bread, rolls and noodles. *Food Technology*, **26**, 66–74.

Lorenz, K.L. (1991) Rye, in *Handbook of Cereal Science and Technology* (eds K.J. Lorenz and K. Kulp), Marcel Dekker, New York, pp. 331–71.

Macri, L.J., Ballance, G.M. and Larter, E.N. (1986) Factors affecting the breadmaking potential of four secondary hexaploid triticales. *Cereal Chemistry*, **63**, 263–6.

McMullen, M.S. (1991) Oats, in *Handbook of Cereal Science and Technology* (eds K.J. Lorenz and K. Kulp), Marcel Dekker, New York, pp. 199–232.

Meuser, F., Brummer, J.-M. and Seibel, W. (1994) Bread varieties in Central Europe. *Cereal Foods World*, April, 222–30.

Petersen, M.A., Hansen, A., Venskaiyte, A., Juodeikiene, G. and Svenickaite (2005) Flavour of rye bread made with scalded flour in, *Using Cereal Science and Technology for the Benefit of Consumers* (eds S.P. Cauvain, S.E. Salmon and L.S. Young), Woodhead Publishing Ltd, Cambridge, pp. 69–73.

Rozsa, T.A. (1976) Rye Milling, in *Rye: Production, Chemistry and Technology* (ed. W. Bushuk), American Association of Cereal Chemists, St Paul, Minnesota, USA, pp. 111–26.

Satin, M. (1988) Bread without wheat. *New Scientist*, 28 April, 56–9.

Satin, M. (1989) Wheatless breads. FAO leaflet, FAO, Rome, Italy.

Taylor, J.R.N., Hugo, L.F. and Yetnerberk, S. (2005) Development in sorghum baking, in *Using Cereal Science and Technology for the Benefit of Consumers* (eds S.P. Cauvain, S.E. Salmon and L.S.Young), Woodhead Publishing Ltd, Cambridge, pp. 51–6.

Tsen, C.C., Hoover, W.J. and Farrell, E.P. (1973) Baking quality of triticale flours. *Cereal Chemistry*, **50**, 16–26.

Varughese, G., Pfeiffer, W.H. and Pena, R.J. (1996) Triticale: A successful alternative crop (part 2). *Cereal Foods World*, **41**(7), 637–45.

Weberpals, F. (1950) Fundamentals of rye bread production. *Bakers' Digest*, **24**, 84–7, 93.

# Index

Page numbers in bold refer to figures; those in *italic* refer to tables.

Acetic acid, *67*, 86
Acrylamide, 74
Activated dough development, 36, 65, 276
Additives, 36, 57–68, 353–354
  in bread, 16, 52, 64, 232
  to flour, 300, 354
Air
  flow in provers, 149, 150
  incorporation, 22, 318–319
  velocity in ovens, 142
Alcohol, *see* Ethanol
*Alpha*-amylase, 68–70
  bacterial, 71, 288–289
  in CBP, 70
  cereal, 31, 69, 70, 71, 156, 183, 288, 289,
    338, 339, 354, 361, 378
  fungal, 31, 69, 70, 71, 183, 200, 289
  heat stability, 69, 71
  maltogenic, 72
  in straight dough bulk fermentation, 26,
    28, 31
  in triticale, 375–376
Alveograph, 320, 322, 328, 362, 364
Ambient part-baked, 170; *see also* Milton
  Keynes Process
AMF Pan-O-Mat, 251
Amino acid composition, 307
Amylopectin, 68, 71, 72, 284, 285,
  286, 301
Amylose, 59, 60, 68, 284, 285, 286, 301
Appearance of wheat, 336
Ascorbic acid, 35, 45, 46, 56, 61–65, 358
  in CBP, 45, 46, 56, 64, 98
Ash, 226, 304–305
Asparaginase, 74
Aspiration, 342
  definition, 367

Bacterial *alpha*-amylase
  effect on staling, 289
Bacterial spoilage, 277
Baguette, 4
  moulding, 125, 131, 234
  part-baked, 255
  use of lecithin, 61
Bakers yeast, 2, 7, 76–89
Baking
  hamburger buns, 253
  part-baked breads, 257–258
  retarded doughs, 184, 185
  rye breads, 235, 374
  triticale bread, 375
Baking nets, 257
Balady bread, 384, 385
*Beta*-amylase, 68, *69*, 70, 71, 354, 355
Biological contaminants in flour, 345–351
Black spots in retarded doughs, 200
Blast freezing, 182, 191, 194, 294
Blending
  in dough, 312–313
  of flours, 352–353
  of wheat, 352
Blisters in retarded and frozen doughs, 196,
  199, 200
Bloomer
  moulding, 125
Bran, 6, 15, 30, 226, 333, 334, 377
Bran finishers, 344
Branscan, 360
Bread
  American rye, 374–375
  in Australia, 225, 226, 232, 241
  with cassava, 380, 381
  character, 208, 277, 319
  composition, 15
  consumption, 15, 16, 235

crustless, 239
eating quality, 14–15
enrichment, 15, 16, 178, 198
fault diagnosis by expert system, 208
fault diagnosis by Retarding Advisor,
    215–217
faults in baking frozen dough retarding,
    212, 213
flavour, 6, 47
in France, 8, 11, 16, 55
freezing, 260, 292–294
freshness, 6, 275
gluten-free, 379
in Japan, 66, 102, 235
with maize, 357, 381
in the Middle East, 1, 2, 4, 8, 224, 225
multi-grain, 376–377
in North America, 4, 16, 112, 115, 223,
    229, 371
nutrition, 15–17, 355–356
quality, 8, 10, 47–48, 54, **184**, **185**
quality assessment, 11–13
with rice, 351, 381
with rye, 3, 67, 346, 371, 374–375
scoring, 11
softness, 4, 14, 228, 287, 289
with sorghum, 381, **382**, 383
spoilage, 275–294
staling, 275–294
texture, 14, 283
triticale, 375, 379
types, **8**, 225
in the UK, 4, 52, 57–58, 60, 61
unleavened, 141, 233, 384, 385, 386
volume measurement, 12, 23, 44, 58
wheatless, 379, 380
Bread Advisor, 209–210, 212, 214
Bread faults expert system, 209,
    210, 215
Breadmaking in
    New Zealand, 230, 232, 233, 234, 237
    South Africa, 98, 231
Breadmaking processes, 21–48, 176–177
Breakdown of dough, 39, 131, 317
Break system, 343–344
Brown flour, 347, 356
Bun divider moulders, **136**, 137
Butter in laminated products, 264, 268


Calcium propionate
    in pan-baked breads, 66, 68, 86, 259, 278,
        279
Calcium stearoyl lactylate, 59

Carbon dioxide
    in frozen proved doughs, 196
    in modified atmosphere packaging, 279
Cassava bread, 381
CBP-compatible mixers, 45, 46, 95–99, 104
Cell creation and control, 25
Cereal *alpha*-amylase
    heat stability, 71
    in straight dough bulk fermentation, 31
Chapatti, 3, *225*, 228, 232, 233, 333, 385
Chemical contaminants in flour, 349
Chorleywood Bread Process
    effects of atmospheric pressure during
        mixing, 94
    in New Zealand, 40, 230, 232, 235, 241
    production of hamburger buns, 65,
        247, 249
    requirements for wheat variety, 209, 229
    role of energy during mixing, 40
Commercial yeast production, 81
Compressed yeast, 77, 85
Condensation spotting in retarded dough,
    197, 200
Conditioning of wheat, 342
    definition, 367
Conical rounders, 117, **188**
Continuous mixers, 106–109, 382
Contributing factors, 211, **213**, 213
Controlling flour quality and
    specification, 351–352
Control of microbiological spoilage, 278
Conveyorized first provers, 125
Conveyorized oven, **163**
Cooler
    design, 167
    spiral, 168
Cooling
    hamburger buns, 253–254
    part-baked breads, 258–259
Cream yeast, 77–78
Croissant
    frozen proved, 196
    part-baked, 255
    recipes, 264–266
Cross-grain moulding, 132
Crumb cell structure
    changes during baking, 156
    creation in CBP, 44, 48, 65
    effect of bacterial *alpha*-amylase,
        71, 288
    effect of fungal *alpha*-amylase, 70, 289
    firmness, 14, 259, 262
    grain, 13, 318, 319
    straight dough bulk fermentation, 26, 28

Crust
  colour, 13, *32*, 183, 200
  crispness, 160
  crumb-type, 159
  fissures in retarded and frozen dough,
    198–199
  formation, 156–158, 257
  gelation, 159
  paste-type, 159
  ragged breaks in retarded and frozen
    dough, 199
  staling, 283
  temperature, **160**, 165, 169
Curling, 129, **130**, 270
  chains, **130, 131**
Cylindrical rounders, 118
Cysteine, 36, **62**; *see also* L-cysteine

Danish pastry
  frozen proved, 196
  part-baked, 255
  recipes, 264, 267, 268, 270
DATA esters, 57
  effect on staling, 59
  in high volume bread, 58
  in long-proof breads, 58
  in UK white bread, 58
  in wholemeal breads, 58–59
  in yeasted laminated
    products, 266
Deactivated yeasts, 79
Defrosting cycle in retarder–prover, 204
Defrosting frozen dough, 194
Degassing flour brews, 247
Dehydrated sours in rye bread, 373
Delayed salt, 27, 87
Depanning
  bread, 258
  hamburger buns, 253
Diastatic malt flour, 71
Dielectric heating, 164
Distilled monoglyceride, 59, 60; *see also*
    Glycerol monostearate
Disulphide bonds, 40, 41, 65
Dividers
  extrusion, 113–115
  model K, 250
  two-stage oil suction, 113, **114**
  vacuum, 115
Dividing
  in Australia, 233
  hamburger buns, 250
  in New Zealand, 233
Do-flow unit, 250

Dough
  breakdown, 131, 317
  conditioners, 51–52, 57, 59
  cutting, **132**
  damage during dividing, 112
  development, 22, 24, 26, 36, 38,
    228–231, 362
  elasticity, 301, 303, **323**
  extensibility, 328
  formation, 28, 299–329
  freezing, 190–194, 193
  hydration, 312
  mixing, 24, 31, 55, 93, 261, 269–270
  moulding, 132
  resistance to deformation, 23, 31, 42,
    46, 47
  retarding, 175–204
  rheology, 46–47, 57, 120, 320–323, *321*
    in ADD, 36
    in CBP, 46
  stickiness, 23, 38, 278
  temperature, 29, 37, 42, 109–110, 193
    in CBP, 35, 41, 97
    in no-time doughs, 37
    in sponge and dough, 33, 35
    in straight dough fermentation, 28, 257
    in yeasted laminated products, 263–272
  transfer systems, 100–111
  unmixing, 317–318
  volume during retarding, 182, 199
Dried pellet yeast, 78
Dried wheat gluten, 377
  in CBP, 43
Dusting, 251, 343, 344, 346
Dutch green dough process, 36, 37, 121

Elasticity, 326
Electrophoresis, 304, 308, 309, 339–340
Emulsifiers
  in Australia, 225, 226, 232
  in CBP, 39
  effect on staling, 59, 61, 232, 289
  in New Zealand, 232
  in straight dough fermentation, 26, 28
Encapsulated yeast, 78
Endosperm of wheat, 72
Energy calculator, 217, **218**
Enzyme activity during
  baking, 156, 288
  retarding, 288
Enzymes
  effect on staling, 71, 74
Equilibrium relative humidity, 183, 277

Ethanol
   effect on staling, 279, 291
Expert systems, 208, 210, 215
Extensibility, 23, 328–329
Extensograph, **364**, 364
Extrusion bun divider, 245, 250–251
Extrusion dividers, 113, 138, 250

Falling Number, 31, 35, 232, 372; *see
      also* Hagberg
Falling Number
Farinograph, 360, 362–**363**, 363
Fat
   in CBP, 53, 54, 380
   laminating, 266–269, 270–271
   melting point, 39, 53, 54, 55, 57, 60, 61, 268
   in straight dough fermentation, 32
   in wholemeal, 52–55, **54**
   in yeasted laminated products, 265–266
Final moulding, 129–131
Final proof, 235, 258
First proof, 116, 120–121
First prover
   conveyorized, 125
   pallet in-feed, 124
   pusher in-feed, 123–124, **124**
Floor time, 26, 28, 33, 35, 51, 199, 299, 327
Flour
   additives, 57
   amino acid composition, 307–308
   ash content, 304–305, 342, 351, 359, 375
   biological contaminants, 349–351
   blending, 313–314, 353
   brew, 246–249, **249**
   in CBP, 95, 230
   chemical contaminants, 349
   components, 299–300, 304, 305, 312, 366
   definition, 30, 352
   dressing, 346–347
   foreign bodies, 348–349
   grade colour figure, 359–360
   infestation, 337, 347, 350
   milling, 333, 335–347, 351
   moisture content, 338, 346, 357–359
   nutritional additions, 355–356
   packing, 347
   protein content, 264–265, 357–359
   quality, 43, 351–352
   in rapid processing, 36–38
   rheology, 362
   specification, 351–352
   in sponge and dough, 33–36
   storage, 292–293, 340, 347

   in straight dough bulk fermentation, 26,
      28–33, *29*, **30**
   taints, 336, 349
   testing, 357, 365, 366
   for unleavened breads, 384–387
   water absorption, 305–307, 360–361
   in yeasted laminated products, 263–265
Folic acid, 15, 356
Food safety and product protection, 347–351
Form-fill sealing, 281
Formulation, *see* Recipes
Four-piecing, 125, 131–132
Freezing
   bread, 260, 292–294
   fermented doughs, 187–195, 193
   laminated doughs, 271
   part-baked breads, 260
   proved doughs, 195–196
   rate, 182, 186, 191, **192**, *193*, 294
Frozen dough
   breadmaking processes, 176–177, 187–190
   defrosting, 194, 197
   effect of storage temperature, 178,
      **181**, 181
   recipe, 187–190
   volume losses, 188, 199
   yeast level, 28, 178
Frozen yeast, 78, **87**, 87
Functional ingredients, 51–90
Functions of mixing, 93–94
Functions of the breadmaking process, 21–25
Fungal *alpha*-amylase
   in CBP, 70
   effect on staling, 71–72
   heat stability, 69
   in straight dough bulk fermentation, 31

Gas
   in CBP, **84–85**
   in frozen dough, 37, 187, 191, 196
   production, 24, 84, 177, 201
   retention, 9, 23, 24–25, 59, 197, 198, 199,
      200, 213, 376, 377, 378
   in straight dough fermentation, 77, 270
Germ, 6, 7, 15, 226, 334, 344
Glass transition temperature, 193, 291, 292–294
Glassy state, 293, 294
Gliadin, 308, **309**, 327
Gloss formation, 158–160
Glucose oxidase, 74
Gluten
   development, 21–22, 307, 309, 312,
      313–316, **314**
   elasticity, **326**, 326

matrix, 319
oxidation by soya flour, 56
structure, 323–**324**, 324
viscosity, 326–328
Gluten-free bread, 379
Glutenin, 308–311
Glycemic index, 17
Glycemic load, 17
Glycerol monostearate
effect on staling, 266
in yeasted laminated products, 266
Grading, *see* Dusting; Scalping
Granular yeast, 77
Gristing of wheat
definition, 352–353
Guidelines for retarded dough production, 186

Hagberg Falling Number, 338–339, 361; *see
also* Falling Number
Hamburger buns
baking, 253
using CBP, 210
cooling, 253–254
depanning, 253–254
flour recovery, 251–252
formulations, 246
liquid brews and ferments, 246–248
mixing, 249
production rates, 245–246
proving, 252
seed application, 252, **253**
Hardness in wheat, 339
Hemicellulase, 72–73, 355
High speed mixers, 100–102
High velocity convection, 164
Horizontal mixers, 108, 312
Hydration, 43, 312, 314

Image analysis, 11, 12, 13, 360
Improvers
in part-baked breads, 257
in rapid processing, 36–38
in rye bread, 374
in sponge and dough Inactivated yeasts, 154
Indexing conveyors, 123
Infestation in flour, 337, 347, 349, 350
Infrared radiation, 164, 282
Instant yeast, 78, 232
Intermediate proof
with frozen doughs, 190, 191
Irradiation, 282

Kneading
action in Wendal mixer, 100, **101**

Knock-back, 28, 32
Knowledge-based systems, 208, 210, 215

Laminated products, *see* Croissant
Danish pastry, 263–271, **264**, *265*
Laminating fat, 266–269
Lamination, 271
L-Cysteine, 36, 65–66
Lecithins
in yeasted laminated products, 266
Life cycle costs, 170–172, **171**
Lipase, 74
Lipids, 303–304
Liquid brews, 35, 36, 246–248, 261
Loaf core temperature, 141, 154, **155**
Low-grade flour
definition, 346
Low speed mixers, 36, 37, 65, 105–106

Maillard reactions, 158
Maize bread, 381, 382, **383**
Malt bread, 71, 168, 346, 378
Malted barley in bread, 378
Malted grain flours, 357
Malt flour
in straight dough fermentation, 30
Maltogenic amylase, 72
Manufacture of part-baked breads, 257
Mechanical dough development, 26, 38–47
Microbiological spoilage of bread, 275–282
Microwave radiation, 282
Mill screenroom, 340–342
Milton Keynes Process, 170, 261–263
Mixers
CBP-compatible mixers, 95–98, **95**, 99
effects of pressure in CBP, 98–99
types, 94–109
Mixing
control of mixer headspace atmosphere,
39, 99
hamburger buns, 249
laminated products, 269, 271
role of energy in CBP, 40–42, **40**, **41**
rye doughs, 374
Model K divider, 250
Modified atmosphere packaging, 279–281
Moulding
cross-grain, 132
Mould inhibitors
in part-baked breads, 259
Mould spoilage, 275–277
Multi-grain breads, 376–377

Naan, **9**, 385
Nitrogen, 25, 99, 196, 280

North America
  sponge and dough, 33, 34, 66

Oats in bread, 377, 378
Oblique axis fork mixer, 105–106
Off-odours in wheat, 336
Oil suction divider, 113, **114**, 137
Oven
  conveyorized, 162, **163**
  design, 162–164, **163**
Oven break, 158, **159**, 160–161
Oxidizing agents in the USA, 66
Oxygen, 25, 55–56, 64, 99

Packaging
  form-fill, 281
  vacuum, 281
Packing, 237, 347
Pallet in-feed systems, 124
Panning, 135, 139, 234
Pans, 162, 234, 251
Papadams, 386
Part-baked breads
  bake-off, 170, 256
  baking, 169, 255, 256, 257, 258, 260
  cooling, 256, 258
  depanning, 253, 258–259, 262
  freezing, 260
  manufacture, 255, 256, 257
  proving, 256, 257
  storage, 256, 259
  uses, 256
Part-baked processes
  ambient, 170, 259
  frozen, 257, 260
Partial vacuum in CBP, 46, 56, 64, 97, 98–99, 104
Patent flour
  definition, 346, 367
Pea fibre in bread, 377
Pearling of wheat, 351
Pentosans
  effect on staling, 287, 290
  in rye, 302
pH in brews, 248
Pizza yeasts, 79
Pocket-type prover, 116, 121
Potassium sorbate
  in part-baked breads, 259
Preservatives, 66, 67, 278
Primary causes, 213
Process economics, 170
Processing frozen doughs, 187

Proof determination by Retarding Advisor, 215–216, 221
Propionic acid, 66, 67, 86, 278–279
Protease, 183, 277, 300, 354, 355
Protein
  bonding, 324
  content in, 337, 338, 353, 360
  flour, 30–31, 58, 264, 265, 360
  wheat, 3, 61, 307, 337–338, 352
Proteinases, 73
Provers
  carrier, 151
  spiral, 152
Proving, see also Intermediate proof
  Final proof, 235
  frozen doughs, 195
  hamburger buns, 252
  part-baked breads, 169, 255
  retarded doughs, 184
Psychrometric chart, 145, 146
Psychrometry, 144
Pumpernickel bread, 372, 375
Purifiers, 343, 345
Pusher in-feed systems, 123

Radio frequency heating, 164, 282
Rapid processing methods, 26, 36
Recipes for
  balady, 385
  cassava bread, 381
  chapatti, 385
  frozen dough, 86–87, 187, 190, 195
  hamburger buns, 245–246, 247
  maize bread, 381, 382–383
  naan, 385–386
  papadams, 386
  retarding, 176
  rice bread, 381
  rye bread, 371
  sorghum bread, 381–382
  sponge and dough, 33, 34, 35
  straight dough bulk fermentation, 28
  tortilla, 386
  UK white bread, 52, 57
  yeasted laminated products, 263
Reciprocating rounders, 120
Reduction system, 345
Refrigerated proving, 151
Refrigeration principles, 201
Relative humidity
  definition, 144
  in retarding, 177

Resistance to deformation, 23, 31, 42, 46–47, 127, 387
Retarder
  construction, 204
Retarder–provers
  construction, 203
  control by Retarding Advisor, 215–216
Retarding
  effect of storage temperature, 179, 181, 189
  effect of storage time, 179, 181, 182, 189
  effect of yeast level, 182, 185, 188–189
  pizza doughs, 186
  production guidelines, 186
  recipes, 176
  suitability of processes, 176
Retarding Advisor, 215–216, 221
Rice bread, 381
Role of energy during mixing, 40
Roles of the sponge, 33
ROLLOUT, 218–219
Roll plants, 135, 137–138, 139
Rope, 86, 275, 277, 278–279, 323
Rounders
  conical, 117–118
  cylindrical, 118
Rounding, 115, 137
Rounding belts, 118
Rye bread
  American, 374
  baking, 374
  dehydrated sours, 373
  doughmaking, 373
  improvers, 374
  keeping qualities, 375
  recipe, 371
  starter, 373
  use of sour dough, 373

Salt
  effect on yeast, 86, 88
Sampling of wheat deliveries, 336–337
Sandwiches
  UK market, 228, 238–239, 287, 378
Scalping, 344
Scratch system, 343, 344
Screenings in wheat
  definition, 336, 367
Seed applications, 252
Self-raising flour, 357
Sheeting, **127–129**, 132, *133*
Shelf life of bread
  effect of ethanol, 290
  of rye bread, 375
Shock freezing, 260

Skinning in retarded doughs, 198
Slicing, 237
Sodium stearoyl-2-lactylate
  effect on staling, 59, 287
  in yeasted laminated products, 263
Sorbic acid, 66, 67, 317
Sorghum bread, 381, **382**
Sour dough in rye bread, 375
Soya flour
  in CBP, 55–56
  as a flour bleaching agent, 55
  gluten oxidation, 56
  in straight dough fermentation, 55
Specific humidity, 146, 148
Spiral mixers
  with no-time doughs, 33, 42, 133, 178
Spoilage by
  bacteria, 277
  mould, 275
  yeasts, 278
Sponge
  effect of pH, 34
  in retarding, 177
  role of temperature, 35
Sponge and dough
  in North America, 33, 34, 66, 112, 137
  in UK, 33
Staling
  role of gluten, 286
  role of starch, 284, 285
Staling inhibitors
  emulsifiers, 289–290
  enzymes, 288–289
  ethanol, 290–291
  pentosans, 290
  sugars, 291–292
Starch
  damage, 155, 306, 345, 353, 357, 360
  gelatinization, 155, 284
  retrogradation, 71, 72, 262, 285–287, 290
Stickiness of dough, 23, 38, 47
Straight dough bulk fermentation
  dough temperature, 28, 29
  effect of fermentation time, 28, **30**, 32, 38, 40, 177
  recipes, 28
  in retarding, 28
Sugar
  effect on staling, 291
  effect on yeast, 69, 70, 87
  level in yeasted laminated products, 263

Taints
  in flour, 336
  in wheat, 336
Temperature
  changes during
    baking, 156
    cooling, **166**, 167, 168, 262
    defrosting frozen dough, 194
    freezing, 293
    proving, **149, 184**, 185
    retarding, 180–182, 183, 186, 197, 198,
      199, 200, 201, 217
    vacuum cooling, 168–169, 170, 262, 378
  dew point, 144, 146, 148, 156, 159, 160
  dry bulb, 144, 148
  wet bulb, 146
Texture profile analysis, 14
Tortillas, 386
Total titratable acidity (TTA) in brews, 247
Training, 220, 260
Transglutaminase, 74
Triticale
  *alpha*-amylase activity, 375
  baking, 375
  milling, 375
Twin-arm mixers, 105, **106**
Twin spiral mixers, 100–102

Ultraviolet irradiation, 282
Unleavened bread, 384–385, 386
Unmixing of dough, 317–318

Vacuum
  cooling, 236
  dividers, 115
  packaging, 115
Vinegar, *see* Acetic acid
Vitamin C, *see* Ascorbic acid
Volume losses
  in retarded and frozen doughs, 198

Water
  absorption, 360
  level in CBP, 353
  straight dough bulk fermentation, 31
  redistribution in staling, 287
  role in dough formation, 28
  splitter, 252
Water-soluble proteins, 304
Water temperature calculator, 42, 109, 217
Waxy patches in retarded doughs, 200
Weight loss during
  baking, 170
  cooling, 170

  processing, 170, **171**
  retarding, 170
Wendal mixers, 100
Wheat
  appearance, 336
  blending, 312–313
  in Canada, 16, 224
  cereal *alpha*-amylase, 71
  conditioning, 342
  delivery, 347
  density, 337
  endosperm, 306, 339, 359
  gluten content, 338
  gluten protein, 307–308
  grain, 15, 333, **334**, 337, 339, 346, 351, 376
  gristing, 352–353
  Hagberg Falling Number, 338, 340, 361
  hardness, 339
  impurities separation, 336–337, 339, 340,
    353
  off-odours, 336
  pearling, 351
  protein content, 337–338
  screenings, 336–337
  storage, 340
  taints, 336, 337
  testing, 336–340
  in the USA, 224
Wheatless breads, 379–383
White spots
  bread composition, 177, 178, 187
  effect of fat in bread, **54**, 55
  effect of yeast, **185**, 185, 197–198
  flour, 178
  in frozen dough, 196–198
  mould, 190, 197
  in retarded dough Wholemeal, 196
Wishbone mixer, 105
Work input in the CBP, 40, 229, 230
Wrapping, 280, 281, 282, 283, 294

Yeast
  activity during baking, 155
  activity during proof, 146
  in the CBP, 84
  cell biology, 79–81, **80**
  commercial production, 81–83
  compressed, 77, 85
  cream, 77–78
  deactivated, 79
  dried pellet, 78
  effects of acidity or alkalinity, 84
  effects of mould and rope inhibitors, 86
  effects of salt, 87

effects of spices, 86
effects of sugar, **88**
effects on white spots in retarded dough
      extracts, 177–178
encapsulated, 78
fermentation, 246–248, 290
fermentative, 278
filamentous, 278
frozen, 78, 87
in frozen dough, 86–87, **87**
granular, 77
in liquid brews, 246–248
in retarded dough, 178, 187, 199, 201
in straight dough bulk fermentation,
      28–30
in yeasted laminated products, 263–271
inactivated, 154
instant, 78

level, 28–30, 177–179
pizza, 79
storage stability, **85**
Yeasted laminated products
      addition of fat, 265–266
      dough mixing, 269–270
      dough temperature, 266, 267
      flour quality, 43, 58
      formulations, 263–264
      laminating fat, 266, 267, 270
      lamination, 271
      sugar level, 266
      use of emulsifiers, 289–290
      yeast levels, 266
Yeast level, 28, 177, 266

Z-blade mixers, 102

Printed in the United States of America.